Chemistry of High-Temperature Superconductors II

ACS SYMPOSIUM SERIES 377

Chemistry of High-Temperature Superconductors II

David L. Nelson, EDITOR
Office of Naval Research

Thomas F. George, EDITOR
State University of New York at Buffalo

Developed from a symposium sponsored by
the Division of Physical Chemistry
at the 195th Meeting
of the American Chemical Society,
Los Angeles, California,
September 25-30, 1988

American Chemical Society, Washington, DC 1988

Library of Congress Cataloging-in-Publication Data
(Revised for vol. 2)

Chemistry of high-temperature superconductors.

(ACS symposium series, 0097-6156; 351, 377)
Vol. 2. edited by David L. Nelson and Thomas F. George.

Developed for symposia sponsored by the Divisions of Inorganic Chemistry and Physical Chemistry at the 194th and 196th Meetings of the American Chemical Society, held Aug. 30–Sept. 4, 1987, New Orleans, La., and Sept. 25–30, 1988, Los Angeles, Calif.

Includes bibliographies and indexes.

1. Superconductors—Chemistry—Congresses. 2. Materials at high temperatures—Congresses.

I. Nelson, David L., 1942– . II. Whittingham, M. Stanley (Michael Stanley), 1941– . III. George, Thomas F., 1947– . IV. American Chemical Society. Meeting (194th: 1987: New Orleans, La.) V. American Chemical Society. Meeting (196th: 1988: Los Angeles, Calif.) VI. American Chemical Society. Division of Inorganic Chemistry. VII. American Chemical Society. Division of Physical Chemistry. VIII. Series: ACS symposium series; 351, etc.

QD473.C488 1987 537.6'23 87-19314
ISBN 0-8412-1431-X (v. 1)
ISBN 0-8412-1541-3 (v. 2)

PRINTED IN THE UNITED STATES OF AMERICA

ACS Symposium Series

M. Joan Comstock, *Series Editor*

Foreword

The ACS SYMPOSIUM SERIES was founded in 1974 to provide a medium for publishing symposia quickly in book form. The format of the Series parallels that of the continuing ADVANCES IN CHEMISTRY SERIES except that, in order to save time, the papers are not typeset but are reproduced as they are submitted by the authors in camera-ready form. Papers are reviewed under the supervision of the Editors with the assistance of the Series Advisory Board and are selected to maintain the integrity of the symposia; however, verbatim reproductions of previously published papers are not accepted. Both reviews and reports of research are acceptable, because symposia may embrace both types of presentation.

Contents

Preface

If the present momentum is maintained, high-temperature superconducting compounds may approach semiconductors, such as silicon, as one of the most intensively studied classes of materials in the history of science. This year marks the second anniversary since the discovery of these remarkable materials. Spectacular progress has been made in areas ranging from the synthesis and characterization of new classes of high-temperature superconductors to new theoretical descriptions of the phenomenon. The full integration of these fascinating compounds into existing and new technologies continues to progress, although somewhat more slowly. A number of applications that utilize the new high-temperature superconductors look promising for the near future, while others remain controversial and may not be realized for some time.

While Volume I included a focus on applications, research needs, and opportunities, Volume II reports on the rapid progress that has been made in the following areas of physical chemistry: theory, new materials, surfaces and interfaces, and processing.

Acknowledgments

We thank Darlene S. Miller for her major contributions to the editing process of this book. Such contributions included numerous telephone calls, extensive typing, and an amazingly cheerful attitude throughout the various crises that arose.

David L. Nelson
Office of Naval Research
Arlington, VA 22217–5000

Thomas F. George
State University of New York at Buffalo
Buffalo, NY 14260

July 12, 1988

Chapter 1

Overview of High-Temperature Superconductivity

Theory, Surfaces, Interfaces, and Bulk Systems

D. Sahu[1], A. Langner[1], Thomas F. George[1], J. H. Weaver[2], H. M. Meyer, III[2], David L. Nelson[3], and Aaron Wold[4]

[1]Departments of Chemistry and Physics, 239 Fronczak Hall, State University of New York at Buffalo, Buffalo, NY 14260
[2]Department of Chemical Engineering and Materials Science, University of Minnesota, Minneapolis, MN 55455
[3]Chemistry Division, Office of Naval Research, Arlington, VA 22217–5000
[4]Department of Chemistry, Brown University, Providence, RI 02912

An overview of the theoretical and experimental aspects of the recently-discovered superconducting compounds is presented. This overview is divided into three sections. In the first section a review of some of the theoretical and computational works is presented under the subsections entitled pairing mechanisms, electronic structure calculations and thermophysical properties. In the second section surface and interface chemistry issues related to the fabrication and use of high-temperature superconductors for high-performance applications are presented. Specific issues that are discussed include metallization and the formation of stable ohmic contacts, and chemically-stable overlayers that are suitable for passivation, protection and encapsulation of superconducting material structures that can then be used under a wide range of environmental conditions. Lastly, issues are discussed that are related to each of the bulk high-temperature superconducting ceramic oxides which have received so much attention the past two years. These include $T\ell_2Ba_2Ca_2Cu_3O_{10}$ with a critical temperature of 125 K, which is the current record.

0097–6156/88/0377–0001$06.00/0

Theory

Since the discovery of high-temperature superconductivity in the CuO ceramics in the year 1986, there has been an explosion of theoretical works relating to these systems. It is not possible, in a short review of theoretical work like this, to examine the essential ideas of even a small fraction of this large body of literature. We select, quite arbitrarily, some of the work we are aware of and discuss the essential ideas of these. We apologize, in advance, to all the research workers whose work we have not reviewed.

This theoretical overview is divided into three sections: (1) pairing mechanism, (2) electronic structure calculations and finally, (3) thermophysical properties. In the section on pairing mechanism we review the basic microscopic mechanisms that might be responsible for producing superconducting states. This is a fundamental question and the success of any theoretical model addressing it should be judged in relation to the experiments it explains and the phenomena it predicts. The second part, dealing with electronic structure calculations, is quite important in that such calculations give detailed information about conduction processes, density of states and anisotropies in $\vec{k}$-space. Finally, in the section on thermophysical properties we review works relating to phonon dispersion and soft modes, dynamics of tetragonal-to-orthorhombic phase transitions and the temperature and concentration dependence of the stability of these phases.

(1) Pairing Mechanism

In the conventional superconductors a pair of electrons of opposite spin and momentum form a bound state which leads to a coherent and highly correlated many-body superconducting state. The attractive interaction between the electrons of the pair is mediated by lattice vibrations (phonons) and the electrons overcome their strong Coulomb repulsion by staying away from each other in time (retardation). The basic question in the new superconductors is - What is the pairing mechanism? For a review of some of the early pairing models proposed for the high-T_c materials we refer the reader to the work by Rice [1].

One of the leading theories for the high-T_c superconductors is the resonating-valence-band (RVB) model proposed by Anderson [2] and his coworkers [3]. According to this model, the pairing mechanism is magnetic in origin and not of the conventional BCS type. The starting point of the RVB theory is a two-dimensional Hubbard model at half-filling with strong on-site Coulomb repulsion U and an attractive inter-site hopping energy t. Without oxygen doping (i.e., dopant concentration $\delta = 0$), the ground state of the above model is expected to be a long-range anti-ferromagnetic (AF) state [4], but Anderson argues that frustration might favor a RVB state over an AF ground state. The basic idea of the RVB theory is that strong electron-electron correlations result in a separation of charge degrees of freedom from spin degrees of freedom. At low doping ($\delta \neq 0$) and temperature, the quasi-particle excitations are believed to be "holons" (i.e., charge-carrying spinless particles) and "spinons" (i.e., spin-½ chargeless particles).

Superconductivity is due to formation of a condensate consisting of holon pairs. Since Bose-Einstein condensation is not possible in a strictly two-dimensional system, the interplanar couplings are important in giving rise to superconductivity.

The RVB theory has been amplified or modified in several ways [5,6,7], the details of which we won't go into here. Rice and Wang [8] have proposed a model in which superconductivity is due to condensation of a pair of bosons which gives rise to quasi-particle excitation energies which are similar to those of RVB theory, but different from BCS theory. Rice and Wang, however, favor a phonon interaction which mediates the attraction between the boson pair in question. Coffey and Cox [9] have given a nice summary of the essential points of the RVB theory in Section II of their paper.

Another theory that has been proposed is the spin-bag mechanism of Schrieffer's group [10]. The starting point of their theory is the strong AF spin order in the neighborhood of the superconducting transition (T_c) and two-dimensional spin-correlations over a length L (~200 Å) in the neighborhood of the Neel transition ($T_N >> T_c$). Doping creates extra holes in the system and these are assumed to be localized over a length ℓ (<<L). The spin of the hole couples to the spin density of the AF background through an exchange interaction. Within a given domain a hole produces an effective potential well or bag in its vicinity in which the hole is self-consistently trapped. The pairing is due to an effective attractive interaction between two holes which overcomes the short range Coulomb repulsion.

The importance of spins in producing a superconducting state has also been emphasized in a model proposed by Emery [11]. He has proposed that oxygen doping in the 214-material creates holes at the oxygen sites and that there is a narrow band of oxygen holes which couple strongly to the local spin configuration of the Cu-sites. This strong coupling is responsible for an attractive interaction between the oxygen holes.

We mention in passing that other mechanisms such as plasmons [12] and excitons [13] have also been proposed as possible candidates for being the condensate of the superconducting state.

Recently doubts have been expressed about the importance of the magnetic origin of the pairing interaction. The bismuth superconductors [14,15] ($BaKBiO_3$) which are closely related in structure to the earlier superconductors and which are free of copper provide counter examples [16] to the magnetic origin of superconductivity. Another counter example [16] is the 123-material in which Cu is substituted 100% by Ag bringing T_c down to 40 K. This has led some people to propose that local structure might play a role in producing a superconducting state. A recent double-well model [17] of oxygen motion indicates trends in this direction; it has been shown that a strong T-dependent electron-phonon coupling parameter could produce $T_c \approx 100$ K.

(2) Electronic Structure Calculations

The electronic structure [18-25] of the undoped compounds is of immense interest in getting a clue to the origin of superconductivity. These parent compounds are La_2CuO_4 (denoted by 214), $YBa_2Cu_3O_7$ (denoted by 123) and $Bi_2Sr_2CaCu_2O_8$ (denoted by

2212). The important question is: Why is the ground state of these compounds an antiferromagnetic (AF) insulator? The band structure calculations, so far performed within the local density approximation, give results which are in contrast to the experimental situation - they all produce a metallic ground state. The reason for this discrepancy is believed to be the strong electron correlations in the CuO planes which are not adequately taken into account in a band picture. It would be interesting to have electronic structure calculations which take these correlations into account. In this regard, spin-polarized band structure calculations are expected to yield improved results over conventional band structure calculations. However, the results [23-25] of spin-polarized band structure calculations do not seem to be definitive.

We would like to conclude this section by stating some of the important conclusions that have emerged from the band structure calculations. In these calculations [18,19] the importance of the two-dimensional nature of the CuO planes was emphasized. The copper $d(x^2-y^2)$ orbitals and the neighboring oxygen p(x,y) orbitals interact to produce bonding σ and antibonding σ^* orbitals. Similar results [20-22] were also obtained for 123- and 2212-compounds. Another important result is that the antibonding band was positioned closer to the Fermi energy E_f. In the new 2212-compounds, a pair of slightly filled Bi 6p bands provide additional carriers in the Bi-O planes [21]. A remarkable feature in these compounds is the charge separation between the two Bi-O planes [22].

(3) Thermophysical Properties

In this brief discussion of the thermophysical properties of the high-T_c oxide superconductors we restrict ourselves to theoretical investigations of the lattice dynamics of these systems. In particular, we review the work on phonon modes and oxygen vacancy ordering and their influence on the structural transitions of the 214 and 123 superconductors. Knowledge of the phonon spectrum and its dependence on the oxygen distribution may prove to be of prime importance in elucidating the mechanism of high-T_c superconductivity. Several scenerios of how electron-phonon interactions can lead to high transition temperatures have been proposed including oxygen motion in double wells [17], interlayer coupling [26,27] and coupling to soft quasicyclic modes associated with underconstrained nearest-neighbor rearrangements [28].

There have been a limited number of theoretical investigations on phonon frequencies and eigenvectors of the $La_{2-x}(Ba,Sr)_x CuO_4$ [29-31] and $YBa_2Cu_3O_{7-\delta}$ [32-35] superconductors. Unscreened lattice dynamical models [30,35], yielding only the bare phonon frequencies, gave fair agreement with experimentally determined total phonon density of states, mean square atomic vibrational amplitudes and Debye temperatures. Weber [29] has shown that the effect of screening, spectrum renormalization, due to conduction electrons gives rise to large Kohn anomalies near the Brillouin-zone boundary involving oxygen breathing modes. It was pointed out by Chaplot [35] that an additional effect of renormalization is to hybridize the high-frequency modes, dominated by oxygen, with the medium-frequency, heavier-nuclei modes thus making necessary an effective-mass description for the analysis of the isotope effect. Cohen, et al. [31] employing the potential-induced breathing model were able to predict the tetragonal-to-orthorhombic phase transition in La_2CuO_4 as arising from an instability, softening, in the B_{3g} tilting mode of the I4/mmm tetragonal structure. The onset of this transition has been recently derived from symmetry principles.[36]

As is now well established, $YBa_2Cu_3O_{7-\delta}$ undergoes a tetragonal-to-orthorhombic phase transition for $1 > \delta > 0$ which is a consequence of oxygen vacancy ordering in the Cu-O basal plane. Several approaches have been undertaken to model this transition and describe the phase diagram in (T,δ)-space. Most treatments are of the 2-D lattice gas type, employing first and second nearest-neighbor interactions for oxygen and vacancies and solved by mean field techniques [37-45]. Another approach, which has been successful in predicting the microstructure of the multiphase region, utilizes the method of concentration waves [46]; here the oxide is treated as an interstitial compound of ordered oxygen atoms and vacancies on a simple lattice [47]. The model of Mattis [48] deserves mention since it specifically accounts for the copper-oxygen bonds and thus enables predictions to be made concerning the dielectric response of the various phases. The picture that emerges from these studies is that at high temperatures, $T > 750$ K, and/or stoichiometries $\delta < 0.5$ a tetragonal phase exists with random ordering of oxygen and vacancies in the Cu-O basal plane. For material with $\delta = 0$ the superconducting orthorhombic phase is stable with oxygen (atoms) and vacancies forming alternate chains along the crystal b-direction. At low temperatures and intermediate stoichiometries phase separation occurs with micro regions of tetragonal phase, orthorhombic phase and a second cell-doubled orthorhombic phase [45]. Phase transitions between the tetragonal and orthorhombic phases are of a second-order disorder-order type.

As a final note we mention the work on establishing the temperature domain in which thermal fluctuations and critical behavior dominate. Many of the theories described in this review are based on mean field techniques which become invalid when fluctuations are important. Estimates of the critical region, Ginzburg criterion, are in the range $\xi^G \equiv (T_c-T)/T_c \sim 0.1 - 0.7$ [49-51]. However, it has been pointed out that the breakdown of mean-field behavior is progressive [51]; the ability of mean-field theory to predict non-universal quantities (prefactors, GL-parameters and T_c) is lost within a region $\xi^B \sim (\xi^G)^{1/2}$, Brout criterion [52], which may cover most of the superconducting regime.

Surfaces and Interfaces

The full integration of bulk and thin film high-temperature superconductors into existing and new technologies of high commercial value appears limited by a number of surface and interface materials issues. Many of these issues can be stated in general terms because they are shared by each type of ceramic superconductor (2-1-4, 1-2-3, or 2-1-2-2). Others are more specific to the material under study, e.g., the toxic character of the Tl 2-1-2-2 compound. As new materials with even higher critical temperatures are developed, analogous problems will be encountered.

Hence, the knowledge base developed for one class of ceramic material may also apply to others.

First, there are issues related to materials synthesis so that structures can be fabricated with predetermined shapes, sizes, and current carrying ability. These range from macroscopic to microscopic. Second, there are challenges related to the fabrication of superconducting thin films on a variety of substrates, with Si being an obvious choice from the perspective of microelectronic devices. Third, there are issues related to the formation of stable ohmic contacts, particularly for small samples and thin films. Fourth, there are problems related to the passivation, protection or encapsulation of small structures such as fibers or thin films, so that the superconducting oxides can be used under a wide range of environments. These and a great many other issues raise challenging chemical questions. Spectacular progress has been made in the few months that the new superconductors have been in existence, and we can anticipate that rapid progress will be made in the near future. These efforts that address fundamental issues will bring us one step closer to realizing breakthroughs in technologies of high commercial value.

It might be thought that surface and interface issues should be separated from those involved in the synthesis of the superconductors themselves. This is certainly not the case, however, because many of the most exciting opportunities for these materials will be in high performance applications. In these applications, the size of the superconducting elements will be comparable to those of the other components. Scaling down or shrinking the size of a structure exacerbates problems related to interfacial phenomena. Indeed, the challenges of forming, contacting, and protecting a superconducting strip 1 μm wide and 0.1 μm thick point to the intermingling of a wide range of materials issues.

To date, issues related to bulk synthesis and surface or interface characterization have also been intimately tied. This can be seen by noting that the starting point for surface or interface research is a well-characterized bulk material. Unfortunately for the surface scientist, early samples did not spring completely characterized from the firing furnaces, to paraphrase the springing-forth of Minerva completely armed from the head of Jupiter. Instead, early samples were sintered, were 70-90% dense, and individual grains were clad with other phases [53,54]. Difficulties in characterizing these interfaces and identifying their intrinsic properties are reflected in the literature of the last year for the 1-2-3 and 2-1-4 materials, and the last few months for the 2-1-2-2 materials. Only recently have bulk samples of sufficient quality been available so that fracturing could provide a clean surface [53]. Today, it is relatively routine to find single crystals having dimensions of greater than 1 mm, and several small companies are preparing to sell single crystals as large a 1 cm × 1 cm so that full characterization can be achieved.

Early studies indicated that polycrystalline, sintered samples of the 1-2-3 and 2-1-4 materials degraded rapidly upon exposure to a range of environments, including H_2O, CO, CO_2, O_2 and solvents [55-57]. More recent work suggests that degradation is much slower and that some of the early problems were related to the intergranular phases (e.g. carbonates or cuprates). Early work also showed that exposure of the 1-2-3's to high-intensity ultraviolet and X-ray photon beams produces substantial changes in the surfaces [58]. Again, recent work has indicated that most of the changes are related to the presence of second phases and were not intrinsic to the superconductors. At the same time, studies by Rosenberg and coworkers [59] pointed out some of the details of photon stimulated desorption. Electron beams of high current or high energy also induce damage. Exposure of these materials to energetic ions, such as used in Ar sputtering, leads to surface modification, structural changes, and the loss of superconductivity [53]. This damage points to the fragile character of these superconductors but also indicates that it may be possible to selectively alter portions of a thin film, for example writing nonsuperconducting lines on a superconducting template.

An issue that has arisen repeatedly has been the possiblity of oxygen loss through the surface at room temperature. To our knowledge, there is no clear evidence that oxygen is lost under static vacuum conditions. Instead, the exposure of freshly prepared surfaces to ultrahigh vacuum leads to the chemisorption of residual gases from the vacuum system. Recent work with single crystals has provided evidence for the rearrangement of surface atoms after cleaving in vacuum [60]. This has been attributed to transgranular fracture and the exposure of energetically unfavorable surfaces. Indeed, such effects can be understood in terms of the highly anisotropic unit cell, but the restructuring does not necessarily result in oxygen loss.

Attempts to form contacts to surfaces or to investigate the electrical properties of the high-temperature superconductors have often been complicated by nonreproducibility. This can be related to the chemical processes that occur at these surfaces. As detailed studies of representative interfaces have shown, there is a strong tendency for reactive metals to leach oxygen from the superconductor to form new metal oxide bonding configurations [53,61]. The result of this interfacial chemistry is a heterogeneous transition region between the buried superconductor and the metal film. In particular, a cross section through an interface based on the 1-2-3, 2-1-4 or 2-1-2-2 superconductors would show the superconductor; a region where oxygen has been removed, where the structure is likely to be disrupted, and which is not superconducting; a region where the metal adatoms have formed electrically-resistive oxides; and the metal overlayer, possibly containing oxygen and dissociated superconductor atoms in solution and at the surface. Such an interface is shown schematically in Fig. 1. These interfaces are metastable because thermal processing will enhance oxygen transport to the metal layer and will increase the amount of substrate disruption. Certainly, these interfaces would not form the ohmic contacts desired in device applications. Likewise, the cladding of superconducting filaments with copper sheaths does not seem propitious since there will be interfacial interactions and the formation of a nonsuperconducting layer. The scale of this disrupted region is at least 50 Å for room temperature metal deposition and is likely to be much larger if the interface is processed at a higher temperature.

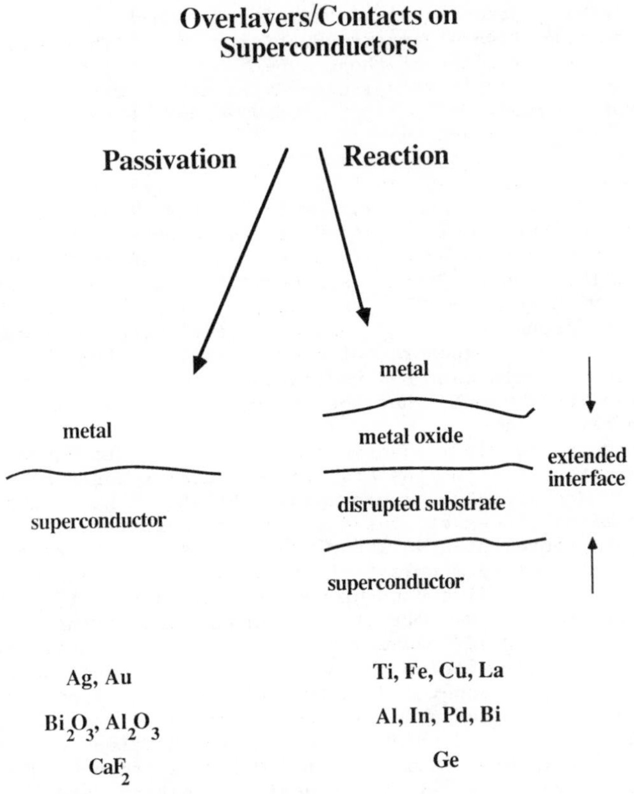

Fig. 1 Schematic showing passivation of high-T_c superconductor surface with Ag, Au, and composite materials (left) compared to disruptive reaction for chemically active overlayers (right).

Ohmic contacts to these superconductors can be formed with Ag and Au overlayers [62]. These metals show minimal tendency to form oxides, and their deposition does not seriously disrupt the superconductor. For contacts, then, these metals appear to be the materials of choice. Two caveats should be noted, however. First, processing may lead to Au clustering, as has been observed for Au layers on many substrates. Second, if the surface to be metallized has been exposed to the air, it will be coated with hydrocarbons, water vapor and the like, and its superconducting properties will most likely be compromised. Attempts using Ar-ion sputtering to remove these contaminants prior to metallization have been successful, but almost certainly at the expense of the structural integrity of the surface. It remains to be seen whether the amount of surface damage (and loss of superconductivity) compromises the effective use of this approach to the fabrication of ohmic contacts for devices.

The protection of these superconductors is also crucial if they are to be integrated with other technologies. Much less is known about such passivation issues, although attempts are presently under way to develop protective overlayers. Two types of materials that have been examined so far and appear promising are metal oxides and non-metallic insulators [63]. Metal oxides have been formed by the deposition of metal atoms in an activated oxygen atmosphere to form a metal oxide precursor that does not leach oxygen from the substrate. The oxide layer then serves as a diffusion barrier against oxygen loss or atomic intermixing. To date, efforts have been successful with the oxides of Al and Bi, and it is likely that other oxides will prove to be effective as passivation layers and/or diffusion barriers. Those investigated so far are insulators, but future work may lead to conducting oxides that can serve as both contacts and passivators.

The second type of material that shows promise for passivation and electrical isolation is CaF_2 [63]. This large-bandgap ionic insulator has a high static dielectric constant, can be readily evaporated from thermal sources in molecular form, and shows no tendency to modify the superconductor. As such, it may be useful as a dielectric layer in advanced devices.

We also note that both organic and inorganic polymers may be useful as encapsulants. In fact, fully protective coatings for high temperture superconductors may need to be developed to meet the specific demands of large- and small-scale applications. These protective coatings will likely involve a multilayer structure that would consist of thin- or thick-films of metals, low dielectric ceramics, and possibly organic and inorganic polymeric films. These multifunctional coatings would be effective as a diffusion barrier, would withstand environmental stress and temperature cycling, and would maintain strong adhesion. Specific opportunities exist for the synthesis of precursor molecules, such as organometallics, for chemical vapor deposition (CVD), or sol/gel approaches for low temperature fabrication of these multilayer protective coatings.

Bulk Superconducting Ceramic Oxides

Superconducting oxides have been known since 1964, but until recently the intermetallic compounds showed higher superconducting temperatures. In 1975 research scientists at E. I. DuPont de Nemours [64] discovered superconductivity in the system $BaPb_{1-x}Bi_xO_3$ with a T_c of 13 K. The structure for the superconducting compositions in this system is only slightly distorted from the ideal cubic perovskite structure. It is generally accepted that a disproportionation of the Bi(IV) occurs, namely, $2Bi(IV)(6s^1) \rightarrow Bi(III)(6s^2) + Bi(V)(6s^0)$ at approximately 30 percent Bi. Sleight found that the best superconductors were single phases prepared by quenching from a rather restricted single-phase region, and hence these phases are actually metastable materials. At equilibrium conditions, two phases with different values of x would exist; the phase with a lower value of x would be metallic and with a higher value of x would be a semiconductor. It is important to keep in mind that the actual assignment of formal valence states is a convenient way of electron accounting; the actual states include appreciable admixing of anion functions. The system $BaPb_{1-x}Bi_xO_3$ should be studied further since it contains compositions showing the highest T_c for any superconductor not containing a transition element. Recently, for example, Cava and Batlogg [14] have shown that $Ba_{.6}K_{.4}BiO_3$ gave a T_c of almost 30 K, which is considerably higher than the 13 K reported for $BaPb_{.75}Bi_{.25}O_3$.

La_2CuO_4 was reported by Longo and Raccah [65] to show an orthorhombic distortion of the K_2NiF_4 structure with a = 5.363 Å, b = 5.409 Å and c = 13.17 Å. It was also reported [66,67] that La_2CuO_4 has a variable concentration of anion vacancies which may be represented as La_2CuO_{4-x}. Superconductivity has been reported for some preparations of La_2CuO_4. However, there appears to be some question as to the stoichiometry of these products since only a small portion of the material seems to exhibit superconductivity [68].

The extent of the anion vacancies has been recently reexamined [69], and the magnitude of this deficiency is less than can be unambiguously ascertained by direct thermogravimetric analysis which has a limit of accuracy in x of 0.01 for the composition La_2CuO_{4-x}. However, significant shifts in the Néel temperatures confirm a small variation in anion vacancy concentrations.

In the $La_{2-x}A_xCuO_4$ phases (A = Ca,Sr,Ba) the substitution of the alkaline earth cation for the rare earth depresses the tetragonal-to-orthorhombic transition temperature. The transition disappears completely at x > 0.2, which is about the composition for which superconductivity is no longer observed. Compositions of $La_{2-x}A_xCuO_4$ can also be prepared [70,71] where A is Cd(II) or Pb(II). However, these phases are not superconducting, and it appears that the more basic divalent cations are necessary to allow Cu(III) to coexist with O^{2-}. The existence of Cu(I) and Cu(III) in $La_{1.8}Sr_{.2}CuO_4$ is consistent with the ESR spectra, which shows the absence of square planar Cu(II), and the Pauli-paramagnetic behavior over the temperature range from 77 to 300 K. Since the Pauli-paramagnetic behavior of $La_{1.8}Sr_{.2}CuO_4$ is consistent with delocalized electrons, this would also indicate a high probability for the existence of Cu(I)-Cu(III) formed as a result of disproportionation of Cu(II) [72]. Subramanian et al. [73] have recently substituted both sodium and potassium into La_2CuO_4, giving rise to the compositions $La_{2-x}A_xCuO_4$ (A = Na,K). However, only the sodium substituted samples exhibited superconducting behavior.

The compound $Ba_2YCu_3O_7$ shows a superconducting transition of ~ 93 K and crystallizes as a defect perovskite. The unit cell of $Ba_2YCu_3O_7$ is orthorhombic (Pmmm) with a = 3.8198(1) Å, b = 3.8849(1) Å and c = 11.6762(3) Å. The structure may be considered as an oxygen-deficient perovskite with tripled unit cells due to Ba-Y ordering along the c-axis. For $Ba_2YCu_3O_7$, the oxygens occupy 7/9 of the anion sites. One third of the copper is in 4-fold coordination and 2/3 are five-fold coordinated. A reversible structural transformation occurs with changing oxygen stoichiometry going from orthorhombic at x = 7.0 to tetragonal at x = 6.0 [74]. The value x = 7.0 is achieved by annealing in oxygen at 400-500°C, and this composition shows the sharpest superconducting transition. It was shown by Davison et al. [72] that these materials are readily attacked by water and carbon dioxide in air to produce carbonates.

Recently Maeda et al. [75] reported that a superconducting transition of 120 K was obtained in the Bi/Sr/Ca/Cu/O system. The structure was determined for the composition $Bi_2Sr_2CaCu_2O_8$ by several laboratories [76-78].

In most of the studies reported to date on the Bi/Sr/Ca/Cu/O system measurements were made on single crystals selected from multiphase products. The group at DuPont selected platy crystals having the composition $Bi_2Sr_{3-x}Ca_xCu_2O_{8+y}$ (0.9 > x > 0.4) which showed a T_c ~ 95 K. Crystals of $Bi_2Sr_{3-x}Ca_xCu_2O_{8+y}$ for x = 0.5 gave orthorhombic cell constants a = 5.399 Å, b = 5.414 Å, c = 30.904 Å [76]. These cell dimensions are consistent with the results of other investigators [77,78]. The structure consists of pairs of CuO_2 sheets interleaved by Ca(Sr), alternating with double bismuth-oxide layers. Sunshine et al. [78] have indicated that the addition of Pb to this system raises the T_c above 100 K. There are now three groups of superconducting oxides which contain the mixed Cu(II)-Cu(III) oxidation states, namly $La_{2-x}A^{II}_xCuO_4$ where A^{II} = Ba,Sr,Ca; $RBa_2Cu_3O_7$ where R is almost any lanthanide; and $Bi_2Sr_{3-x}Ca_xCu_2O_{8+y}$.

Sheng and Herman have recently reported [79] on a high-temperature superconducting phase in the system Tl/Ba/Ca/Cu/O. Two phases were identified by Hazen et al. [80], namely $Tl_2Ba_2CaCu_2O_9$ and $Tl_2Ba_2Ca_2Cu_3O_{10}$. Sleight et al. [76,81] have also reported on the structure of $Tl_2Ba_2CaCu_2O_8$ as well as $Tl_2Ba_2CuO_6$. In addition, the superconductor $Tl_2Ba_2Ca_2Cu_3O_{10}$ has been prepared by the DuPont group [82] and shows the highest T_c of any known bulk superconductor, namely ~ 125 K.

A series of oxides with high T_c values has now been studied for the type $(A^{III}O)_2A^{II}_2Ca_{n-1}Cu_nO_{2+2n}$, where A(III) is Bi or Tl, A(II) is Ba or Sr, and n is the number of Cu-O sheets stacked. To date, n = 3 is the maximum number of stacked Cu-O sheets examined consecutively. There appears to be a general trend whereby T_c increases as n increases. Unfortunately, these phases involve rather complex ordering, crystals of the phases are grown in sealed gold tubes, and excess reactants are always present. The toxicity as well as volatility of thallium, coupled with problems in obtaining reasonable quantities of homogeneous single-phase material, presents a challenge to the synthetic chemist. It will also be interesting to see if these materials are truly more stable over time than the $La_{2-x}A_xCuO_4$ or $RBa_2Cu_3O_7$ phases.

Acknowledgments

Support by the Office of Naval Research is gratefully acknowledged.

Literature Cited

1. Rice, T. M. Z. Phys. B 1987, 67, 141.
2. Anderson, P. W. Mater. Res. Bull. 1973, 8, 153; Fazekas, P.; Anderson, P. W. Phil. Mag. 1974, 30, 432; Anderson, P. W., Science, 1987, 235, 1196.
3. Baskaran, G.; Zou, Z.; Anderson, P. W. Solid State Commun. 1987, 63, 973; Anderson, P. W.; Baskaran, G.; Zou, Z.; Hsu, T. C. Phys. Rev. Lett. 1987, 58, 2790; Anderson, P. W.; Zou, Z. Phys. Rev. Lett. 1988, 60, 132; Wheatly, J. M.; Hsu, T. C.; Anderson, P. W. Phys. Rev. B 1988, 37, 5897.
4. Hirsch, J. E. Phys. Rev. B 1985, 31, 4403.
5. Kivelson, S. A.; Rokhsar, D. S.; Sethna, J. P. Phys. Rev. B, 1987, 35, 857.
6. Ruckenstein, A. E., Hirschfeld, P. J.; Appel, J. Phys. Rev. B. 1987, 36, 857.
7. Kotliar, G. Phys. Rev. B 1988, 37, 3664.
8. Rice, M. J.; Wang, Y. R. Phys. Rev. B, 1988, 37, 5893.
9. Coffey, L.; Cox, D. L. Phys. Rev. B, 1988, 37, 3389.
10. Schrieffer, J. R.; Wen, X.-G.; Zhang, S.-C. Phys. Rev. Lett. 1988, 60, 944.
11. Emergy, V. Phys. Rev. Lett. 1987, 58, 2794.
12. Varma, C. M.; Schmitt-Rink, S; Abrahams, E. Solid State Commun. 1987, 62, 681.
13. Ruvalds, J. Phys. Rev. B 1987, 36, 8869.
14. Cava, R. J.; Batlogg, A. B. Nature 1988, 332, 814.
15. Rice, T. M. Nature 1988, 332, 780.
16. Kao, Y. H. colloquium talk at SUNY-Buffalo on May 24, 1988.
17. Hardy, J. R.; Flochen, J. W. Phys.Rev. Lett. 1988, 60, 2191.
18. Mattheiss, L. F. Phys. Rev. Lett. 1987, 58, 1028.
19. Yu, J.; Freeman, A. J.; Xu, J.-H. Phys. Rev. Lett. 1987, 58, 1035.
20. Mattheiss, L. F.; Hamann, D. R. Solid State Commun. 1987, 63, 395.
21. Hybersten, M. S.; Mattheiss, L. F. Phys. Rev. Lett. 1988, 60, 1661.
22. Krakauer, H.; Pickett, W. E. Phys. Rev. Lett. 1988, 60, 1665.
23. Guo, G. Y.; Temmerman, W.; Stocks, G. J. Phys. C 1988, 21, L103.
24. Leung, T. C.; Wang, X. W.; Harmon, B. N. Phys. Rev. B 1988, 37, 384.
25. Sterne, P. A.; Wang, C. S. Phys. Rev B 1988, 37, 7472.
26. Arnold, G. B. In Novel Superconductivity; Kresin, V. Z.; Wolf, S. A., Eds.; Plenum: New York, 1987; p. 323.
27. Gulasci, Zs.; Gulasci, M.; Pop, I. Phys. Rev. B 1988, 37, 2247.
28. Phillips, J. C. Phys. Rev. Lett. 1987, 59, 1856.
29. Weber, W. Phys. Rev. Lett. 1987, 58, 1371; 1987, 58, 2154(E).
30. Prade, J.; Kulkarni, A. D.; de Wette, F. W.; Kress, W.; Cardona, M.; Reiger, R.; Schröder, U. Solid State Commun. 1987, 64, 1267.

31. Cohen, R. E.; Pickett, W. E.; Boyer, L. L.; Krakauer, H. Phys. Rev. Lett. 1988, 60, 817.
32. Thomsen, C.; Cardona, M.; Kress, W.; Liu, R.; Genzel, L.; Bauer, M.; Schönhers, E.; Schröder, U. Solid State Commun. 1987, 65, 1139.
33. Bates, F. E.; Eldridge, J. E. Solid State Commun. 1987, 64, 1435.
34. Liu, R.; Thomsen, C.; Kress, W.; Cardona, M.; Gegenheimer, B.; de Wette, F. W.; Prade, J.; Kulkarni, A. D.; Schröder, U. Phys. Rev. B 1988, 37, 7971.
35. Chaplot, S. L. Phys. Rev. B 1988, 37, 7435.
36. Sahu, D.; George, T. F. Solid State Commun. 1988, 65, 1371.
37. Bakker, H.; Welch, D.; Lazareth, O. Solid State Commun. 1987, 64, 237.
38. Shi-jie, X. J. Phys. 1988, C21, L69.
39. Kubo, Y.; Igarashi, H. Jpn. J. Appl. Phys. 1987, 26, L1988.
40. Inouie, M.; Takemori, T.; Sakudo, T. Jpn. J. Appl. Phys. 1987, 26, L2015.
41. Bell, J. M. Phys. Rev. B 1988, 37, 541.
42. Varea, C.; Robledo, A. Mod. Phys. Lett. B 1988, 1, in press.
43. Sanchez, J. M.; Mejia-Lia, F.; Moran-Lopez, J. L. Phys. Rev. B 1988, 37, 3678.
44. Khachaturyan, A. G.; Semenovskaya, S. V.; Morris, Jr., J. W. Phys. Rev. B, 1988, 37, 2243.
45. Wille, L. T.; Berera, A.; de Fontaine, D. Phys. Rev. Lett. 1988, 60, 1065.
46. Khachaturyan, A. G.; Morris, Jr., J. W. Phys. Rev. Lett. 1987, 59, 2776.
47. Pokrovskii, B. I.; Khachaturyan, A. G. J. Solid State Chem. 1986, 61, 137, 154.
48. Mattis, D. C. Int. J. Mod. Phys. Lett., in press.
49. Deutscher, G. In Novel Superconductivity; Kresin, V. Z.; Wolf, S. A., Eds.; Plenum: New York, 1987; p. 293.
50. Lobb, C. J. Phys. Rev. B 1987, 36, 3930.
51. Kapitulnik, A.; Beasley, M. R.; Castellani, C.; DiCastro, C. Phys. Rev. B 1988, 37, 537.
52. Hohenberg, P. C. In Proceedings of Fluctuations in Superconductors; Goree, W. S.; Chilton, F., Eds.; Stanford Research Institute, Stanford, 1968.
53. Meyer, H. M.; Wagener, T. J.; Hill, D. M.;Gao, Y.; Weaver, J. H.; Capone, D. W.; Goretta, K. C. Phys. Rev. B 1988, 38, xxx.
54. Verhoven, J. D.; Bevolo, A. J.; McCullum, R. W.; Gibson, E. D.; Noack, M. A. Appl. Phys. Lett. 1987, 52, 745.
55. Yan, M. F.; Barnes, R. L.; O'Bryan, Jr., H. M.; Gallagher, P. K.; Sherwood, P. K.; Jim, S. Appl. Phys. Lett. 1987, 51, 532.
56. Qui, S. L.; Ruckman, M. W.; Brookes, N. B.; Johnson, P. D.; Chen, J.; Lin, C. L.; Strongin, M.; Sinkovic, B.; Crow, J. E.; Jee, Chou-Soo Phys. Rev. B 1988, 37, 3747.
57. Kurtz, R. L.; Stockbauer, R. A.; Madey,T. E.; Mueller, D.; Shih, A.; Toth, L. Phys. Rev. B 1988, 37, 7936.
58. Chang, Y.; Onellion, M.; Niles,D. W.; Margaritondo, G. Phys. Rev. B 1987, 36, 3986.
59. Rosenberg, R. A.; Wen, C.-R. Phys. Rev. B 1988, 37, 5841.
60. Weaver, J. H.; Meyer, H. M.; Wagener, T. J.; Hill, D. M.; Peterson, D.; Fisk, Z.; Arko, A. J. Phys Rev. B (in press).

61. Gao, Y.; Wagener, T. J.; Hill, D. M.; Meyer, H. M.; Weaver, J. H.; Arko, A. J.; Flandermeyer, B.; Capone, D. W. In Chemistry of High-Temperature Superconductors; American Chemical Society Symposium Series 351, Nelson, D. L.; Whittingham, M. Stanley; George, Thomas F., Eds.; Washington, D.C., 1987, p. 212; Gao, Y.; Meyer, H. M.; Wagener, T. J.; Hill, D. M.; Anderson, S. G.; Weaver, J. H.; Flandermeyer, B.; Capone, D. W. In Thin Film Processing and Characterization of High-Temperature Superconductors; American Institute of Physics Conference Proceedings No. 165, Harper, J. M. E.; Colton, R. J.; Feldman, Leonard C.; Eds.: New York, NY, 1987; p. 358.
62. Wagener, T. J.; Vitomirov, I. M.; Aldao, C. M.; Joyce, J. J.; Capasso, C.; Weaver, J. H.; Capone, D. W. Pnys. Rev. B 1988, 38, xxx; Meyer, H. M.; Wagener, T. J.; Hill, D. M.; Gao, Y.; Anderson, S. G.; Krhan, S. D.; Weaver, J. H.; Flandermeyer, B.; Capone, D. W. Appl. Phys. Lett. 1987, 51, 1118.
63. See the chapter in this book by Meyer et al.
64. Sleight, A. W.; Gillson, J. L.; Bierstedt, P. E. Solid State Commun. 1975, 17, 299.
65. Longo, J. M.; Raccah, P. M. J. Solid State Chem. 1973 6, 526.
66. Johnston, D. C.; Stokes, J. P.; Goshorn, D. P.; Lewandowski, J. T. Phys. Rev. B 1987, 36, 4007.
67. Mitsuda, S.; Shirane, G.; Sinha, S. K.; Johnston, D. C.; Alvarez, M. S.; Vaknin, D.; Moncton, D. E. Phys. Rev. B, 1987 36, 822.
68. Grant, P. M.; Parkin, S. S. P.; Lee, V. Y.; Engler, E. M.; Ramirez, M. L.; Vazquez, J. E.; Lim, G.; Jacowitz, R. D.; Greene, R. L. Phys. Rev. Lett. 1987, 58, 2482.
69. DiCarlo, J.; Niu, C. M.; Dwight, K.; Wold, A. (see this volume).
70. Shaplygin, I. S.; Kakhan, B. G.; Lazareo, Russ. V. B. J. Inorg. Chem. 1979, 24, 820.
71. Gopalaknshnan, J.; Subramanian, M. A.; Sleight, A. W. to be published.
72. Davison, S.; Smith, K.; Zhang, Y.-C.; Liu, J-H.; Kershaw, R.; Dwight, K.; Reiger, P. H.; Wold, A. ACS Symposium Series No. 351, 1987, 65.
73. Subramanian, M. A.; Gopalaknshnan, J.; Torardi, C. C. Askew, T. R.; Flippen, R. B.; Sleight, A. W. Science, submitted.
74. Gallagher, P. K.; O'Bryan, H. M.; Sunshine, S. A.; Murphy, D. W. Mat. Res. Bull. 1987, 22, 995.
75. Maeda, H.; Tanaka, Y.; Fukutomi, M.; and Asano, T. Jpn. J. Appl. Phys. 1988, 27, L209.
76. Subramanian, M. A.; Torardi, C. C.; Calabrese, J. C.; Gopalaknshnan, J.; Morrissey, K. J.; Askew, J. R.; Flippen, R. B.; Chowdry, U.; Sleight, A. W. Science, 1988, 239, 1015.
77. Tarascon, J. M.; LaPage, Y.; Barboux, P.; Bagley, B. G.; Greene, L. H.; McKinnon, W. R.; Hull, G. W.; Giroud, M.; Hwang, D. M. Phys. Rev. B communicated.
78. Sunshine, S. A.; Siegrist, T.; Schneemeyer, L. F.; Murphy, D. W.; Cava, R. J.; Batloggg, B.; van Dover, R.B.; Fleming, R. M.; Olanim, S. H.; Nakahara, S.; Farrow, R.; Krajewski, J. J.; Zahurak, S. M.; Wasczak, J. V.; Marshall, J. H.; Marsh, P.; Rupp, Jr., L. W.; Peck, W. F. Phys Rev. Lett. communicated.

79. Sheng, Z. Z.; Hermann, A. M. Nature 1988, 332, 55; 1988, 332, 138.
80. Hazen, R. M.; Finger, L. W.; Angel, R. J.; Prewitt, C. T.; Ross, N. L.; Hadidiacos, C. G.; Heaney, P. J.; Veblen, D. R.; Sheng, Z. Z.; El Ali, A.; Hermann, A. M. Phys. Rev. Lett. submitted.
81. Torardi, C. C.; Subramanian, M. A.; Calabrese, J. C.; Gopalaknshnan, J.; McCarron, E. M.; Morrissey, K. J.; Askew, T.R.; Flippen, R. B.; Chowdry, U.; Sleight, A. W. Phys. Rev. B submitted.
82. Torardi, C. C.; Subramanian, M. A. Calbabrese; Gopalaknshnan, J.; Morrissey, K. J.; Askew, T. R.; Flippen, R. P.; Chowdry, U.; Sleight, A. W. Science submitted.

RECEIVED July 6, 1988

THEORY

Chapter 2

Analysis of Thermodynamic and Transport Properties of $La_{2-x}M_xCuO_4$ and $YBa_2Cu_3O_{7-\delta}$ Superconductors

A. Langner, D. Sahu, and Thomas F. George

Departments of Chemistry and Physics, 239 Fronczak Hall, State University of New York at Buffalo, Buffalo, NY 14260

Anisotropic Ginzburg-Landau theory for coupled s-wave and d-wave order parameters is used to analyze the unique thermodynamic and transport properties of the new $La_{2-x}(Ba,Sr)_xCuO_4$ and $YBa_2Cu_3O_{7-\delta}$ superconductors. This simple phenomenological approach is used to explain the prevalence of the large Sommerfeld coefficients of the specific heat, the existence of multiple specific heat anomalies, the ultrasonic attenuation peak, and model the anisotropic critical field data as observed in oriented samples.

Following the discovery by Bednorz and Müller [1] of "high-temperature" superconductivity in the rare-earth copper oxides, there have been numerous investigations of the anisotropic electronic [2,3] and magnetic [3-7] properties of these materials. It is now well recognized that any successful theory of superconductivity for the high-T_c oxides must include the quasi-two-dimensional nature of the Cu-O planes; the theory must provide, in addition, for a coupling between the planes [8,9]. One of the best known theories of the new superconductors is the resonating-valence-bond (RVB) model of Anderson [10] which describes the onset of superconductivity as a Bose condensation of quasi-particle pairs within a large-U Hubbard model. It has been shown by Kotliar [11] and Inui, et al [12] that the superconducting order parameter of this model possesses s-wave and d-wave components, the latter being favored at large U and near half-filling. At low temperatures the mixed (s+d)-state is favored, similar to that found in the heavy-fermion superconductor $U_{1-x}Th_xBe_{13}$ [13-15]. It is interesting to note that the low-temperature behavior of the penetration depth, $\lambda(T)$ [16], the large Sommerfeld coefficients of the specific heat, γ [17,18], the enhancement of the sound velocity and ultrasonic attenuation [19,20], and the thermopowers [17] of the $La_{2-x}(Sr,Ba)_xCuO_4$ (called 214) and $YBa_2Cu_3O_{7-\delta}$ (called 123)

0097-6156/88/0377-0018$06.00/0

materials are very similar to the heavy-fermion systems. This leads us to believe, as has been suggested on the basis of high-resolution X-ray scattering experiments, [21] that s- and d-wave coupling may exist in the high-T_c superconductors.

Model

In this work we apply anisotropic Ginzburg-Landau (GL) theory [22], previously extended by us to include coupled s-wave and d-wave superconducting order parameters [23], to qualitatively analyze the single-crystal and oriented-film data on the 214- and 123-materials. In particular we think that the large Sommerfeld coefficients γ = 5 mJ/mol K^2 [4,24,25] and 9 mJ/mol K^2 [18,20] for the 40 K and 90 K superconductors, respectively, the anomalous peak in the ultrasonic attenuation at $T \sim 0.9\, T_c$ [19,20], the upturn in the $Hc_2(T)$ curve [6,7], and the anisotropy in the magnetic properties of these materials can be explained in the context of coupled (s+d)-wave states. A brief investigation of the (s+d)-wave state on a square lattice has been reported previously [26] and will be compared with the full three-dimensional results. We are aware that the limitations on any mean-field-theory description of the high-T_c materials, namely the Brout condition, due to critical fluctuations is very restrictive [27]; however, the qualitative agreement of the GL thoery with experiment deserves mention.

As is done in the GL-theory for a single even-parity order parameter, we write the free energy density difference between the superconducting state and the normal state as an expansion in even powers of the complex gap function $\Delta(\vec{k})$, which is related to the anomalous thermal average $<c_{\vec{k}\uparrow} c_{-\vec{k}\downarrow}>$ of the microscopic theory [28], where $c_{\vec{k}\uparrow}$ is the electron annihilation operator with wave vector $\vec{k}$ and spin $\uparrow$. However, for the multiple-order parameter case we must expand $\Delta(\vec{k})$ as a linear combination of the angular momentum basis functions $\{Y_j(\hat{k})\}$,

$$\Delta(\vec{k}) = \sum_{j=0}^{2} \eta_j(k)\, Y_j(\hat{k}) = \sum_{j=0}^{2} \Delta_j(k)\, \exp(i\theta_j) Y_j(\hat{k}) \tag{1}$$

where Y_0, Y_1 and Y_2 are analogous to the s, $d_{x^2-y^2}$ and d_{z^2} atomic orbitals. Y_0 and Y_2 both belong to the A_{1g} irreducible representations of the D_{4h} (tetragonal) and the D_{2h} (orthorhombic) point groups, while Y_1 degenerates from a B_{1g} to an A_{1g} representation in going over from D_{4h}- to D_{2h}-symmetry. The consequence of this is to induce some low-angular-momentum s-$d_{x^2-y^2}$-coupling as described below. Generating the invariant terms of the free-energy density, as previously described [29], we can write the free-energy difference between the superconducting and normal state for a tetragonal lattice as

$$F_s - F_n = \int d^3r\, [\, \mathcal{F}_{sq} + \mathcal{F}_T + \mathcal{F}_{GS} + \mathcal{F}_{GT} + b^2/(8/\pi)] \tag{2a}$$

$$L\cdot\, \mathcal{F}_{sq} = \sum_{j=0}^{1} (\alpha_j \Delta_j^2 + \beta_j \Delta_j^4) + \Delta_0^2 \Delta_1^2\, (\gamma_1 + \delta_1 \cos 2\theta_1) \tag{2b}$$

$$\mathcal{F}_T = \alpha_2\Delta_2^2 + \beta_2\Delta_2^4 + \Delta_0^2\Delta_2^2(\gamma_2+\delta_2\cos 2\theta_2)$$
$$+ \Delta_0\Delta_2\cos\theta_2\ (\lambda_2 + \mu_{20}\Delta_0^2 + \mu_{22}\Delta_2^2) \qquad (2c)$$

$$\mathcal{F}_{GS} = \sum_{j=0}^{1} |\alpha_j|\xi_{jp}^2[|D_x\eta_j|^2 + |D_y\eta_j|^2]$$
$$+ M_{01}[(D_x\eta_0)(D_x\eta_1)^* - (D_y\eta_0)(D_y\eta_1)^* + cc] \qquad (2d)$$

$$\mathcal{F}_{GT} = \sum_{j=0}^{2} |\alpha_j|\xi_{jz}^2|D_z\eta_j|^2 + |\alpha_2|\xi_{2p}^2[|D_x\eta_2|^2 + |D_y\eta_2|^2]$$
$$+ \sum_{j=0}^{1} M_{j2}[(D_x\eta_j)(D_x\eta_2)^* + (-1)^j(D_y\eta_j)(D_y\eta_2)^* + cc]$$
$$+ M_z[(D_z\eta_0)(D_z\eta_2)^* + cc] \qquad (2e)$$

Here we define the coherence lengths, $\xi_{j\ell}$, as $\xi_{j\ell}^2 = \hbar^2/[2m_{j\ell}|\alpha_j|]$, $\alpha_j = A_j(T-T_j)$, where j refers to the species and ℓ the orientation (p refers to the xy-plane), as is done in GL-theory for axial symmetry. The gauge-invariant differential operators are defined as $D_\zeta = (\partial/\partial\zeta - i\phi_{inv}A_\zeta)$ $(\zeta = x,y,z)$, with vector potential $\vec{A}$ and $\phi_{inv} = 2\pi/\phi_0$, $\phi_0 = hc/(2e)$ being the flux quantum. The coupling terms in the gradient expressions are characterized by reciprocal effective masses, $M_{ij} = \hbar^2/4m_{ij}$. The phase angles θ_1 and θ_2 are taken relative to θ_0, the phase of η_0, thus ensuring the gauge invariance of Eq. (2). We use $b^2/8\pi$ to represent the internal magnetic field energy density.

Equation (2) has been subdivided into terms arising from a two-dimensional analysis of the square xy-planes, $\mathcal{F}_{sq}$ and $\mathcal{F}_{GS}$, and the additional terms required to analyze tetragonal systems, $\mathcal{F}_T$ and $\mathcal{F}_{GT}$. Reduction of the symmetry to an orthorhombic point group adds an additional term of the form $\Delta_0\Delta_1\cos\theta_1$ to Eq. (2c) and destroys the axial symmetry of the gradient terms. In light of the smallness of the orthorhombic distortion and prevalence of twinning in the copper oxide superconductors [30], we assume the $\Delta_0\Delta_1$ term to act as a perturbation on the free energy of the tetragonal lattice and ignore the effect of reduced symmetry on the gradient terms. It is interesting to note that the two d-wave states above do not couple directly to each other up to order $\ell = 2$ of the relative orbital angular momentum of the Cooper pairs.

Results

We have performed a full minimization of the free energy with respect to the Δ_i's, θ_i's and the vector potential $\vec{A}$ to obtain a self-consistent picture of the thermodynamics and spatial variation of the order parameters which reproduces the dominant features of the single-crystal data of the high-T_c oxides. Even though many parameters appear in Eq. (2), we understand the basic physics in simple qualitative terms. The simplest scenario is that of the coexistence of a highly anisotropic $d_{x^2-y^2}$-state, Δ_1, responsible for the quasi-two-dimensional character of these materials, with a nearly isotropic, mixed ($s+d_{z^2}$)-state, possibly characterizing the "holon"-pair hopping within the RVB picture [31]. As determined by Kotliar [11], the transition temperature, T_1, of the d-state is higher than that of the mixed state. A schematic picture of the relative magnitudes of the order parameters is given in Fig. 1. The relative phases are $\theta_1 = \pi/2$ and $\theta_2 = \pi$ near the transition temperatures. The small amount of $\Delta = \Delta_0 + \Delta_2$ state persisting above the onset temperature, $\tilde{T}$, is a consequence of the small perturbation to Eq. (2) caused by a shift from tetragonal to orthorhombic symmetry. Perhaps in a naive way, this may be viewed as adding the three-dimensional character necessary for the onset of superconductivity [9]. The existence of d-wave states, consequently gapless superconductivity, would explain the large observed Sommerfeld coefficients, while the multiple transitions of these states would explain the two specific heat anomalies observed near T_c [32,33].

We feel that the peak in the ultrasound attenuation results from the oscillations of the relative phases θ_1 and θ_2 about their equilibrium values $\theta_1^0 = \pi/2$ and $\theta_2^0 = \pi$, as suggested by Kumar and Wolfle [13] in a different context. Defining $\omega_j^2 = \partial^2 \mathfrak{F}_L/\partial\theta_j^2$ ($j = 1,2$), where $\mathfrak{F}_L = \mathfrak{F}_{sq} + \mathfrak{F}_T$, the oscillation frequencies are given by

$$\omega_1^2 = 4\Delta_0^2\Delta_1^2\delta_1 \tag{3a}$$

and

$$\omega_2^2 = \Delta_0\Delta_2(\lambda_2 - 8\Delta_0\Delta_2\delta_2 + \mu_{20}\Delta_0^2 + \mu_{22}\Delta_2^2) \tag{3b}$$

There will be a sharp onset of these oscillations at $\tilde{T}$ which will correspond to the attenuation peak at $T = 0.9\ T_c$.

We next consider the variation of the upper critical field, H_{c2} with orientation and temperature. Using a straightforward variational approach on the linearized form of Eq. (2), we have derived the differential GL equations, the full details of which will be presented elsewhere. For the sake of simplicity we assume a ($s+d_{z^2}$)-wave mixed state with $\Delta_0 = \Delta_2 = \Delta_m$ and $\xi_{0p} = \xi_{2p} = \xi_m$ and write differential equations for fields parallel, $H^{\parallel}$, and perpendicular, $H^{\perp}$ to the xy-plane. For $H^{\parallel} = (H,0,0)$ and $\vec{A} = (0,-zH,0]$, we have,

$$\Delta_m - (\xi_m\phi_{inv}Hz)^2\ \Delta_m + \xi_m^2(d^2\Delta_m/dz^2) = 0 \tag{4a}$$

$$\Delta_1 - (\xi_{1p}\phi_{inv}Hz)^2\ \Delta_1 + \xi_{1z}^2(d^2\Delta_1/dz^2) = 0 \tag{4b}$$

Similarly, for $H^{\perp} = (0,0,H)$ and $\vec{A} = (0,xH,0)$, we have

$$(\alpha_m-\lambda_2)\Delta_m - (\alpha_m\xi_m^2+2M_{02})(\phi_{inv}Hx)^2\Delta_m + (\alpha_m\xi_m^2+2M_{01})(d^2\Delta_m/dx^2) = 0, \tag{4c}$$

$$\Delta_1 - (\xi_{1p}\phi_{inv}Hx)^2\,\Delta_1 + \xi_{1p}^2(d^2\Delta_1/dx^2) = 0 \qquad (4d)$$

These equations are decoupled and can readily be solved for H_{c2} within the harmonic oscillator approximation to yield

$$H_{c2}^{\parallel} = (\phi_{inv}\xi_{1p}\xi_{1z})^{-1}\ , \quad H_{c2}^{\perp} = (\phi_{inv}\xi_{1p}^2)^{-1}$$

$$H_{c2} = (1 - \lambda_2/\alpha_m)[\phi_{inv}(\xi_m^2 + 2M_{02}/\alpha_m)]^{-1} \qquad (5)$$

Figure 2 gives the variation of the critical fields with temperature for $\xi_{1z} < \xi_m < \xi_{1p}$. For $H^{\parallel}$ the upper critical field is always determined by the smallest coherence length $\xi_{1z}(0\ K) \sim 7$Å. For $H^{\perp}$ the upper critical field becomes the larger between $H_{c2}^{\perp}$ and H_{c2} as given above. This may explain the discrepancy in the reported 0 K values of the in-plane coherence length ($\xi_{1p}(0) \sim 34$Å, $\xi_m(0) \sim 22$Å), as well as the kink in the $H_{c2}^{\perp}$ data.

The variation of the lower critical field, H_{c1}, with orientation and temperture for the mixed state can be approximated by the expression $H_{c1} = (\Phi_0/4\pi\,\lambda_{eff}^2)\ell n\,(\kappa_{eff})$ [34], which is valid for large values of the GL-parameter, $\kappa_{eff} = \lambda_{eff}/\xi_{eff}$. For this case the variation of the internal field occurs mainly in a region where the order parameters exhibit their maximum values. One can therefore obtain the penetration depth, λ_{eff}, by casting the current relations into the form of the London equation, $\vec{\nabla} \times \vec{b} = -\lambda_{eff}^{-2}\vec{A}$. The results for $H_{c1}^{\parallel}$ and $H_{c1}^{\perp}$ are,

$$H_{c1}^{\parallel}:\quad \lambda_{\parallel}^{-2} = 2\lambda_m^{-2} + \lambda_{1z}^{-2} + \lambda_z^{-2} \qquad (6a)$$

$$H_{c1}^{\perp}:\quad \lambda_{\perp}^{-2} = 2\lambda_m^{-2} + \lambda_{1p}^{-2} + \lambda_p^{-2} \qquad (6b)$$

where the same assumptions on Δ_0 and Δ_2 were made as for the calculation of H_{c2}. At tempertaures near $T_c = T_1$ the lower critical field should behave as λ_1^{-2} since it is proportional to the square of the order parameter. Consequently the anisotropy of H_{c1} should go as the square of the anisotropy of H_{c2}. At lower temperatures the influence of the coupling terms λ_p^{-2} and λ_z^{-2} makes predictions more difficult. The anticipated behavior of H_{c1} for several values of the coupling terms is given in Fig. 3. We are at present not aware of any single-crystal H_{c1} studies over the entire temperature range 0 - T_c.

Summary

We have analyzed the thermodynamic, magnetic and ultrasound attenuation data on oriented samples of the high-T_c superconductors within the context of anisotropic Ginzburg-Landau theory for coupled, even-parity superconducting states. We are able to present a consistent interpretation of the data in terms of the coexistence of a quasi-two-dimensional d-wave state, with critical temperature $T_1 = T_c$ and a more isotropic mixed (s+d)-wave state with critical tempertaure $T_m < T_c$. We predict the possibility of a "kink" in the temperature dependence of the lower critical field near $0{\cdot}9T_c$, which should be tested by experiments on single crystals.

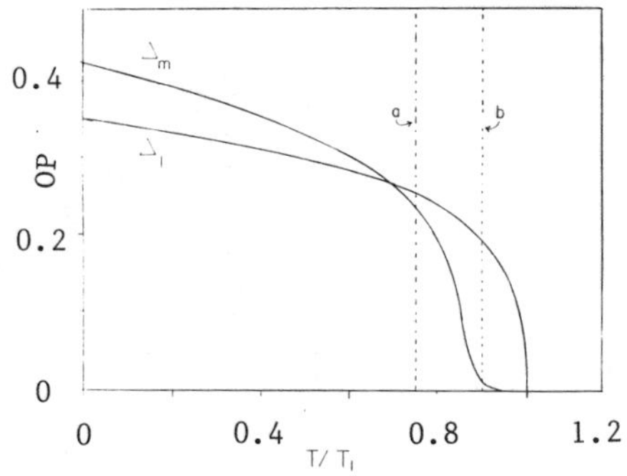

Figure 1. Schematic temperature dependence of the superconducting order parameters, where Δ_m is for the mixed (s+d)-state and Δ_1 for the pure $d_{x^2-y^2}$-state. T_1 and T = a = T_m are the critical temperatures of the mixed and pure states, respectively, and T = b = $\tilde{T}$ is the onset temperature.

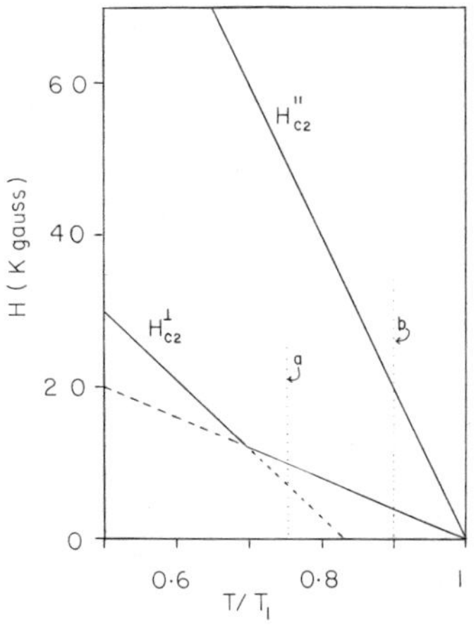

Figure 2. Schematic temperature dependence of the upper critical field, H_{c2}. The dashed curves are not experimentally observable. $H^{\parallel}_{c2}$ is the field parallel to the ab-plane, and $H^{\perp}_{c2}$ is the field parallel to c-axis. T = a = T_m and T = b = $\tilde{T}$.

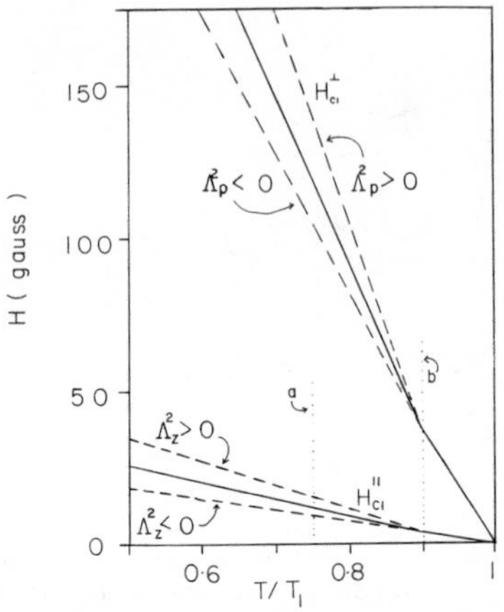

Figure 3. Schematic temperature dependence of the lower critical field, H_{c1}. The dashed curves represent the effect of the coupling terms λ_p^{-2} and λ_z^{-2}. $T = a = T_m$ and $T = b = \tilde{T}$.

Acknowledgments

This research was supported by the Office of Naval Research.

Literature Cited

1. Bednorz, J. G.; Müller, K. A. Z. Phys. B 1986, 64, 188.
2. Tozer, S. W.; Kleinsasser, A. W.; Penney, T.; Kaiser, D.; Holtzberg, F. Phys. Rev. Lett. 1987, 59, 1768.
3. Dinger, T. R.; Worthington, T. K.; Gallagher, W. J.; Sandstrom, R. L. Phys. Rev. Lett. 1987, 58, 2687.
4. Batlogg, B.; Ramirez, A. P.; Cava, R. J.; von Dover, R. B.; Rietman, E. A. Phys. Rev. B 1987, 35, 5340.
5. Hidaka, Y.; Enomoto, Y.; Sukuki, M.; Oda, M.; Murakami, T. Jpn. J. Appl. Phys. 1987, 26, L377.
6. Worthington, T. K.; Gallagher, W. J.; Dinger, T. R. Phys. Rev. Lett. 1987, 59, 1160.
7. Moodera, J. S.; Meservey, R.; Tkaczyk, J. E.; Hao, C. X.; Gibson, G. A.; Tedrow, P. W. Phys. Rev. B 1988, 37, 619.
8. Gulácsi, Zs.; Gulácsi, M.; Pop, I. Phys. Rev. B 1988, 37, 2247.
9. Wen, X.-G.; Kan, R. Phys. Rev. B 1988, 37, 595.
10. Anderson, P. W. Science 1987, 235, 1196.
11. Kotliar, G Phys. Rev. B 1987, 37, 3664.
12. Inui, M.; Doniach, S.; Hirschfeld, P. J.; Ruckenstein, A. E. Phys. Rev. B 1988, 37, 2320.
13. Kumar, P.; Wolfle, P. Phys. Rev. Lett. 1987, 59, 1954.

14. Rauchschwalbe, U.; Steglich, F.; Stewart, G. R.; Giorgi, A. L.; Fulde, P.; Maki, K. Europhys. Lett. 1987, 3, 751.
15. Rauchschwalbe, U.; Bredl, C. D.; Steglich, F.; Maki, K.; Fulde, P. Europhys. Lett. 1987, 3, 757.
16. Cooper, J. R.; Chu, C. T.; Zhou, L. W.; Dunn, B.; Grüner, G. Phys. Rev. B 1988, 37, 638.
17. Cheong, S. W.; Brown, S. E.; Fisk, Z.; Kwok, R. S.; Thompson, J. D.; Zirngiebl, E.; Grüner, G.; Peterson, D. E.; Wells, G. L.; Schwarz, R. B.; Cooper, J. R. Phys. Rev. B 1987, 36, 3913.
18. von Molnár, S.; Torressen, A.; Kaiser, D.; Holtzberg, F.; Penney, T. Phys. Rev. B 1988, 37, 3762.
19. Xu, M-F.; Baum, H-P.; Schenstrom, A.; Sarma, B. K.; Levy, M.; Sun, K. J.; Toth, L. E.; Wolf, S. A.; Gubser, D. U. Phys. Rev. B 1988, 37, 3675.
20. Bhattacharya, S.; Higgins, M. J.; Johnston, D. C.; Jacobson, A. J.; Stokes, J. P.; Lewandowski, J. T.; Goshorn, D. P. Phys. Rev. B 1988, 37, 5901.
21. Horn, P. M.; Keane, D. T.; Held, G. A.; Jordan-Sweet, J. L.; Kaiser, D. L.; Holtzberg, F.; Rice, T. M. Phys. Rev. Lett. 1987, 59, 2772.
22. Morris, R. C.; Coleman, R. V.; Bhandari, R. Phys. Rev. B 1972, 8, 895. Spatial and not $\vec{k}$-space anisotropy is implied here.
23. Sahu, D.; Langner, A.; George, T. F. unpublished.
24. Ferreira, J. M.; Lee, B. W.; Dalichaouch, Y.; Torikachvili, M. S.; Yang, K. N.; Maple, M. B. Phys. Rev. B 1988, 37, 1580.
25. Kumagai, K.; Nakamichi, Y.; Watanabe, I.; Nakamura, Y.; Nakajima, H.; Wada, N.; Lederer, P. Phys. Rev. Lett. 1988, 60, 724.
26. Langner, A.; Sahu, D.; George, T. F. Proceedings of the Conference on Superconductivity and Applications, ed. by H. S. Kwok (Elsevier, New York, 1988).
27. Kapitulnik, A.; Beasley, M. R.; Castellani, C.; DiCastro, C. Phys. Rev.B 1988, 37, 537.
28. Baskaran, G.; Zou, Z.; Anderson, P. W. Solid State Commun. 1987, 63, 973.
29. The term $\Delta_0\Delta_2\cos\theta_2[\lambda_2 + \mu_2(\Delta_0^2+\Delta_2^2)]$ of Ref. [23] is too restrictive since Δ_2^2 and Δ_0^2 need not have the same coefficients to be invariant terms.
30. You, H; Axe, J. D.; Kan, X. B.; Moss, S. C.; Lin, J. Z.; Lam, D. J. Phys. Rev. B 1988, 37, 2361.
31. Wheatly, J. W.; Hsu, T. C.; Anderson, P. W. Phys. Rev. B 1988, 37, 5897.
32. Inderhees, S. E.; Salamon, M. B.; Goldenfeld, N.; Rice, J. P.; Pazol, B. G.; Ginsberg, D. M.; Liu, J. Z.; Crabtree, G. W. Phys. Rev. Lett. 1988, 60, 1178.
33. Butera, R. A. Phys. Rev. B 1988, 37, 5909.
34. Fetter, A. L.; Hohenberg, P. C. In Superconductivity; Parks, R. D., Ed.; Marcel Dekker, New York, 1969; p. 817 ff.

RECEIVED July 13, 1988

Chapter 3

Electronic Structure and High Critical Temperature in Oxide Superconductors

B. M. Klein[1], W. E. Pickett[1], R. E. Cohen[1], H. Krakauer[2], D. A. Papaconstantopoulos[1], P. B. Allen[1,3], and L. L. Boyer[1]

[1]Condensed Matter Physics Branch, Naval Research Laboratory, Washington, DC 20375–5000

[2]Department of Physics, College of William and Mary, Williamsburg, VA 23185

Results of studies of the electronic structure of several of the new high T_c oxides are presented. These are determined within the framework of the local density approximation for exchange and correlation. By comparing with results of the *ab initio* model of potential-induced-breathing (which uses overlapping ionic charge densities) the ionicity of the components of these compounds is determined. Specifically, we argue against an interpretation of Cu^{2+} in the "planes" and Cu^{3+} in the "chains" in the 1-2-3 materials. We find, in fact, all Cu sites are best described as being closer to the 2+ valence state consistent with the O ions being close to a fully 2- state. Calculations of the electron-phonon interaction using "standard" approaches yields values of T_c and resistivity significantly too small compared with experiment, but we argue that "standard" approximations are no longer valid in these ionic metals. We emphasize two spectacular successes of electronic structure theory: (1) the predicted anisotropies and carrier signs of the Hall and thermopower tensors are in full agreement with single-crystal data; and (2) we show that the temperature dependence of the resistivity, in particular the linear dependence of resistivity on temperature well below the Debye temperature, which has been termed termed "anomalous" by many, arises in a straightforward manner from standard electron-phonon theory.

[3]Permanent address: Department of Physics, SUNY Stony Brook, Stony Brook, NY 11794

0097–6156/88/0377–0026$06.00/0

The excitement and mystery of attempting to generate a comprehensive explanation for high-temperature superconductivity (1,2) in the ceramic oxides is nearly indescribable, but we will try anyhow, at least from the perspective of a group experienced in the fields of electronic structure and superconductivity theory. The wealth of new physics uncovered experimentally presents an unprecedented challenge to theoreticians, with many new ideas and "exotic" approaches and mechanisms having been elucidated in a relatively short time. To do justice to all of the theoretical research in a short article is impossible, so we will be content with emphasizing and explaining the status of results and interpretations of electronic structure theory, in particular those that follow from applications of density-functional theory-based approaches. A number of successes of this theory will be emphasized, and a number of limitations that have been uncovered will be enumerated, too. The present status is, we believe, that "school is still out" on whether these approaches and the generalized electron-phonon scattering mechanism, for which we have generated a number of results, are key ingredients to the high T_c in these materials. The recent discovery of T_c of nearly 30 K in a perovskite of Bi-K-Ba-O (3), containing no Cu, has weakened the argument for an exclusively magnetic mechanism. Our view is that the combination of ionic and metallic behavior is very important in these materials, leading to the possibility that modifications to "traditional" approaches are needed to explain the chemistry and physics in the oxides. Time, and further experiments and computational results, will tell.

First-Principles Electronic Structure Results

Since the bulk of our discussions will involve the La_2CuO_4 and $YBa_2Cu_3O_7$ classes of compounds, we will often use the notation 2-1-4 and 1-2-3, respectively, for these materials. The crystal structures (4,5) are shown in Figures 1 and 2. The first-principles electronic structure method used in the NRL calculations of the high-T_c oxides is the highly accurate linearized-augmented-plane-wave (LAPW) method as implemented by Wei and Krakauer (6). This all-electron method uses the local (spin) density approximation L(S)DA for exchange and correlation which, together with the Hohenberg-Kohn-Sham formalism [reviewed recently in (7)], leads to a solution of the electronic structure problem in terms of single-particle eigenvalues(functions). This approach, and other equivalent implementations of LSDA theory (e.g. all-electron or pseudopotential methods), have given good descriptions of a number of ground state properties of condensed matter systems, especially for non-magnetic states. These include structural ground state properties (lattice constant and internal coordinates), elastic constants, cohesive energies, and phase transitions, including superconductivity. Even excited state properties (such as transport) are often successfully described. For magnetic systems, recent work has shown that the LSDA method has some quantitative inaccuracies, predicting the incorrect ground state structure of Fe, for instance, as shown by Wang, *et al.* (8). Even before the oxide superconductors were discovered, it was known that improvements beyond the LSDA were needed for magnetic systems. The apparent failure of LSDA in predicting the magnetic structures of the oxides is further impetus for seeking such improvements, although the role of oxygen vacancies in determining the magnetic behavior in these materials is not yet clear.

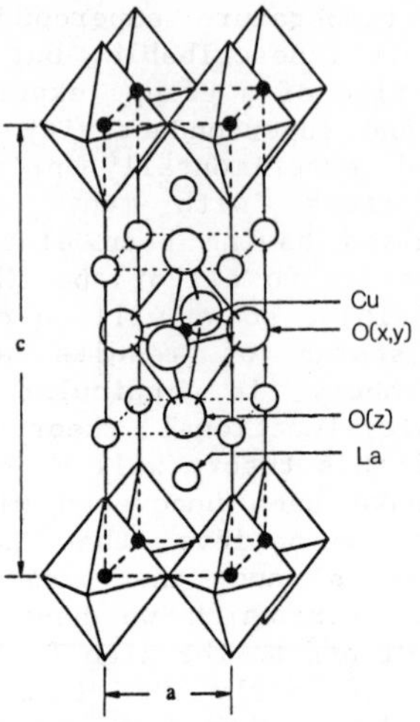

Figure 1. Atomic structure of body-centered tetragonal La_2CuO_4 with inequivalent sites labelled.

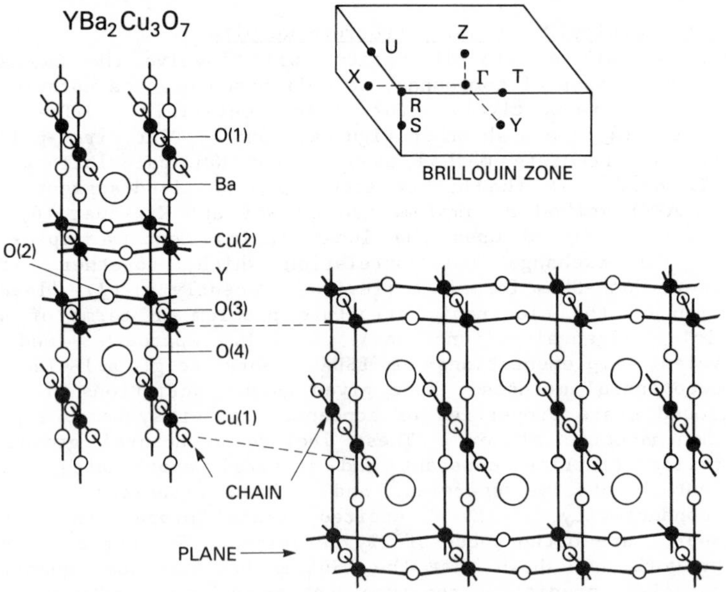

Figure 2. Atomic structure of orthorhombic Y-Ba-Cu-O from (5) with inequivalent sites labelled. Solid bars denote Cu-O coordinates and illustrate the plane and chain configurations.

La_2CuO_4. This material has the I4/mmm body-centered tetragonal space group with a single formula unit in the primitive cell shown in Figure 1. There are two equivalent out-of-(Cu)plane O_z atoms, and two equivalent in-(Cu)plane O_{xy} atoms. There are a total of four inequivalent sites in the unit cell. The La 2-1-4 compound distorts to an orthorhombic structure, arrived at by the tilting of the Cu-O octahedra along a <110> direction (9) in a classic perovskite soft mode kind of behavior. The soft (110) zone boundary "tilt" mode freezes into the lattice below about T_s= 500 K, resulting in a doubling of the primitive unit cell. The resulting lower symmetry structure is orthorhombic with space group Abma. Below the structural transition temperature, long-range antiferromagnetic order is observed (10) below about T_N = 250 K, with the moment primarily residing on the Cu site, and with a strong sensitivity to the oxygen stoichiometry. Early reports indicated O vacancies of y ~0.03 in La_2CuO_{4-y} were needed for obtaining the magnetic, insulating ground state (11). This appears not to be absolutely confirmed, and it is commonly accepted (whether true or not) that stoichiometric 2-1-4 is an antiferromagnet.

The paramagnetic energy bands of tetragonal 2-1-4, which must be metallic due to the odd number of electrons in the unit cell, are shown in Figure 3, and the density of states (DOS) is shown in Figure 4, and tabulated in Table I (12). Referring first to the tetragonal structure, we note that the Cu(d)-O(p) interactions are strong, leading to a strongly hybridized complex of bands with an approximate 10 eV band width. In most transition metals oxides the O(p) bands are split-off below the d bands, although calculations at NRL for hypothetical rocksalt structure CuO gives Cu(d) and O(p) bands centered at approximately the same energy, with strong hybridization.

Table I. Total, $N(E_F)$, and muffin-tin component densities of states (in states/eV cell). Also shown for $La_{2-x}(Sr,Ba)_xCuO_4$ is the x=0.14 values using a rigid-band model

$La_{2-x}(Sr,Ba)_xCuO_4$						
E_F	x	Electrons	$N(E_F)$	Cu	O(z)	O(x,y)
0.000	0.00	0.00	1.24	0.54	0.04	0.13
-0.062	0.14	-0.14	2.03	1.05	0.11	0.18

$YBa_2Cu_3O_7$						
$N(E_F)$	Cu(1)	Cu(2)	O(1)	O(4)	O(2)	O(3)
5.54	0.43	0.70	0.64	0.60	0.20	0.20

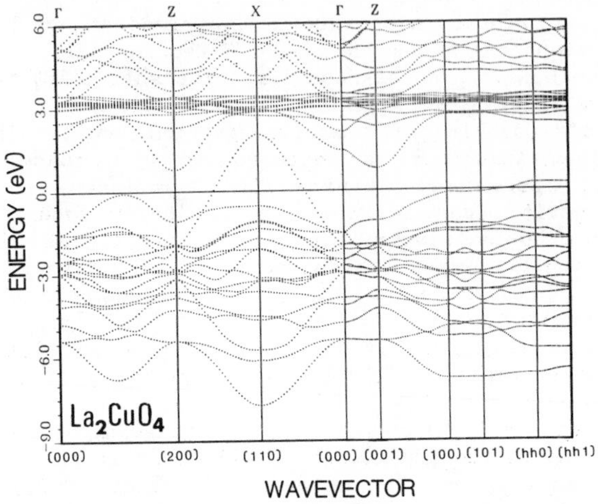

Figure 3. Energy bands of tetragonal La_2CuO_4 plotted along several directions in the Brillouin Zone. The Fermi energy is 0.0.

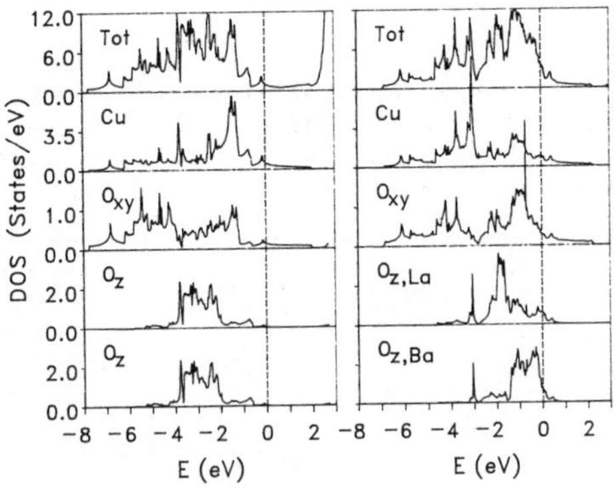

Figure 4. Densities of states (DOS) for La_2CuO_4 and $LaBaCuO_4$. The total DOS is shown in the upper panels, and the site-decomposed muffin-tin components are shown in the remaining panels. The Fermi energy is 0.0. Refer to Figure 1 for site definitions.

The Fermi level, E_F, crosses a single band with antibonding pdσ character. The highly nested Fermi surface can be characterized roughly as a square cylinder with rounded edges (13,14). Since the O_z states (see Figure 3) are completely filled, we see that the Cu(d)-O_{xy}(p) band crossing E_F is exactly half-filled for stoichiometric 2-1-4, an essential attribute for some of the "new mechanism" models being proposed for the oxides. We find that N(E_F), the DOS at E_F, is 1.24 states/eV/unit cell, a small value compared to previous "high" T_c compounds such as the A15's.

In addition, since superconductivity occurs in this material only for approximately 7% doping of divalent Sr or Ba replacing La, it is important to determine the effects of these substitutions. To ascertain the applicability of applying rigid-band types of arguments to the 2-1-4 system we have also done calculations for the stoichiometric compound $LaBaCuO_4$, and have concluded that even in this extreme doping limit there is a qualitative rigid-band behavior near the Fermi energy for these cation substitutions (12). This is important for our superconductivity arguments and transport calculations to be discussed below.

Several spin-polarized first-principles LSDA calculations have been performed for the 2-1-4 system in the tetragonal structure with conflicting conclusions (15-17), although there are indications that the system is "close" to a magnetic instability. Our preliminary results for the orthorhombic band structure shows distinct features compared with the tetragonal case which may make the true orthorhombic structure more amenable to spin splitting. This should be explored by explicit spin-polarized calculations. It may also be possible that theoretical improvements, especially to the LSDA exchange-correlation functional, may lead to a stable AF state from electronic structure calculations. However, there is also still some experimental question (11) of the role of dilute O vacancies on the magnetic properties of 2-1-4, and the antiferromagnetism in particular, and dilute O vacancies are difficult to treat theoretically. Further theoretical work in this area is going on at NRL and elsewhere.

$YBa_2Cu_3O_7$. The 1-2-3 material crystallizes in an orthorhombic structure with space group Pmmm containing one formula unit per primitive cell shown in Figure 2. This oxide exhibits both planar Cu-O bonding as in the 2-1-4 material, as well as planes containing Cu-O chains with missing oxygens along two of the four cube edges. The structure can be viewed as a defect perovskite structure with a superlattice ordering of Y and Ba along the c direction. There are no oxygen atoms in the Y layer. In the unit cell, which contains eight inequivalent sites, there is one Y and one Ba site, and two inequivalent Cu sites: Cu(1) in the chains, and Cu(2) in the planes. Above each Cu(2) there is an O(4) site (in a Ba plane); and in the Cu(2) planes there are O(2) and O(3) sites along the a and b axes, respectively. The oxygen atom site in the chains is labeled O(1). It was believed, early on, that the occurrence of Cu-O chain layers was crucial for obtaining 90 K and above transition temperatures, but the recent discovery of the new Bi and Tl layered oxides, which do not have chains, have shown that this is not likely to be the key, in general.

The energy bands of the 1-2-3 material, shown in Figure 5, are much more complicated than for the 2-1-4 compound, with a quite distinct Fermi surface as well, discussed fully in (18). There are four bands crossing E_F, two of which are chain related, and two from the Cu-O planes. These latter bands are similar to those in 2-1-4, but are not even approximately half-filled. The density of states results are shown in Figure 6, with a tabulation given in Table I. The DOS shows clear metallic character with major contributions to the Fermi level values from all of the coppers and oxygens, and very minor Y or Ba components. Further details may be found in (18). The $N(E_F)$ value is 5.54 states/eV/unit cell, much larger than the 2-1-4 compound, although the value per copper atom is similar to the 2-1-4 material. Due to the additional complexity of the 1-2-3 crystal structure, there have not been accurate LSDA calculations of the magnetic state of $YBa_2Cu_3O_6$, so that a theoretical comparison with the observed AF ground state for the O_{6+x} system, with x ~0.1 cannot be made. There are several groups attempting to perform these highly computer intensive, important calculations.

Oxides Based on Bi and Tl. Very recently, newly discovered layered ceramic oxides based on Bi (19) or Tl (20) instead of a rare earth element, have been fabricated with T_c's above 100 K (up to ~125 K for Tl). These compounds have neither rare earths nor Cu-O chains adding further evidence that we are dealing with very subtle physical/chemical effects which have eluded a simple understanding. Recently, we have performed LAPW calculations on a prototype of these new structures, $Bi_2Sr_2CaCu_2O_8$, which showed a quite distinct band structure from the previous materials (21). Most notably, there are Bi-O hybrid bands which cross E_F in addition to Cu-O bands. Some features of the bonding in these materials seem quite different from 1-2-3 or 2-1-4. It has been suggested that, qualitatively, the higher T_c's in these new materials are related to larger numbers of Cu-O planes per unit cell, the only remaining common feature in all the 90 K and above superconductors, but a real theory remains to be elucidated. Calculation of the electronic structure of the Tl compounds is currently being pursued by several of us.

Ab Initio Ionic Models. The potential-induced-breathing (PIB) model (22,23) is a generalization of the Gordon-Kim approach which uses overlapping ionic charge densities and a Thomas-Fermi model for the kinetic energy of the electrons in forming the total energy of the electron-ion system. In PIB, radial relaxation of the overlapped charge densities in response to the crystal field are allowed, adding an important improvement to the Gordon-Kim method. This approach is particularly applicable to closed shell or ionic systems, and has been highly successful in a number of applications including predicting relative phase stability, elastic behavior, and thermal equations-of-state for a number of systems, including oxides (24). Although O^{2-} is unstable in the free atom state, the doubly charged anion is stabilized in a solid by the crystal field. Although PIB neglects covalent or metallic polarization effects, it may be applicable to the "ionic" high-T_c oxides where ionic behavior is likely to be an important component of the bonding. The PIB

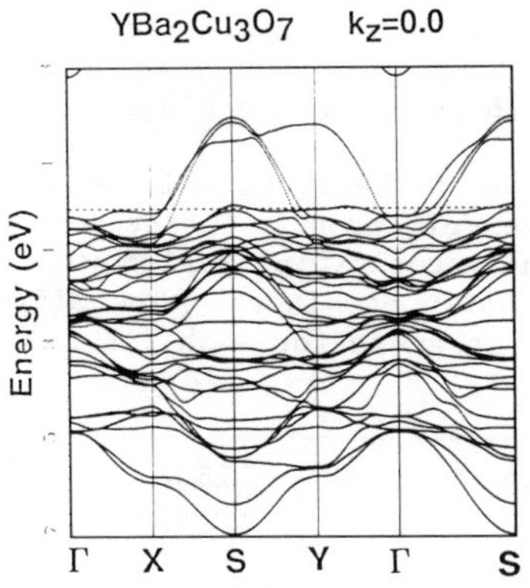

Figure 5. Energy bands of orthorhombic $YBa_2Cu_3O_7$ plotted along several directions in the Brillouin Zone. The Fermi energy is 0.0.

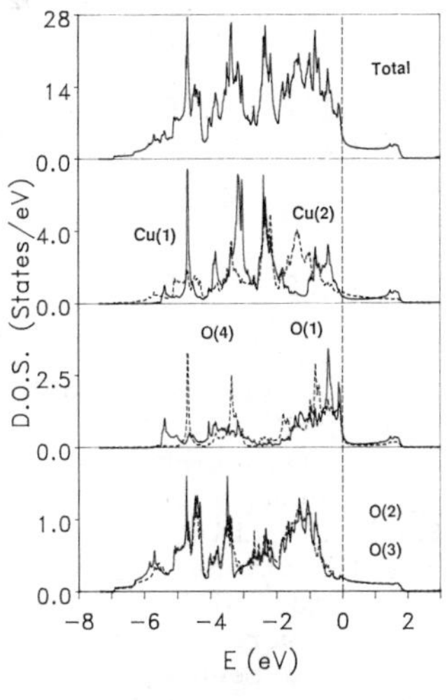

Figure 6. Densities of states (DOS) for the $YBa_2Cu_3O_7$ compound. The total DOS is in the top panel, and the site-decomposed muffin-tin components are shown in the remaining panels. The Fermi energy is 0.0. Refer to Figure 2 for site definitions. Solid lines are for the Cu(1), O(1), and O(2) sites. Dashed lines are for the Cu(2), O(4), and O(3) sites.

approach is attractive on two levels: (a) it is six orders of magnitude faster on the computer than an equivalent LAPW calculation; and (b) one can vary the ionic configurations of the overlapping charge densities to compare (see Table II) with the LAPW results and hence obtain insights into the ionicity of these materials (18,25). Both the LAPW and PIB methods use the LDA.

Table II. Comparison of the total self-consistent electronic charge (number of electrons) inside the various muffin-tin spheres and in the interstitial region with that obtained from overlapping PIB charge densities. The difference (Δ) from the LAPW results is also given for each PIB calculation

Atom	LAPW	PIB $Cu^{+2\ 1/3}$ (O^{-2})	Δ	PIB Cu^{+2} $(O^{-1.86})$	Δ	PIB $Cu^{+1.62}$ $(O^{-1.69})$	Δ
Y	37.639	37.469	0.170	37.401	0.238	37.329	0.310
Ba	53.934	53.935	-0.001	53.919	0.015	53.905	0.029
Cu(1)	27.074	26.730	0.344	26.891	0.183	27.026	0.048
Cu(2)	27.077	26.679	0.398	26.873	0.204	27.061	0.016
O(1)	6.858	6.905	-0.047	6.879	-0.021	6.852	-0.006
O(4)	6.898	6.919	-0.021	6.897	-0.100	6.875	-0.023
O(2)	6.945	7.072	-0.127	7.048	-0.103	7.022	-0.077
O(3)	6.942	7.064	-0.122	7.042	-0.100	7.002	-0.060
Interstitial:							
	18.837	19.559	-0.722	19.270	-0.433	19.029	-0.192

The first applications of PIB to the oxides was work at NRL on La_2CuO_4, where it was predicted that the Cu-O "tilt" mode, observed to be unstable in neutron diffraction experiments, was unstable in the tetragonal phase (25). Equally important was the PIB prediction that the in-plane oxygen breathing mode was stable. This is contrary to earlier calculations (26) which were based on approximations appropriate for high DOS metals --- essentially, important ionic contributions were neglected (see below).

The PIB approach has also been applied to 1-2-3 (18,25). The extant ionic configurations of Cu and O in the oxides is clearly of great importance for obtaining a theory of high T_c in the oxides. A common interpretation for 1-2-3 is that the Cu(1) chain coppers are in the 3+ state, while the Cu(2) plane coppers are in the 2+ state. Charge balance in the unit cell results if all of the oxygens are in the 2- state and Y and Ba are 3+ and 2+, respectively. Guided by the LAPW results which show that to a very good approximation Y^{3+} and Ba^{2+} are the correct configurations for these elements, and that the muffin-tin charges of Cu(1) and Cu(2) and of all the O sites are very nearly identical, there is a copper charge and an oxygen charge to choose in implementing PIB. The procedure that was used was to perform separate PIB calculations for different choices of Cu ionic charges: +3, +2 1/3, +2, and +1.62, and to choose the O charge to neutralize the unit cell in each case. These results were compared with the LAPW charge densities and, by performing DOS calculations using overlapping charge densities from the PIB picture, compared with the LAPW DOS. The fits with $Cu^{+1.62}$, and the resulting $O^{-1.69}$, give the best overall agreement with the LAPW results. Comparisons with XANES measurements, discussed fully in (18), show a consistency with our interpretation in that it is unlikely that there is any Cu^{+3}, and that the copper valence is more nearly +2. However, it must be kept in mind that PIB is a spherical model and there certainly are non-spherical components to the true LAPW charge density. There is also a tendency for the LDA to underestimate the on-site Coulomb repulsion, and additional correlation may increase the Cu valence.

Electron-Phonon Interaction

In this Section we discuss the electron-phonon interaction, and the possibility that it plays a significant role in BCS pairing in the oxide materials. Explicit calculations using a standard model have been done for the 2-1-4 and 1-2-3 materials, and are discussed in detail. We begin with some relevant background information, for which further details may be found in several fairly recent review articles (27,28).

In the strong coupling version of BCS theory, knowledge of the Eliashberg spectral function $\alpha^2F(\omega)$ determines most of the superconducting properties, including T_c. $\alpha^2F(\omega)$ is often approximated as being proportional to the phonon density of states $G(\omega)$. Experimental comparisons of $\alpha^2F(\omega)$ and $G(\omega)$ for a number of metals shows this to be a good approximation in many cases. Central to the determination of $\alpha^2F(\omega)$ is the quantity $\delta_aV(\mathbf{r})$, which is the change in crystal potential when atom a is displaced from equilibrium, all other atoms being held fixed. In principle, since the atomic displacements break periodicity, the spatial integrals of $\delta_aV(\mathbf{r})$ are over the whole crystal. However, on physical grounds, for good metals it is expected that the valence/conduction electrons will effectively screen the perturbation limiting any important contributions to regions very close to the displaced atom. This fact has been exploited in most approaches to calculations of electron-phonon scattering (see below), but as we shall show, this approximation may be qualitatively inappropriate for the oxide superconductors, where longer-range Coulomb forces are present, due to the ionic nature of the constituents.

Even in today's supercomputer generation fully first-principles calculations of $\alpha^2F(\omega)$ are rare due to the numerical complexity of having to determine all of the electronic and phononic properties and their couplings. Therefore, simplifying approximations for determining the key quantity, $\delta_aV(\mathbf{r})$, have been made which have proven to be quite accurate for several elemental transition metals, and at least adequate for a variety of superconducting compounds. One of these approaches is the rigid-muffin-tin-approximation (29)

(RMTA) whereby $\delta_a V(r)$ is replaced by the gradient of the muffin-tin potential (preferably self-consistent) obtained from an electronic structure calculation (e.g., APW, LAPW, KKR, LMTO, etc.). The muffin-tin potential is assumed to move rigidly with the atom, and $\delta_a V(r)$ is assumed to vanish outside of the muffin-tin of the displaced atom, a physically reasonable approximation for high carrier density metals, as discussed above. Tight-binding theory based approximations, as pioneered by W. Weber and co-workers (30) make other approximations which, equivalently, neglect long-range effects in $\delta_a V(r)$, and obtain very similar results.

To obtain further insights into the transition temperature in particular, McMillan showed that it is worthwhile to define the electron-phonon coupling constant λ by,

$$\lambda = 2 \int \omega^{-1} \alpha^2 F(\omega) d\omega \qquad = \sum_a \frac{\eta_a}{M_a} \langle\omega^2\rangle^{-1} \tag{1}$$

$$\eta_a = N(E_F) I_a^2 \tag{2}$$

M_a is the mass of atom a and I_a^2 the square of an electron-phonon matrix element. Expressions for T_c have been given by McMillan (31) and by Allen and Dynes (32) which are functions of λ and μ^*, a repulsive Coulomb pseudo-potential parameter usually taken as ~0.1, and moments of $\alpha^2 F(\omega)$, usually replaced in practice by moments over the phonon density of states ($\langle\omega^2\rangle$ is one such moment in Equation 1). We note the following: large η derives from either a strong electron-phonon matrix element (I^2) (strong scattering by moving atoms), or a large $N(E_F)$ (a large number of electrons available to pair, and condense), these two quantities entering as a product. In addition, "soft" phonons, viz., $\langle\omega^2\rangle$ ---> 0, will also raise λ and enhance T_c. This becomes apparent when it is realized that the phonon spectrum also depends on the product $N(E_F)I_a^2$, since electron-phonon coupling not only is responsible for Cooper pairs in conventional superconductivity theory, but also is fundamental to the phonons themselves since this is the mechanism by which the "bare" ion vibrational frequencies are screened. The stronger the electron-phonon scattering, the lower (more highly screened) are the phonon frequencies.

The last point is particularly important leading to the following qualitative argument: as the electron-phonon interaction increases, phonons will soften, leading to the possibility of phonon instabilities (imaginary frequencies) and, hence, structural instabilities that inhibit arbitrary increases in λ and lead to a practical limit on the value of T_c from electron-phonon scattering within a given class of materials. This "maximum" value of T_c for a solid due to the accompanying lattice instability is actually potentially quite high --- say up to the melting temperature of the material. It has been argued recently (33) that the fundamental limit is ~40K, without any proof other than the observation that previous "conventional" superconductors had been locked into T_c's lower than this value.

It is important to recognize that it has been shown that there is <u>no limitation</u> on T_c from the Eliashberg theory itself, with T_c scaling as,

$$T_c \Rightarrow 0.18 \left[\sum_a \frac{\eta_a}{M_a} \right]^{1/2} \tag{3}$$

in the large λ limit (32). The only limitation is the practical one of lattice (in)stability. We conclude that there is no fundamental intrinsic reason why electron-phonon coupling cannot play a major role in the oxide superconductors with T_c's of 100K or more. But the numerical results we now discuss show that it is not "business as usual" in the high T_c oxides.

Using the electronic structure results discussed above, we have followed the approach of Gaspari and Gyorffy (29) to calculate η_a defined in Equation 2. We emphasize that these are RMTA calculations. The results are presented in Table III. The η_a values are rather small, especially for the $YBa_2Cu_3O_7$, when compared with "conventional" high T_c materials. Previous "high" T_c materials such as the A15 family (e.g. Nb_3Sn, Nb_3Ge, V_3Si) derive their relatively large transition metal values of η_a from a large $N(E_F)$ (34), generally more than twice the value of the parent transition metal itself, with $I_a{}^2$ being typical of the transition metal parent. To the contrary, refractory carbonitride materials (e.g. NbC or NbN) have unusually large $I_a{}^2$ values and modest values of $N(E_F)$ (35,36). In both cases strong electron-phonon coupling can result.

Table III. McMillan-Hopfield parameters, η (eV/Å^2) for 2-1-4 and 1-2-3 materials

	La	Ba	Cu	O(x,y)	$O(z)_{La}$	$O(z)_{Ba}$
$\eta(La_2CuO_4)$	0.00	....	0.62	0.72	0.04	0.04
$\eta(LaBaCuO_4)$	0.08	0.16	1.08	0.32	0.15	0.33

$$\sum_a \frac{\eta_a}{M_a} = 0.149 \text{ (eV/Å}^2\text{ amu) } La_{1.85}(Sr,Ba)_{0.15}CuO_4$$

$YBa_2Cu_3O_7$

Y	Ba	Cu(1)	Cu(2)	O(1)	O(2)	O(3)	O(4)
0.003	0.011	0.267	0.199	0.834	0.153	0.144	0.127

$$\Sigma \frac{\eta_a}{M_a} = 0.116 \text{ (eV/Å}^2\text{ amu)}$$

Analyzing the results in Table III in more detail we see that for the 2-1-4 compounds, the La or Ba atoms always have very small values of η_a; the values for copper are moderate, and enhanced by Ba doping; the O_z atom has very small η_a in La_2CuO4, with a strong enhancement by Ba doping; the $O_{x,y}$ η_a are substantial, though smaller in the Ba doped material. Note that while $N(E_F)$ nearly doubles in the Ba-doped material, the η_a values do not increase proportionally. In the 1-2-3 material, the Y or Ba contributions to η are negligible, and the total Cu and O contributions are comparable, but somewhat smaller than in the 2-1-4 material. We can use the measured phonon densities of states to get an estimate of $<\omega^2>$, and T_c for these materials, finding λ values of 0.65 and 0.32, and T_c values of 8 K and 0.4 K, respectively, for 2-1-4 and 1-2-3, values which are much too low. Another estimate is obtained for the maximum T_c values possible from Equation 3 where we find 53 K and 44 K, respectively, for 2-1-4 and 1-2-3, using the η_a in Table III. This estimate for 2-1-4 is the "large λ limit" and shows a possibility of an electron-phonon explanation for 2-1-4 in agreement with our earlier conclusions (12), but the 1-2-3 value is still much too low. We are left with the question --- does the electron-phonon interaction fail?

Before abandoning electron-phonon scattering, it is important to examine the approximations that go into the RMTA and other calculations of λ and T_c to date. In the RMTA there is a so-called spherical approximation discussed at length by W. Butler and collaborators (37). Given the fairly low symmetry of the oxide crystalline sites, one might expect a substantial correction, although this has not been the case in other (cubic) materials where these corrections have been calculated (37). However, this will almost certainly not make up for the factor of 5-10 needed for the 1-2-3 material. A more important approximation to examine is that of treating $\delta_a V(\mathbf{r})$ in the RMTA, which assumes that there is sufficient screening so that no contribution outside of the displaced atom's muffin-tin need be considered. This turns out to be a bad approximation in La_2CuO_4 as seen in Figure 7 where we illustrate that a displacement of the O_z atoms in the axial breathing mode, where the two O_z atoms move out of phase (breathe) along the c-axis, induces substantial charge transfer between the in-plane Cu-O atoms. This calculation was done using a fully self-consistent frozen displacement LAPW calculation and casts doubt on the applicability of the RMTA to the oxides. The non-RMTA behavior is caused by the ionic character which, together with the rather low carrier densities leads to long-range contributions to the electron-ion matrix element. Further evidence that this may be an important effect is that recent frozen phonon LAPW calculations of in-plane oxygen breathing modes performed by our group shows them to be stable, contrary to Weber's (26) tight binding phonon results which neglected certain ionic contributions to the electron-phonon scattering similar to what is neglected in the RMTA. These recent LAPW results confirm our earlier PIB results (25), discussed above, which early on indicated that the in-plane oxygen breathing mode appears to be stable, with one of the most unstable modes being the tilt mode which drives the tetragonal to orthorhombic transition in the 2-1-4 material. Soft in-plane breathing modes have not been observed experimentally, a fact which has been used to cast doubt upon band theory applicability to the oxides. However, a proper calculation of these modes shows no instability, and further emphasizes that ionicity in the oxides leads to unusual electron-phonon effects. At this writing, elaborate calculations of η

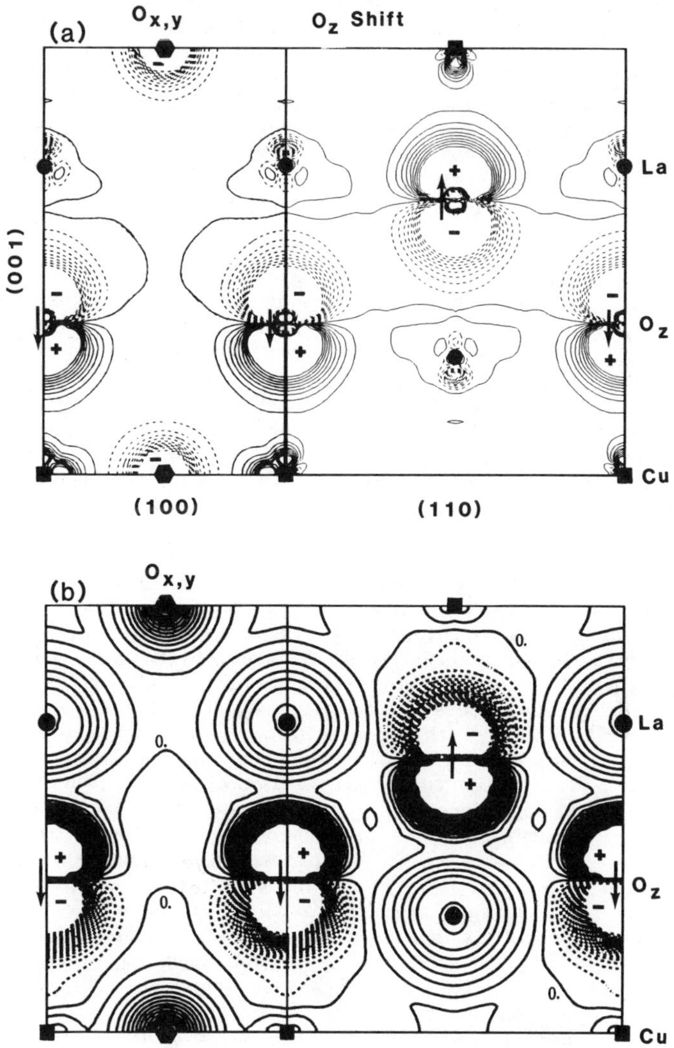

Figure 7. Shift in the valence charge density (top) and the one-electron potential (bottom) in La_2CuO_4 due to a displacement of O_z atoms in the axial breathing normal mode.

without using the RMTA are in progress at NRL to see the quantitative effects on the electron-phonon pairing.

Finally we mention some of our recent work using the coherent potential approximation (CPA) to study the effects of cation alloying on T_c in the 2-1-4 material, in particular Ba and Sr substitutions for La. The CPA approach, a much more rigorous method than the rigid-band approach, allows a determination of the $N(E_F)$ variation with alloying from which an estimate of λ and T_c can be obtained. The calculations confirm a rigid-band-like behavior with cation alloying, but oxygen vacancies lead to more complex behavior. In both cases the calculated values of $N(E_F)$ are strongly correlated with the measured values of T_c.

Transport Properties

A thorough discussion of several transport properties, viz., the electrical resistivity, Hall coefficient, and thermopower tensors, has been given in a recent paper by several of us (38,39). These results, derived from our LAPW electronic structure calculations, show that the transport properties are extremely anisotropic, so that comparisons with experiments on single crystals is crucial. A number of predictions from the LDA LAPW calculations agree quite well with experiment, lending support to the idea of a conventional Fermi-liquid ground state, a picture that has been discounted by some. Further single-crystal transport experiments will be highly valuable.

For the resistivity, the results show out-of-plane components nearly a factor of ten larger than the in-plane components, exhibiting the highly anisotropic conduction in these materials. This seems to agree with experiments to date (40,41). However, the overall magnitudes of ρ based on RMTA calculations are much smaller than experiment. Another interesting result is shown in Figure 8 where we plot the calculated temperature dependence of the electron-phonon contribution to the resistivity [see (39) for details] compared with experiment. The observed and calculated linear behavior of ρ with T is to be particularly noted. The "folklore" that this linear dependence of ρ with T cannot be explained by the electron-phonon interaction is without basis. In fact, many metals show linear behavior down to 10-20% of the Debye temperature.

For both 1-2-3 and 2-1-4 the Hall coefficient is predicted to be hole-like for electrons orbiting in the Cu-O planes, and electron like if the **B** field is rotated 90° into the plane, contrary to simple expectations. This has been verified experimentally by Tozer, *et al.* (42) for the 1-2-3 compound, with reasonable quantitative agreement with theory. Single-crystal data for 2-1-4 for $x \cong 0.06$ Sr doping have been given by Suzuki and Murakami (43). For **B** perpendicular to the planes, the Hall voltage is hole-like and within a factor of three of a rigid-band treatment of the band calculations.

Thermopower calculations show results opposite to that of the Hall coefficient: electron-like for in-plane thermal gradients and a change of sign (or nearly so, depending on the energy dependence of the scattering) for a 90^0 rotation of ∇T. These rather unusual

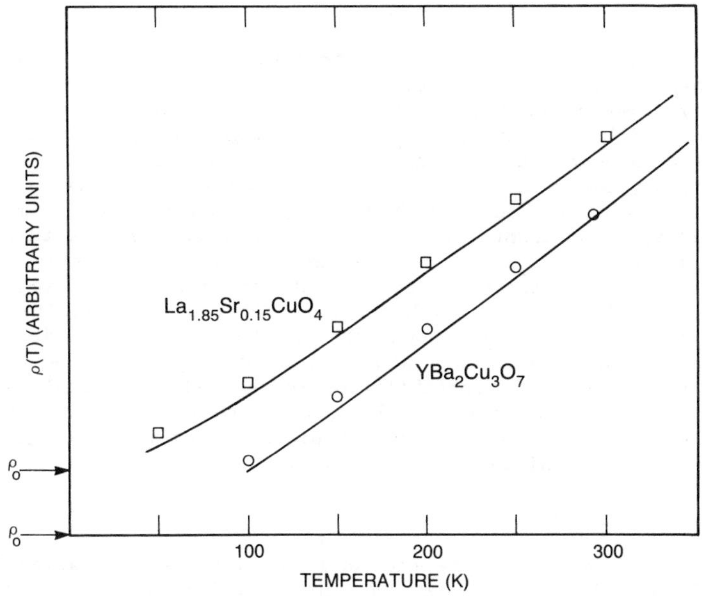

Figure 8. Expected shape of the electron-phonon part of ρ(T) according to Boltzmann theory. The curve of $La_{1.85}Sr_{0.15}CuO_4$ is displaced upward by an arbitrary amount ρ_o. See (39) for details.

thermopower signs are in agreement with recent single crystal measurements on 1-2-3 by R. C. Yu, *et al.* (44).

The predictions (38,39) of the correct temperature dependence of the resistivity and the correct signs for both the Hall and thermopower tensors is a remarkable accomplishment for the band structure approaches. The discrepancies in the magnitudes of the calculated transport coefficients compared with experiment seem to us relatively minor. We therefore wish to oppose strongly the view suggested by some that band theory is completely irrelevant. It is very important that accurate single crystal measurements on these and other materials be performed to stimulate further theoretical progress.

High T_c Without Copper

In April of 1988, a group at AT & T Bell Labs reported achieving superconductivity at nearly 30 K in the perovskite $Ba_{0.6}K_{0.4}BiO_3$, a material without any Cu at all (3). This work, a spinoff of earlier work on superconductivity at much lower temperatures in the Ba-Pb-Bi-O system, has the potential for setting the theoretical world on its ear. For, unless experimental evidence for magnetic correlations in these systems is found, it would imply either a common, non-magnetic mechanism, or that two different mechanisms for high T_c occur in systems that are closely related, a slightly implausible situation. Mother Nature is having a Field Day!

Conclusions

The status of the applicability of "traditional" band structure and electron-phonon scattering mechanisms to the high-T_c oxides is currently somewhat unclear. The transport calculations show major agreements with experiment but, quantitatively, the electron-phonon scattering appears too weak. The same is true for our T_c calculations, but we argue that there is a good chance that non-RMTA types of corrections, driven by the ionic nature of these metals, will lead to major quantitative corrections which may move things in the right direction. Our view is that when done properly, there will be strong electron-lattice interactions in these ionic materials which, together with their albeit weak metallicity, can account for many of their unusual chemical/physical properties. The discrepancies between theory and experiment regarding the lack of any calculated magnetic instabilities in the 2-1-4 material is also deserving of further work. Such studies are underway at NRL, and elsewhere. Furthermore, the recent discovery of non-Cu containing high T_c materials thickens the plot.

Acknowledgments

This work was supported by the National Science Foundation Grant Nos. DMR-84-16046 (H. Krakauer) and DMR-84-20308 (P. B. Allen), and by the Office of Naval Research. Computations were carried out under the auspices of the National Science Foundation at the Cornell National Supercomputer Facility and at the Pittsburgh Supercomputer Center, as well as on the Naval Research Laboratory Cray X-MP.

Literature Cited

1. Bednorz, J. G.; Müller, K. A. Z. Phys. B 1986, 64, 189.
2. Wu, M. K.; Ashburn, J. R.; Torng, C. J.; Hor, P. H.; Meng, R. L.; Gao, L.; Huang, J.; Wang, Y. Q.; Chu, C. W. Phys. Rev. Lett. 1987, 58, 908.
3. Cava, R. J.; *et al.* Nature 1988, 332, 814.
4. Beno, M. A.; *et al.* Appl. Phys. Lett. 1987, 51, 57.
5. Hazen, R. M.; *et al.* Phys. Rev. B 1987, 35, 7238.
6. Wei, S.-H.; Krakauer, H. Phys. Rev. Lett. 1985, 55, 1200.
7. Pickett, W. E. Comm. Solid State Phys. 1985, 12, 1.
8. Wang, C. S.; Klein, B. M.; Krakauer, H. Phys. Rev. Lett. 1985, 54, 1852.
9. Birgeneau, R. J.; *et al.* Phys. Rev. Lett. 1987, 59, 1329.
10. Mitsuda, S.; *et al.* Phys. Rev. B 1987, 36, 822.
11. Johnston, D. C.; Stokes, J. P.; Goshorn, D. P.; Lewandowski, J. T. Phys. Rev. B 1988, 36, 4007.
12. Pickett, W. E.; Krakauer, H.; Papaconstantopoulos, D. A.; Boyer, L. L. Phys. Rev. B 1987, 35, 7252.
13. Mattheiss, L. F. Phys. Rev. Lett. 1987, 58, 1028.
14. Xu, J.-H.; *et al.* Phys. Lett. A 1987, 120, 489.
15. Leung, T. C.; Wang, X. W.; Harmon, B. N. Phys. Rev. B 1988, 37, 384.
16. Guo, G. Y.; Temmerman, W. M.; Stocks, G. M. J. Phys. C 1988, 21, L101-L108.
17. Sterne, P. A.; Wang, C. S. Phys. Rev. B 1988, 37, 7472.

18. Krakauer, H.; Pickett, W. E.; Cohen, R. E. J. Supercond. 1988, 1, 111.
19. Chu, C. W.; *et al*. Phys. Rev. Lett. 1988, 60, 941.
20. Sheng, Z. Z.; *et al*. Phys. Rev. Lett. 1988, 60, 937.
21. Krakauer, H.; Pickett, W. E. Phys. Rev. Lett. 1988, 60, 1665.
22. Boyer, L. L.; *et al*. Phys. Rev. Lett. 1985, 57, 2331.
23. Cohen, R. E.; Boyer, L. L.; Mehl, M. J. Phys. Rev. B 1987, 35, 5749.
24. Cohen, R. E.; Boyer, L. L.; Mehl, M. J.; Pickett, W. E.; Krakauer, H. Proc. Chapman Conf. on Perovskites (AGU) 1988, (in press).
25. Cohen, R. E.; Pickett, W. E.; Boyer, L. L.; Krakauer, H. Phys. Rev. Lett. 1988, 60, 817.
26. Weber, W. Phys. Rev. Lett. 1987, 58, 1371. Erratum, ibid 1987, 58, 2154.
27. Klein, B. M.; Pickett, W. E. In Superconductivity in d- and f-Band Metals; Buckel, W.; Weber, W.; Eds.; Kernforschungszentrum: Karlsruhe, 1982, p 97.
28. Butler, W. H. In Treatise on Materials Science and Technology; Fradin. F. Y., Ed.; Academic: New York, 1981; Vol. 21, p 165.
29. Gaspari, G. D.; Gyorffy, B. L. Phys. Rev. Lett. 1972, 29, 801.
30. Varma, C. M.; Weber, W. Phys. Rev. B 1979, 19, 6142.
31. McMillan, W. E. Phys. Rev. 1968, 167, 331.
32. Allen, P. B.; Dynes, R. C. Phys. Rev. B 1975, 12, 905.
33. Anderson, P. W. and Abrahams, E. Nature 1987, 327, 363, and references therein.
34. Klein, B. M.; Boyer, L. L.; Papaconstantopoulos, D. A. Phys. Rev. Lett. 1979, 42, 530.
35. Klein, B. M.; Papaconstantopoulos, D. A. Phys. Rev. Lett. 1974, 32, 1193.
36. Papaconstantopoulos, D. A.; Pickett, W. E.; Klein, B. M.; Boyer, L. L. Nature 1984, 308, 494.
37. Butler, W. H.; Olsen, J. J.; Faulkner, J. S.; Gyorffy, B. L. Phys. Rev. B 1979, 19, 3708.
38. Allen, P. B.; Pickett, W. E.; Krakauer, H. Phys. Rev. B 1987, 36, 3926.
39. Allen, P. B.; Pickett, W. E.; Krakauer, H. Phys. Rev. B 1988, 37, 7482.
40. Schlesinger, Z.; Collins, R. T.; Kaiser, D. L.; Holtzberg, F. Phys. Rev. Lett. 1987, 59, 1958.
41. Worthington, T. K.; Gallagher, W. J.; Dinger, T. R. Phys. Rev. Lett. 1987, 59, 1160.
42. Tozer, S. W; *et al*. Phys. Rev. Lett. 1987, 59, 1768.
43. Suzuki, M.; Murakami, T. Jpn. J. Appl. Phys. 1987, 26, L524.
44. Yu, R. C.; *et al*. Phys. Rev. B 1988, 26, 7963.

RECEIVED July 13, 1988

Chapter 4

High-Temperature Superconductivity from Charged Boson Pairing

M. J. Rice and Y. R. Wang

Xerox Webster Research Center, Webster, NY 14580

The mean field theory of the paired holon superconductor and its predictions are reviewed for the case in which charged holons on different magnetic sublattices in a doped CuO_2 layer interact via a weak attractive pairing potential V. The physical properties of this superconductor often reflect the essential gaplessness of its excitation spectrum.

In the resonating-valence-bond (RVB) model (1) of the cupric oxide superconductors holes doped into a half-filled Mott-Hubbard insulator on a square lattice are thought to delocalize as spinless charged bosons (2), or holons. It has recently been suggested by us (3) and by the Princeton group (4) that in this model high temperature superconductivity might arise from pair condensation of the holon gas, rather than from some kind of single particle (Bose-Einstein) condensation. The possibility of holon pairing stems from our observation (3) that the strictly two-dimensional (2D) bose gas becomes unstable toward pair condensation as arbitrarily weak attractive interactions (V) between its particles are switched on. This property, which does not arise for an isotropic 3D bose gas at zero temperature (T=0), is reminiscent of the Cooper-pair instability in a Fermi gas. In our work, the origin of V is left as an open question and we concentrate on the calculation of the mean-field properties of the resulting paired holon superconductor. In the Princeton work V is identified as an effective in plane attractive interaction between holons driven by interlayer tunneling of holon pairs. In this paper we review the predicted mean field properties of the paired holon superconductor (3,5,6). Specifically, we shall discuss the temperature dependences of the pair potential $\Delta(T)$, the specific heat $C_v(T)$, the critical field $H_c(T)$ and the London penetration depth $\lambda(T)$. We also discuss the microwave absorption (6) which is found to be quite different from

0097–6156/88/0377–0044$06.00/0

Bardeen-Cooper-Schrieffer (7) (BCS) Mattis-Bardeen behavior (8) and which, therefore, provides the possibility of an experimental means of identifying the paired holon superconductor.

The model we introduce to describe the interacting holon system in a doped CuO_2 layer is appropriate for the nearest-neighbour RVB state in which charged holons delocalized on different magnetic sublattices (a and b) interact via a weak attractive pairing potential V. Interactions between holons on the same sublattice are described by a constant repulsive Hartree-Fock potential U. The Hamiltonian describing this model is

$$H = \sum_{\kappa} \{a^+_{\kappa}a_{\kappa}(\epsilon_{\kappa}-\mu_a+\rho_a U)+b^+_{\kappa}b_{\kappa}(\epsilon_{\kappa}-\mu_b+\rho_b U)\}-(V/A)\sum_{\kappa}{}' a^+_{\kappa}b^+_{-\kappa}b_{-\kappa}a_{\kappa} \quad (1)$$

In (1) $a^+_{\kappa}, b^+_{\kappa}, a_{\kappa}$ and b_{κ} are boson creation and annihilation operators for the a and b Hartree-Fock particle states with momentum $\hbar\kappa$ and kinetic energy $\epsilon_{\kappa}=\hbar^2k^2/2m$. ρ_a and ρ_b denote the densities of the component holon gases while μ_a and μ_b denote their respective chemical potentials. In equilibrium, $\mu_a=\mu_b=\mu$, where μ is determined by the condition that the statistical average of A^{-1} $\sum_{\kappa}(a^+_{\kappa}a_{\kappa}+b^+_{\kappa}b_{\kappa})$ be equal to $\rho=\rho_a+\rho_b=N/A$, the total number of holons per unit area ($N\to\infty$, $A\to\infty$). V is taken to satisfy $V<<U$, so that the compressibility of the holon gas, $\chi=2\rho^2U$, is not drastically affected by the presence of the second term in (1) which describes the pairing interaction. Finally, V is restricted to operate between holons with $\epsilon_{\kappa}<\epsilon_A$, a restriction that is denoted by the prime over the second summation symbol in Eq. (1).

To describe the mean field behavior we follow BCS and replace (1) by the bilinear form

$$H=\sum_{\kappa}\{(\epsilon_{\kappa}-\mu)\ (a_{\kappa}^{\dagger}a_{\kappa}+b^{\dagger}_{\kappa}b_{\kappa})+\Delta_{\kappa}a_{\kappa}^{\dagger}b_{-\kappa}^{\dagger}+\Delta_{\kappa}^{*}a_{-\kappa}b_{\kappa}\} \quad (2)$$

where $\Delta_{\kappa}=-V<a_{-\kappa}b_{\kappa}>=\Delta\ \exp(iS)$ is the mean-field pair potential. The brackets denote a Gibbs statistical average taken over the eigenstates of (4) and both Δ and S are real. $\mu\equiv\mu_a-\rho_a(U-V)\equiv\mu_b-\rho_b(U-V)$ is the effective chemical potential. H may be brought to the diagonal form

$$H=\sum_{\kappa}E_{\kappa}(A^{\dagger}_{\kappa}A_{\kappa}+B^{\dagger}_{\kappa}B_{\kappa}) + \sum_{\kappa}[E_{\kappa}-(\epsilon_{\kappa} + |\mu|)] \quad (3)$$

by means of the unitary Bogoliubov transformation $a_{\kappa}=u_{\kappa}A_{\kappa}-v^{*}_{\kappa}B_{-\kappa}^{\dagger}, b_{\kappa} = u_{\kappa}B_{\kappa}-v_{\kappa}^{*}A_{-\kappa}^{\dagger}$, where A_{κ} and B_{κ} are new boson operators which destroy quasi-particle (Q.P.) excitations with energies

$$E_{\kappa} = [(\epsilon_{\kappa} + |\mu|)^2-\Delta^2]^{\frac{1}{2}} \quad (4)$$

relative to the ground state value of H, which is given by the second sum in (4). The amplitudes $u_{\kappa}= |u| \exp(iS/2)$ and $v_{\kappa}= |v_{\kappa}| \exp(-iS/2)$ satisfy $|u_{\kappa}|^2-|v_{\kappa}|^2=1$ and their moduli are given by $2|u_{\kappa}|^2 = (\gamma_{\kappa}+1)$, $2|v_{\kappa}|^2=(\gamma_{\kappa}-1)$, where $\gamma_{\kappa}=(\epsilon_{\kappa}+|\mu|)/E_{\kappa}$. The many-particle wavefunction describing the ground state is the Valatin-Butler (9) wavefunction

$$\Phi_O = \Pi_K |u_K|^{-1} \exp\{-(v_K/u_K) a^\dagger_K b^\dagger_{-K}\} |0> \quad (5)$$

and satisfies $A_K\Phi_O=0$, $B_K\Phi_O=0$ and $\Phi_O^\dagger\Phi_O=1$, where $|0>$ denotes the vacuum. It may be interpreted as describing clusters of n = 1,2,---∞ identical time-reversed pairs. The probability that the particular cluster of n occurs is $P_{K,n}=|v_K|^{2n}/|\mu_K|^{2n+2}$ and satisfies $\Sigma_n P_{K,n}=1$. The equations determining the self-consistent values of $\Delta(T)$ and $\mu(T)$ are

$$A^{-1}\Sigma'_K E_K^{-1}(2n_K+1) = (2/V) \quad (6)$$

$$\rho = (A)^{-1} \Sigma_K E_K^{-1}[(\epsilon_K + |\mu|)(2n_K+1) - E_K] \quad (7)$$

where $n_K = [\exp(E_K/k_BT)-1]^{-1}$ is the number of Q.P's excited with momentum $\hbar K$ at temperature T. Eq. (7) can be evaluated exactly to give the following relationship between μ and Δ;

$$(\mu^2-\Delta^2)^{\frac{1}{2}} = -2k_BT\ell n[\{(x^2+4)^{\frac{1}{2}}-x\}/2] \quad (8)$$

where $x = \exp[(|\mu|-2k_BT_O)/2k_BT]$, where $k_BT_O = \pi\hbar^2\rho/m$ Finally, the thermodynamic properties are determined from the Helmholtz free energy

$$F=<H + \mu N>-k_BT \sum_K [(n_K+1)\ \ell n(n_K+1)-n_K\ell n(n_K)] \quad (9)$$

We now discuss the main consequences of these results.

Transition Temperature T_C

The equation determining the mean-field transition temperature T_C is obtained by setting $\Delta=0$ in Eq. (6); this yields I(T)=1, where I(T) is the integral

$$I(T) = 1/2\ N_OV \int_{|\mu|/k2_BT}^{(\epsilon_A+|\mu|)/2k_BT} (dx/x)\coth(x)$$

Here, $N_O=m/2\pi^2\hbar^2$ is the density of (free holon) states per unit area and $\mu(T)= k_BT\ell n[1-\exp(-T_O/T)]$. In the weak coupling limit, $N_OV\rightarrow 0$, $T_C\sim -T_O/\ell n(N_OV)$, where $T_O=\pi^2\hbar^2\rho/mk_B$ is the degeneracy temperature of one of the component holon gases. Thus, in this limit, T_C is essentially independent of ϵ_A and only weakly dependent on V: the energy scale of T_C is set by T_O. It is important to note that, as also occurs in the fermion case, the pairing instability can survive the presence of a stronger repulsive pair interaction $U_{K,K'}=V_R$ provided that its range, $\epsilon_R>>\epsilon_A$. In this case V in I(T) is to be replaced by $V'=V-\{V_R/[1+N_OV_R\ell n(\epsilon_R/\epsilon_A)]\}$.

Order Parameter and Excitation Spectrum

Eq. (8) shows that the "gap" $E_g=(\mu^2-\Delta^2)^{\frac{1}{2}}$ in the excitation spectrum is non-vanishing at all finite temperatures. For $\lambda=N_OV<\lambda_O=1/\sinh^{-1}$

$[(\epsilon_A/4k_BT_o)^{\frac{1}{2}}]$, however, E_g rapidly diminishes as $T\rightarrow 0$ and is precisely zero at T=0. Fig. 1 shows the magnitudes and temperature dependences of $\Delta(T)$ and $\mu(T)$ as calculated from (6) and (7) for cases corresponding to $\lambda<\lambda_o$(Fig. 1(a)) and $\lambda<\lambda_o$ (Fig.1(b)), where the parameter values employed are indicated in the figures. It is seen that $\Delta(T)$ has a BCS-like T-dependence and the behavior of $\mu(T)$ is quite different from that of the non-interacting gas. For $\lambda<\lambda_o$ and T=0, E_{κ} is phonon-like for $k\xi<<1$, i.e., $E_{\kappa}=\hbar sk$, where $s = (|\mu|/m)^{\frac{1}{2}}=(\Delta/m)^{\frac{1}{2}}$ is the critical superfluid velocity at T=0. In the limit $\lambda\rightarrow 0$, with $\epsilon_A\neq 0$, $\Delta(0)\rightarrow\rho V$.

Critical field

For the charged gas, the critical magnetic field $H_C(T)$ may be calculated from (11) and the relation $F_S-F_n= -(H^2_C/8\pi)$, where F_S and F_n denote the free energies of the superconducting and normal phases, respectively. Fig. 2 shows $H_C(T)/H_C(0)$ versus T/T_C calculated using the same values of ϵ_A that were employed for the calculation of Fig. 1 and λ = 0.6. Within the error of our numerical calculations, the curve shown for $H_C(T)/H_C(0)$ vs. T/T_C coincides with those which we have also calculated for λ = 0.1 and λ =1.2 with the same value of ϵ_A.

Specific heat

The specific heat, $C_v(T)$, of the paired holon superconductor is given by

$$C_v(T) = \sum_{\kappa} E_{\kappa}(dn_{\kappa}/dT) \quad (10)$$

Fig. 3 shows $C_v(T)/Nk_B$ versus T/T_C calculated for the two coupling strengths N_0V=0.1 and N_0V=0.6. These results show that $C_v(T)$ is linear in T for all but the ultra low temperature regime discussed above. In this regime,

$$C_v(T)/Nk_B = 3k_BT^2\zeta(3)/T_0\Delta(0) \quad (11)$$

If in the temperature range in which $C_v(T)$ is linear we represent $C_v(T)$ as $C_v(T) = \gamma T$, an order of magnitude estimate of γ is provided by $\gamma \approx C_v(T_c)/T_c \approx Nk_B/T_c$. For the oxide superconductors this leads to the estimate of $\gamma \approx 20$ mJ/mole(Cu) K^2 on taking (10) $\rho/d = 4\times10^{21}cm^{-3}$ for $YBa_2Cu_3O_{7-\delta}$. The observed linear specific heat (11) is, typically, $\gamma \approx$ 10 mJ/mole K^2. This is consistent with our estimate taking the uncertainty of the carrier density and the coupling into account. This interpretation of the linear specific heat does not necessarily imply a gap E_S to exist in the RVB spinon excitation spectrum. If E_S is zero, spinon excitations will also contribute (12) a specific heat linear in T. If E_S is finite, as it is in a nearest neighbour equal amplitude RVB state (13), the spinon specific heat will be exponentially small at low temperatures.

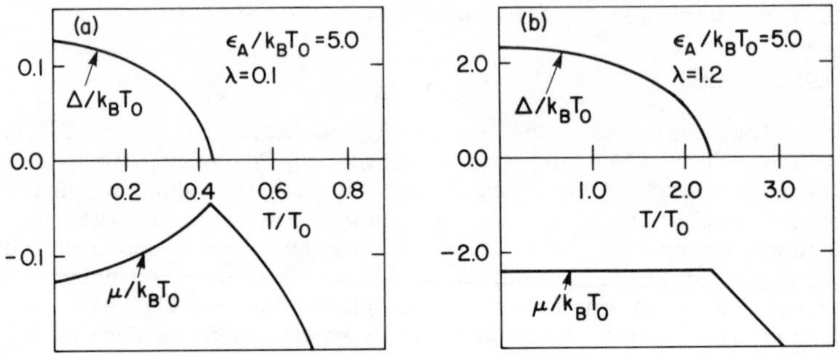

Fig. 1. $\Delta(T)/k_BT_0$ and $\mu(T)/k_BT_0$ versus T/T_0 for (a) λ/λ_0 = 0.096 and (b) λ/λ_0=1.154. In both cases λ_0=1.04.

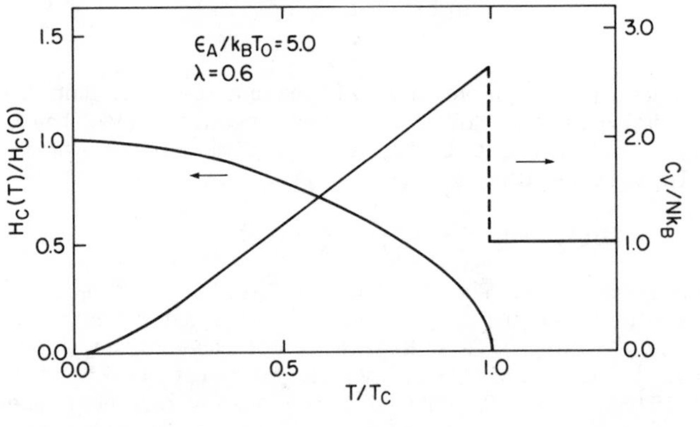

Fig. 2. $C_\nu(T)/Nk_B$ (right hand scale) and $H_C(T)/H_C(0)$ (left hand scale) versus T/T_C.

London penetration depth

From the theory of the diamagnetic response the London penetration depth of the holon superconductor is

$$\lambda(T) = \lambda(0)\ [1-\rho_n(T)/\rho]^{-1/2} \qquad (12)$$

where $\lambda(0) = (mc^2 d/4\pi\rho e^2)^{1/2}$ and

$$\rho_n(T) = -A^{-1} \sum_{\boldsymbol{\kappa}} \varepsilon_{\boldsymbol{\kappa}}(\partial n_{\boldsymbol{\kappa}}/\partial E_{\boldsymbol{\kappa}}) \qquad (13)$$

is the number of "normal" holons per unit area at temperature T. In (12) d is the lattice constant along the c-direction. Eqs. (12) and (13) are valid for the case in which the holon mean free path, l, is large by comparison to the coherence length $\xi(T) \approx \sqrt{\hbar^2/2m\,|\mu|}$. The condition for the London description is then $\lambda(T) >> \xi(T)$. For T = 0, we have $\xi(0) \approx (N_0V\rho)^{-1/2}$, i.e., of order only a few times the mean inter-holon distance $\rho^{-1/2}$ for values of $N_0V \approx 0.1$. By contrast, the observed values of $\lambda(0)$ are of order of 1000Å.

The dependence of $\lambda(T)/\lambda(0)$ on T/T_c as calculated from Eqs. (12) and (13) is shown as the full curve in Figs. 4 and 5. For weak coupling this dependence is practically independent of N_0V. For example, the values of $\lambda(T)/\lambda(0)$ for N_0V as large as 0.6 differ from those for N_0V = 0.1 only in the third decimal place for most values of T/T_c. Thus $\lambda(T)/\lambda(0)$ is essentially a universal function of T/T_c. In Fig. 4 this dependence is compared with the experimental dependence measured by Cooper et al. (14) for $YBa_2Cu_3O_{7-\delta}$ while Fig. 5 shows the comparison with the data of Aeppli et al. (15) obtained for $La_{1.85}Sr_{0.15}CuO_4$. The agreement with the extensive data of Cooper et al. is particularly excellent. In Fig. 6 we have plotted the calculated values of $\lambda(T)/\lambda(0)$ versus $(T/T_c)^2$ in order to check the experimentally reported low temperature T^2 dependence of $\lambda(T)-\lambda(0)$. It is seen that the theoretical dependence is indeed consistent with a T^2 dependence, provided that T is not too low. We note that in the actual $T/T_c \to 0$ limit, Eqs. (12) and (13) predict

$$\lambda(T)/\lambda(0) = 1 + 3\,\zeta(3)k_B^2T^3/2T_0\Delta^2(0) \qquad (14)$$

However, this temperature regime, which reflects the ultra low temperature limit of the excitation spectrum of the holon superconductor, sets in only for $k_BT << \Delta(0)$. Since $\Delta(0) \approx N_0Vk_BT_0$, this temperature criterion is $T/T_c << N_0V\ln(1/N_0V)$. Thus, if N_0V=0.1, say, one has $T/T_c << 0.2$.

Microwave Absorption

The microwave absorption of the paired boson superconductor is calculated in the extremely local (or dirty) limit (6). We find that it is sharply different from that of the Mattis-Bardeen (8) behavior characteristic of a BCS superconductor. In particular, we find that the ratio $\sigma_S(\omega)/\sigma_N$ is strongly temperature dependent and

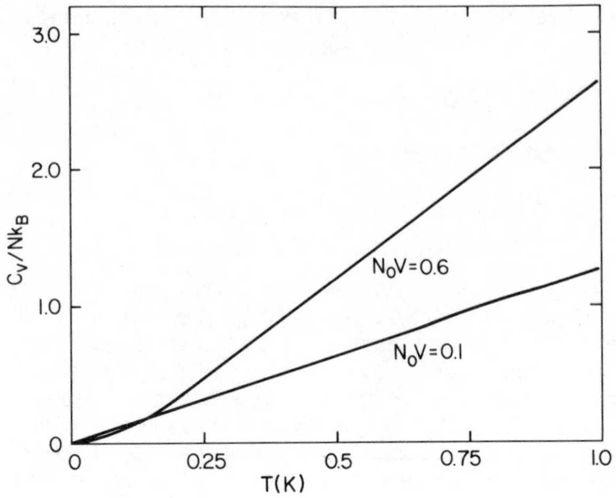

Fig. 3. The temperature dependence of the specific heat of the paired holon superconductor.

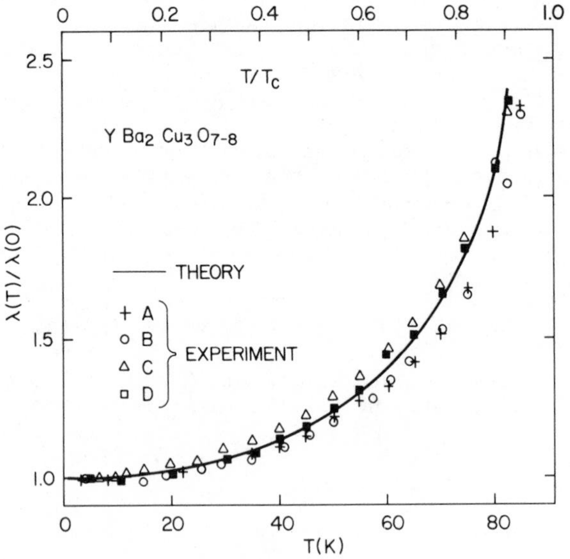

Fig. 4. The temperature dependence of the magnetic penetration depth of the paired holon superconductor (solid line) compared with the experimental data of $YBa_2Cu_3O_{7-\delta}$. The data is from Cooper et al. (Ref. 14).

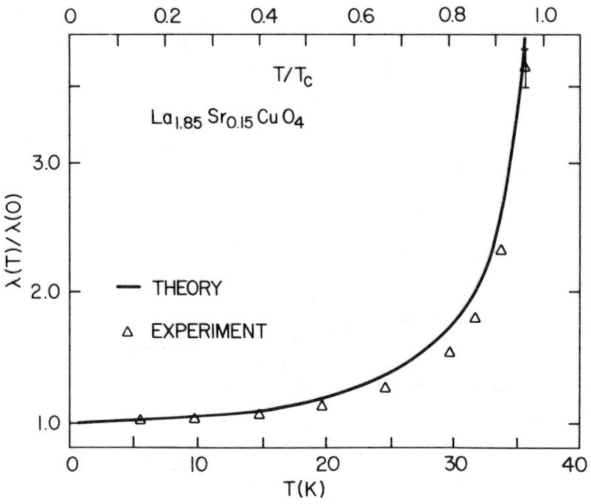

Fig. 5. The temperature dependence of the magnetic penetration depth of the paired holon superconductor (solid line) compared with the experimental data of $La_{1.85}Sr_{0.15}CuO_4$. The data is from Aeppli et al. (Ref. 15).

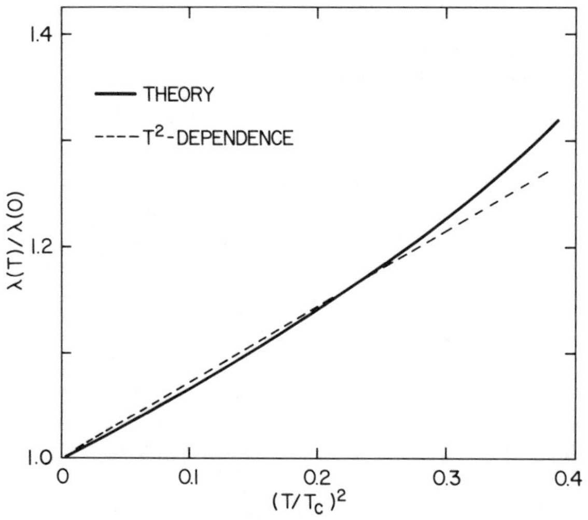

Fig. 6. $\lambda(T)/\lambda(0)$ of the paired holon superconductor as a function of $(T/T_c)^2$ (solid line) compared with the T^2 dependence (the dashed straight line).

only weakly frequency dependent. Here, $\sigma_S(\omega)$ is the absorptive part of the frequency dependent conductivity of the superconducting state and σ_N is the d.c. conductivity of the normal state.

The result is plotted in Fig. 7 for $\sigma_S(\omega)/\sigma_N$ against frequency ω for weak (λ=0.1) interaction. Since the excitation spectrum is gapless at T=0, $\sigma_S(\omega)$ has no threshold. In the limit $\omega \to 0$ and at T=0, we find $\sigma_S(\omega)/\sigma_N \sim (\hbar\omega/|\mu|)^3$, and $\sigma_S(0)/\sigma_N \sim (kT/|\mu|)^4$ when $T \to 0$. Fig. 8 shows a plot of the ultra low frequency absorption, $\sigma_S(0)/\sigma_N$, versus T/T_c.

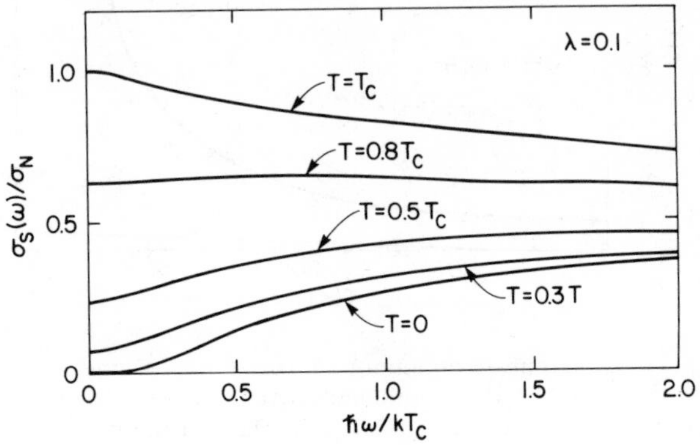

Fig. 7. Frequency dependence of the conductivity of the weakly coupled paired holon superconductor (λ = 0.1). The excitation spectrum is gapless at T = 0.

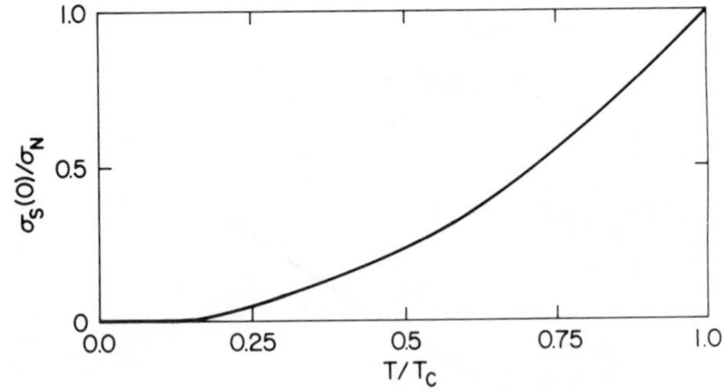

Fig. 8. Temperature dependence of the zero-frequency conductivity of the paired holon superconductor. The curve is indistinguishable for the three coupling strengths λ = 0.1, λ = 0.6 and λ = 1.2.

In summary, we have reviewed the predicted mean field properties of the paired holon superconductor. Mostly, they differ from the BCS case only in the details of their T-dependences. The microwave absorption, however, is sharply different and directly reflects the essential gaplessness of the excitation spectrum.

We thank D. Costenoble for secretarial assistance.

LITERATURE CITED

1. Anderson, P.W. Science, (1987) 235.
2. Kivelson, S.A., Roksar, D.S. and Sethna, J.P. Phys. Rev. (1987) B35, 8865; Kivelson, S. Phys. Rev (1987) B36 7237.
3. Rice, M.J. and Wang, Y.R. Phys. Rev. (1988) B37 5893.
4. Wheatley, J.M., Hsu, T.C. and Anderson, P.W. Phys. Rev. (1988) B37 5897.
5. Wang, Y.R. and Rice, M.J. submitted to Phys. Rev. B.
6. Wang, Y.R., Rice, M.J. and Duke, C.B. Phys. Rev. (1988) B37 9868.
7. Bardeen, J., Cooper, L.N. and Schrieffer, J.R. Phys. Rev. (1957) 106 162.
8. Mattis, D.C. and Bardeen, J Phys. Rev. (1958) 111 412.
9. Valatin, J.G. and Butler,D Nuovo Cimento (1958) 37 10.
10. Panson, A.J., Braginski, A.I., Gavaler, J.R., Hulm, J.K., Janocko, M.A., Pohl, H.C., Stewart, A.M., Talvacchio, J. and Wagner, G.R. Phys. Rev. (1987) B35 8774.
11. Nevitt, M.V., Crabtree,G.W. and Klippert, T.E., Phys. Rev. (1987) B36 2398; von Molnar, S., Torressen,A., Kaiser, D., Holtzberg, F. and Penney, T. Phys. Rev. (1988), B37 3762.
12. Zou, A. and Anderson, P.W. Phys. Rev. (1987) B37 627.
13. Kalmeyer, V. and Laughlin, R.B. Phys. Rev. Lett. (1987) 59 2095; Komoto, M. and Shapir, Y. (preprint, 1988); Mele, E.J. (preprint); Kivelson, S. and Rokhsar, Proc. Inter. Conference on High-temperature Superconductors and Materials and Mechanisms of Superconductivity at Interlaken, Switzerland, 1988.
14. Cooper, J.R., Chu, C.T., Zhou, L.W., Dunn, B. and Gruner, G. Phys. Rev. (1988) B37 638.
15. Aeppli, G., Cava, R.J., Ansaldo, E.J., Brewer, J.H., Krietzman, S.R., Luke, G.M., Noakes, D.R. and Kiefli, R.F. Phys. Rev. (1987) B35 7129.

RECEIVED July 13, 1988

Chapter 5

Topological Aspects of Chemical Bonding in Superconductors

R. B. King

Department of Chemistry, University of Georgia, Athens, GA 30602

The conducting skeletons of the ternary molybdenum chalcogenide (Chevrel phase) $M_nMo_6S_8$ and the ternary lanthanide rhodium boride $LnRh_4B_4$ superconductors consist of infinite three-dimensional networks of metal polyhedra (Mo_6 octahedra or Rh_4 tetrahedra) with the intrapolyhedral metal-metal bonding confined to the polyhedral edges, the interpolyhedral distances short enough for some chemical bonding, and oxidation of the closed shell electron configuration to create the partially filled valence band required for p-type conductivity. The Cu-O conducting planes of the high critical temperature copper oxide superconductors exhibit closely related chemical bonding topologies with metal-oxygen-metal bonds and antiferromagnetic metal-metal interactions rather than direct metal-metal bonds. The much higher critical temperatures of the copper oxide superconductors relative to superconductors based on metal cluster structures can then be related to the much higher ionic character and thus much lower polarizability of metal-oxygen bonds relative to metal-metal bonds. Assuming that T_c scales linearly with the hole concentration leads to a simple formula for estimating T_c's of members of homologous series such as $M^{III}_2M^{II}_2Ca_{n-1}Cu_nO_{5+2n-1+x}$ (M^{III} = Bi, M^{II} = Sr or M^{III} = Tl, M^{II} = Ba). This formula suggests that the maximum T_c's obtainable from copper oxide superconductors will be far below room temperature and probably no more than 180 K.

In 1977 we reported a method based on graph theory for study of the skeletal bonding topology in polyhedral boranes, carboranes, and metal clusters (1). Subsequent work has shown this method to be very effective in relating electron count to cluster shape for diverse metal clusters using a minimum of computation. Discrete metal clusters treated effectively by this method include post-transition metal clusters (2,3), osmium carbonyl clusters (4), gold clusters (5,6), platinum carbonyl clusters (5,7), and

0097–6156/88/0377–0054$06.00/0

rhodium carbonyl clusters having fused polyhedra (8,9). In addition, this graph-theory derived method was shown to be applicable for the study of infinite metal clusters including one-dimensional chains and two-dimensional sheets of fused metal octahedra (10,11) thereby suggesting the application of this method for the study of solid state materials exhibiting interesting electronic properties, particularly solids constructed from metal cluster structural units.

With this background in applying graph-theory derived methods for the study of skeletal bonding topology of both discrete and infinite metal clusters, I undertook the study of the skeletal bonding topology of superconductors in summer 1986. At that time the high T_c copper oxide superconductors had not yet been recognized so that the highest T_c superconductors then known were materials constructed from metal cluster building blocks (12) of which the ternary molybdenum chalcogenides (Chevrel phases) (13) and ternary lanthanide rhodium borides (14) were particularly tractable examples exhibiting relatively high T_c's and H_c's for metal cluster superconductors. Application of the graph-theory derived methods to these metal cluster superconductors did not present any major difficulties and led to the observation that these superconductors exhibit an edge-localized skeletal bonding topology conveniently designated as porous delocalization (15,16). Such a skeletal bonding topology is very different from that of most infinite metal cluster structures (17) thereby suggesting a connection between porous delocalization and superconductivity.

Shortly after acceptance of my manuscripts on the Chevrel phases (15) and the lanthanide rhodium borides (16) the initial scientific reports on the high T_c copper oxide superconductors became available. It soon became apparent that the theoretical approach used for treatment of the skeletal bonding topology in the metal cluster superconductors could be adapted to the copper oxide superconductors (18) to provide a simple rationalization why the T_c's of the copper oxide superconductors are higher than those of any of the known metal cluster superconductors. More recently I have begun to examine factors affecting T_c in the different types of copper oxide superconductors. This paper summarizes the current status of the aspects of our graph-theory derived approach to chemical bonding topology which are relevant to the study of superconductors.

Background

The structures of metal clusters can be described as polyhedra or networks of polyhedra with metal atoms at the vertices. The skeletal bonding topology of such structures can be represented by a graph in which the vertices correspond to atoms or orbitals participating in the bonding and the edges correspond to bonding relationships. The eigenvalues x_k of the adjacency matrix (19) of such a graph are related to Hückel theory molecular orbital energies E_k and the Hückel parameters α, β, and S by the following equation (1,2,9,20-23):

$$E_k = \frac{\alpha + x_k\beta}{1 + x_kS} \quad (1)$$

Positive and negative eigenvalues x_k thus correspond to bonding and antibonding orbitals, respectively.

The two extreme types of skeletal chemical bonding in metal clusters may be called edge-localized and globally delocalized (1,23,24). An edge-localized polyhedron has two-electron two-center bonds along each edge of the polyhedron and is favored when the numbers of internal orbitals of the vertex atoms match the vertex degrees (i.e., number of edges meeting at the vertex in question). The skeletal bonding topology of an edge-localized polyhedral metal cluster having e edges may be depicted by e disconnected K_2 graphs where a K_2 graph consists of two vertices connected by a single edge thus representing a two-electron two-center bond. A globally delocalized polyhedron has a multicenter core bond in the center of the polyhedron and is favored when the numbers of internal orbitals do not match the vertex degrees. The skeletal bonding topology of the multicenter core bond in a globally delocalized polyhedral metal cluster with n vertices may be approximated by a K_n graph where a K_n or complete graph (25) consists of n vertices in which each of the vertices has an edge going to every other vertex for a total of n(n-1)/2 edges.

The skeletal bonding topologies of metal clusters may be described by a skeletal bonding manifold, which represents the space containing the major skeletal bonding electron density (17). The electronic properties of metal cluster structures are closely related to the dimensionality and topology of their skeletal bonding manifolds. The skeletal bonding manifold of an edge-localized metal cluster consists only of the edges of the polyhedron or polyhedral network (i.e., the 1-skeleton (26)) and therefore is porous. The skeletal bonding manifold of a globally delocalized metal cluster consists of the entire volume of the polyhedron or polyhedral network and therefore is dense. Porous skeletal bonding manifolds appear to lead to higher superconducting T_c's and H_c's apparently because the increased localization of the conduction electron wave function leads to an extremely short mean free path and/or a low Fermi velocity corresponding to a small BCS coherence length (27). All of the superconductors discussed in this paper may be considered to have porous rather than dense skeletal bonding manifolds.

The general approach for considering metal cluster bonding models compares the number of available skeletal electrons with the number of electrons required to fill all of the bonding molecular orbitals for various cluster shapes and skeletal bonding topologies, both porous and dense. The number of available skeletal electrons in each of the superconducting structures discussed in this paper is slightly less (typically 1 to 2 electrons for 4 to 6 metal atoms) than the number required to fill all of the bonding molecular orbitals for the applicable skeletal bonding topology thereby corresponding to holes in the valence band required for p-type conductivity.

Metal Cluster Superconductors

A class of metal cluster superconductors of particular interest consists of ternary molybdenum chalcogenides, commonly known as the Chevrel phases (13). These phases were the first type of superconducting ternary system found to have relatively high critical temperatures and exhibited the highest known critical magnetic fields (H_{c2}) before discovery of the copper oxide superconductors.

The most important type of Chevrel phases have the general formulas $M_nMo_6S_8$ and $M_nMo_6Se_8$ (M = Ba, Sn, Pb, Ag, lanthanides, Fe, Co, Ni, etc.) (13). The fundamental building blocks of their structures are Mo_6S_8 (or Mo_6Se_8) units containing a bonded Mo_6 octahedron (Mo-Mo distances in the range 2.67 to 2.78 Å) with a sulfur atom capping each of the 8 faces. This leads to an Mo_6 octahedron within an S_8 cube. Each (neutral) sulfur atom of the S_8 cube functions as a donor of 4 skeletal electrons to the Mo_6 octahedron within that S_8 cube leaving an electron pair from that sulfur to function as a ligand to a molybdenum atom in an adjacent Mo_6 octahedron. Maximizing this sulfur electron pair donation to the appropriate molybdenum atom in the adjacent Mo_6 octahedron results in a tilting of the Mo_6 octahedra by about 20° within the cubic array of the other metal atoms M (28). These other metal atoms M furnish electrons to the Mo_6S_8 units allowing them to approach but not attain the $Mo_6S_8^{4-}$ closed shell electronic configuration. The resulting deficiency in the number of skeletal electrons corresponds to the presence of holes in an otherwise filled valence band thereby providing a mechanism for p-type conductivity. In addition, electronic bridges between individual Mo_6 octahedra are provided by interoctahedral metal-metal interactions with the nearest interoctahedral Mo-Mo distances falling in the 3.08 to 3.49 Å range for Mo_6S_8 and Mo_6Se_8 derivatives (13). The $Mo_6S_8^{4-}$ closed shell electronic configuration for the fundamental Chevrel phase building block is isoelectronic with the molybdenum (II) halide derivatives (29) consisting of discrete $Mo_6X_8L_6^{4+}$ octahedra (X = halogen, L = two-electron donor ligand) remembering that each molybdenum vertex receives an electron pair from a sulfur atom of an adjacent Mo_6S_8 unit and thus may be treated as an LMo vertex. Such LMo vertices are taken to use 4 of their 9 total orbitals as internal orbitals for the skeletal bonding. These considerations lead to the following electron counting scheme for the closed shell $Mo_6S_8^{4-}$ unit (10,11,15):

6 LMo vertices using 4 internal orbitals: (6)[6-(9-4-1)(2)] = (6)(-2) =	-12 electrons
8 μ_3-S bridges: (8)(4) =	32 electrons
-4 charge	4 electrons
Total skeletal electrons	24 electrons

These 24 skeletal electrons are the exact number required for an edge-localized Mo_6 octahedron having two-electron two-center bonds along each of the 12 edges. Furthermore, the 4 internal orbitals used by each LMo vertex match the degree 4 vertices of an octahedron as required for edge-localized skeletal bonding.

Another interesting class of metal cluster superconductors consists of the ternary lanthanide rhodium borides $LnRh_4B_4$ (Ln = certain lanthanides such as Nd, Sm, Er, Tm, Lu) (14), which exhibit significantly higher superconducting transition temperatures than other types of metal borides. The structures of these rhodium borides consist of electronically linked Rh_4 tetrahedra in which each of the 4 triangular faces of each Rh_4 tetrahedron is capped by a boron atom of a B_2 unit (B-B distance 1.86 Å in YRh_4B_4) leading to Rh_4B_4 cubes in which the edges correspond to 12 Rh-B bonds (average length 2.17 Å in YRh_4B_4) and single diagonals of each face (average length 2.71 Å in YRh_4B_4) correspond to 6 Rh-Rh bonds. The ratio between these two lengths, namely 2.71/2.17 = 1.25, is only about 13% less than the $\sqrt{2} = 1.414$ ratio of these lengths in an ideal cube. The Rh-Rh distances of 2.71 Å in these Rh_4B_4 cubes are essentially identical to the mean Rh-Rh distance in the discrete molecular tetrahedral rhodium cluster $Rh_4(CO)_{12}$ (30) regarded as a prototypical example of an edge-localized tetrahedron (1,23).

The boron and rhodium atoms in $LnRh_4B_4$ have 4 and 9 valence orbitals, respectively. All of these valence orbitals are used to form two-center bonds leading to an edge-localized structure. The 4 bonds formed by a boron atom include 3 internal bonds to rhodium atoms in the same Rh_4B_4 cube and 1 external bond to the nearest boron atom in an adjacent Rh_4B_4 cube leading to a discrete B_2 unit. The 9 bonds formed by a rhodium atom include 3 internal bonds to the other rhodium atoms in the same Rh_4 tetrahedron, 3 internal bonds to boron atoms in the same Rh_4B_4 cube, and 3 external bonds to rhodium atoms in adjacent Rh_4B_4 cubes. Each boron atom and each rhodium atom thus use 3 and 6 internal orbitals, respectively, for the skeletal bonding and are therefore donors of 3-1 = 2 and 9-3 = 6 skeletal electrons, respectively, after allowing for the valence electrons required for external bonding. A neutral Rh_4B_4 cube thus has (4)(2) + (4)(6) = 32 skeletal electrons so that the $Rh_4B_4{}^{4-}$ tetraanion corresponds to the closed shell electronic configuration with the 36 skeletal electrons required for an Rh_4B_4 cube with edge-localized bonds along each of the 12 edges (Rh-B bonds) and along a diagonal in each of the 6 faces (Rh-Rh bonds). Since the lanthanides also present in the lattice form tripositive rather than tetrapositive ions, the $LnRh_4B_4$ borides must be $Ln^{3+}Rh_4B_4{}^{3-}$ with the $Rh_4B_4{}^{3-}$ anion having one electron less than the closed shell electronic configuration $Rh_4B_4{}^{4-}$. As in the Chevrel phases discussed above, this electron deficiency of one per Rh_4 tetrahedron corresponds to the presence of holes in an otherwise filled valence band thereby providing a mechanism for p-type conductivity.

The conducting skeletons of both the Chevrel phases MMo_6S_8 and the ternary lanthanide rhodium borides $LnRh_4B_4$ are thus seen to consist of edge-localized discrete metal polyhedra (Mo_6 octahedra and Rh_4 tetrahedra, respectively) linked into a three-dimensional network both through other atoms (sulfur and boron, respectively) and through interpolyhedral metal-metal interactions. The porosity of such edge-localized conducting skeletons coupled with the delocalization of the holes in the valence bands arising from the electron deficiencies with respect to the $Mo_6S_8{}^{4-}$ and $Rh_4B_4{}^{4-}$ closed shell electronic configurations leads naturally to the concept of porous delocalization. This concept is also useful in understanding the topology of the conducting skeletons in the high T_c copper oxide superconductors as discussed in the next section.

Copper Oxide Superconductors

The well-characterized high T_c copper oxide superconductors can be classified into the following four types:

1. The 40 K superconductors $La_{2-x}M_xCuO_{4-y}$ in which the conducting skeleton consists of a single Cu-O plane (31-33);
2. The 90 K superconductors $YBa_2Cu_3O_{7-y}$ in which the conducting skeleton consists of two Cu-O planes braced by a Cu-O chain (33-35);
3. The homologous series of bismuth copper oxide superconductors $Bi_2Sr_2Ca_{n-1}Cu_nO_{5+2n-1-x}$ (n = 1,2,3) consisting of a layer sequence $BiSrCu(CaCu)_{n-1}SrBi$ and exhibiting T_c's as high as 115 K (36-38) for materials isolated in the pure state.
4. The homologous series of thallium copper oxide superconductors $Tl_2Ba_2Ca_{n-1}Cu_nO_{5+2n-1+x}$ (n = 1,2,3) consisting of a layer sequence $TlBaCu(CaCu)_{n-1}BaTl$ and exhibiting T_c's as high as 122 K for materials isolated in the pure state (39,40).

All of these materials appear to consist of layers of two-dimensional Cu-O conducting skeletons separated by layers of the positive counterions, which can be regarded essentially as insulators. Resistivity (41) and critical magnetic field (42) measurements provide experimental evidence for the two-dimensional nature of the conducting skeletons in these materials. Increasing the rigidity of the two-dimensional conducting skeleton by coupling Cu-O layers either by bracing with a Cu-O chain in $YBa_2Cu_3O_{7-y}$ or by close proximity in the higher members of the homologous series $M^{III}_2M^{II}_2Ca_{n-1}Cu_nO_{5+2n-1+x}$ (M^{III} = Bi, M^{II} = Sr or M^{III} = Tl, M^{II} = Ba for n $\geq$ 2) leads to increases in T_c. The presence of bismuth or thallium layers appears to lead to somewhat higher T_c's. This observation coupled with computations of the electronic band structure of $Bi_2Sr_2CaCu_2O_{8+x}$ (43,44) and the low resistivity of Tl_2O_3 (39) suggests that electron transport may occur in the Bi-O or Tl-O layers as well as the Cu-O layers in the $M^{III}_2M^{II}_2Ca_{n-1}$-$Cu_nO_{5+2n-1+x}$ materials. Destruction of the two-dimensional structure of copper oxide derivatives by gaps in the Cu-O planes or by Cu-O chains in the third dimension leads to mixed copper oxides which are metallic but not superconducting (45) such as $La_2SrCu_2O_6$, $La_4BaCu_5O_{13}$, and $La_5SrCu_6O_{15}$.

The two-dimensional conducting skeletons in these copper oxide superconductors may be regarded as porously delocalized and otherwise analogous to those in the Chevrel phases and lanthanide rhodium borides by considering the following points:

1. The conducting skeleton is constructed from Cu-O-Cu bonds rather than direct Cu-Cu bonds. The much higher ionic character and thus much lower polarizability and higher rigidity of metal-oxygen bonds relative to metal-metal bonds can then be related to the persistence of superconductivity in copper oxides to much higher temperatures than that in metal clusters.
2. The required metal-metal interactions are antiferromagnetic interactions between the single unpaired electrons of two d^9 Cu(II) atoms separated by an oxygen bridge similar to antiferromagnetic Cu(II)-Cu(II) interactions in discrete binuclear complexes (46). This idea is closely related to the resonating valence bond model of Anderson (47,48).
3. The positive counterions in the copper oxide superconductors control the negative charge on the Cu-O skeleton and hence the oxidation states of the copper atoms. Positive counterions, which are "hard" in the Pearson sense (49) in preferring to bind to nonpolarizable bases, such as the lanthanides and alkaline earths, do not contribute to the conductivity. However, layers of the positive counterions bismuth and thallium, which are "soft" in the Pearson sense (49) in preferring to bind to polarizable bases, may contribute to the conductivity as noted above.
4. Partial oxidation of some of the Cu(II) to Cu(III) generates holes in the valence band required for conductivity. This is in accord with Hall effect measurements (41,50), which indicate that holes rather than electrons are the current carriers.

Studies on $La_{2-x}Sr_xCuO_{4-y}$ (51) and $YBa_2Cu_3O_{7-y}$ (52) suggest that for a system with a given conducting skeleton, T_c scales approximately linearly to the hole concentration. This idea can be used to estimate the T_c's for members of homologous series such as the $M^{III}_2M^{II}_2Ca_{n-1}Cu_nO_{5+2n-1+x}$ series (M^{III} = Bi, M^{II} = Sr or M^{III} = Tl, M^{II} = Ba) by making the following assumptions:

1. The conductivity occurs only in the Cu-O layers.
2. All of the metal atoms occupy approximately the same volume in the crystal lattice.

These crude assumptions lead to the following formula for estimating T_c:

$$T_c \approx f_{Cu}T_c^{\infty} \qquad (2)$$

In Equation 2 f_{Cu} is the fraction of metal atoms which are copper and T_c^{∞} is a limiting temperature which is characteristic for a given homologous series. Data are now available on enough members of the thallium copper oxide homologous series $Tl_2Ba_2Ca_{n-1}Cu_nO_{5+2n-1+x}$ to test this idea since species with n = 1,2,3 have been characterized and some evidence for the species n = 5 has been observed (39,40). As can be seen from Table I, Equation 2 using T_c^{∞} = 375 K leads to predicted values for T_c within 8 K of the observed values for the members of this homologous series that have been observed. Furthermore, the small deviations of the observed T_c from the T_c calculated using Equation 2 can be rationalized by the occurrence of some conductivity in the Tl-O layers. Similar treatments of the more limited information on the other homologous series suggest $T_c^{\infty} \approx 330$ K for the series $Bi_2Sr_2Ca_{n-1}Cu_nO_{5+2n-1+x}$ and $T_c^{\infty} \approx 180$ K for superconducting copper oxides with only "hard" (49) positive counterions such as the alkaline earths and the lanthanides. This analysis suggests that the highest T_c's for copper oxide superconductors will be found in the $Tl_2Ba_2Ca_{n-1}Cu_nO_{5+2n-1+x}$ homologous series. Furthermore, since $f_{Cu} \rightarrow 0.5$ as $n \rightarrow \infty$, the limiting T_c in the $Tl_2Ba_2Ca_{n-1}Cu_nO_{5+2n-1+x}$ homologous series is $(0.5)(375) \approx 187$ K as illustrated in Table 1 for estimation of T_c's for higher hypothetical members of this homologous series with n = 10 and 20. This analysis suggests that room temperature superconductivity ($T_c \approx 300$K) is far from attainable using structures with any combination of conducting Cu-O planes separated by hard and/or soft positive counterions. I suspect that another breakthrough comparable to the original discovery by Bednorz and Müller (31) of the first high T_c copper oxide will be required before room temperature superconductivity can be achieved.

Table I. The $Tl_2Ba_2Ca_{n-1}Cu_nO_{5+2n-1+x}$ Homologous Series of High Temperature Superconductors

n	Formula	f_{Cu}	T_c, K	
			Calcd.[a]	Found
1	$Tl_2Ba_2CuO_{6+x}$	0.200	75	83
2	$Tl_2Ba_2CaCu_2O_{8+x}$	0.286	107	108
3	$Tl_2Ba_2Ca_2Cu_3O_{10+x}$	0.333	125	122
5	$Tl_2Ba_2Ca_4Cu_5O_{14+x}$	0.385	144	140[b]
10	$Tl_2Ba_2Ca_9Cu_{10}O_{24+x}$	0.435	163	c
20	$Tl_2Ba_2Ca_{19}Cu_{20}O_{44+x}$	0.465	174	c

a. These calculations were performed using Equation 2 with $T_c^{\infty} = 375$ K.

b. Not obtained in the pure state; the T_c (found) = 140 K is based on an observation reported by Torardi at the April, 1988, conference at the University of Alabama.

c. Unknown species.

Acknowledgments

I am indebted to the U.S. Office of Naval Research for partial support of this work.

Literature Cited

1. King, R.B.; Rouvray, D.H. J. Am. Chem. Soc. 1977, 99, 7834.
2. King, R.B. Inorg. Chim. Acta 1982, 57, 79.
3. King, R.B. In The Physics and Chemistry of Small Clusters; Jena, P.; Rao, B.K.; Khanna, S.N., Eds.; Plenum Press: New York, 1987; pp 79-82.
4. King, R.B. Inorg. Chim. Acta 1986, 116, 99.
5. King, R.B. In Mathematics and Computational Concepts in Chemistry; Trinajstić, N., Ed; Ellis Horwood Ltd.: Chichester, England, 1986; pp 146-154.

6. King, R.B. Inorg. Chim. Acta 1986, 116, 109.
7. King, R.B. Inorg. Chim. Acta 1986, 116, 119.
8. King, R.B. Inorg. Chim. Acta 1986, 116, 125.
9. King, R.B. Int. J. Quant. Chem., Quant. Chem. Symp. 1986, S20, 227.
10. King, R.B. Inorg. Chim. Acta 1987, 129, 91.
11. King, R.B. In Graph Theory and Topology in Chemistry; King, R.B.; Rouvray, D.H., Eds.; Elsevier: Amsterdam, 1987, pp 325-343.
12. Vandenberg, J.M.; Matthias, B.T. Science 1977, 198, 194.
13. Fischer, Ø. Appl. Phys. 1978, 16, 1.
14. Woolf, L.D.; Johnston, D.C.; MacKay, H.B.; McCallum, R.W.; Maple, M.B. J. Low Temp. Phys. 1979, 35, 651.
15. King, R.B. J. Solid State Chem. 1987, 71, 224.
16. King, R.B. J. Solid State Chem. 1987, 71, 233.
17. King, R.B. J. Math. Chem. 1987, 1, 249.
18. King, R.B. Inorg. Chim. Acta 1988, 143, 15.
19. Biggs, N.L. Algebraic Graph Theory; Cambridge University Press: London, 1974; p 9.
20. Ruedenberg, K. J. Chem. Phys. 1954, 22, 1878.
21. Schmidtke, H.H. J. Chem. Phys. 1966, 45, 3920.
22. Gutman, I.; Trinajstić, N. Topics Curr. Chem. 1973, 42, 49.
23. King, R.B. In Chemical Applications of Topology and Graph Theory; King, R.B., Ed.; Elsevier: Amsterdam, 1983; pp 99-123.
24. King, R.B. In Molecular Structure and Energetics; Liebman, J.F.; Greenberg, A., Eds.; Verlag Chemie: Deerfield Beach, Florida, 1986; pp 123-148.
25. Wilson, R.J. Introduction to Graph Theory; Oliver and Boyd: Edinburgh, 1972; p 16.
26. Grünbaum, B. Convex Polytopes; Interscience: New York, 1967; p 138.
27. Fischer, Ø.; Decroux, M.; Chevrel, R.; Sergent, M. In Superconductivity in d- and f-Band Metals; Douglas, D.H., Ed.; Plenum Press: New York, 1976, pp 176-177.
28. Burdett, J.K.; Lin, J.-H. Inorg. Chem. 1982, 21, 5.
29. McCarley, R.E. Philos. Trans. R. Soc. London A 1982, 308, 141.
30. Carré, F.H.; Cotton, F.A.; Frenz, B.A. Inorg. Chem. 1976, 15, 380.
31. Bednorz, J.G.; Müller, K.A. Z. Phys. B 1986, 64, 189.
32. Wang, H.H.; Geiser, U.; Thorn, R.J.; Carlson, K.D.; Beno, M.A.; Monaghan, M.R.; Allen, T.J.; Proksch, R.B.; Stupka, D.L. Kwok, W.K.; Crabtree, G.W.; Williams, J.M. Inorg. Chem. 1987, 26, 1190.
33. Williams, J.M.; Beno, M.A.; Carlson, K.D.; Geiser, U.; Ivy Kao, H.C.; Kini, A.M.; Porter, L.C.; Schultz, A.J.; Thorn, R.J.; Wang, H.H.; Whangbo, M.-H.; Evain, M. Accts. Chem. Res. 1988, 21, 1.
34. Wu, M.K.; Ashburn, J.R.; Torng, C.J.; Hor, P.H.; Meng, R.L.; Gao, L.; Huang, Z.J.; Wang, Y.Q.; Chu, C.W. Phys. Rev. Lett. 1987, 58, 908.
35. Whangbo, M.-H.; Evain, M.; Beno, M.A.; Williams, J.M. Inorg. Chem. 1987, 26, 1831.

36. Hazen, R.M.; Prewitt, C.T.; Angel, R.J.; Ross, N.L.; Finger, L.W.; Hadidacos, C.G.; Veblen, D.R.; Heaney, P.J.; Hor, P.H.; Meng, R.L.; Sun, Y.Y.; Wang, Y.Q.; Xue, Y.Y.; Huang, Z.J.; Gao, L.; Bechtold, J.; Chu, C.W. Phys. Rev. Lett. 1988, 60, 1174.
37. Subramanian, M.A.; Torardi, C.C.; Calabrese, J.C.; Gopalakrishnan, J.; Morrissey, K.J.; Askew, T.R.; Flippen, R.B.; Chowdhry, U.; Sleight, A.W. Science 1988, 239, 1015.
38. Raveau, B.; Hasegawa, T.; Wu, M.-K.; Tarascon, J.-M.; Torardi, C.C. papers presented at the International Conference on the First Two Years of High-Temperature Superconductivity, Tuscaloosa, Alabama, April, 1988.
39. Hazen, R.M.; Finger, L.W.; Angel, R.J.; Prewitt, C.T.; Ross, N.L.; Hadidiacos, C.G.; Heaney, P.J.; Veblen, D.R.; Sheng, Z.Z.; El Ali, A.; Hermann, A.M. Phys. Rev. Lett. 1988, 60, 1657.
40. Raveau, B.; Hermann, A.M.; Torardi, C.C. papers presented at the International Conference on the First Two Years of High-Temperature Superconductivity, Tuscaloosa, Alabama, April, 1988.
41. Cheong, S.-W.; Fisk, Z.; Kwok, R.S.; Remeika, J.P.; Thompson, J.D.; Gruner, G. Phys. Rev. B 1988, 37, 5916.
42. Moodera, J.S.; Meservey, R.; Tkaczyk, J.E.; Hao, C.X.; Gibson, G.A.; Tedrow, P.M. Phys. Rev. B 1988, 37, 619.
43. Hybertsen, M.S.; Mattheiss, L.F. Phys. Rev. Lett. 1988, 60, 1661.
44. Krakauer, H.; Pickett, W.E. Phys. Rev. Lett. 1988, 60, 1665.
45. Torrance, J.B.; Tokura, Y.; Nazzal, A.; Parkin, S.S.P. Phys. Rev. Lett. 1988, 60, 542.
46. Cairns, C.J.; Busch, D.H. Coord. Chem. Rev. 1986, 69, 1.
47. Anderson, P.W. Science 1987, 235, 1196.
48. Anderson, P.W.; Baskaran, G.; Zou, Z.; Hsu, T. Phys. Rev. Lett. 1987, 58, 2790.
49. Pearson, R.G. J. Am. Chem. Soc. 1963, 85, 3533.
50. Ong, N.P.; Wang, Z.Z.; Clayhold, J.; Tarascon, J.M.; Greene, L.H.; McKinnon, W.R. Phys. Rev. B 1987, 35, 8807.
51. Shafer, M.W.; Penney, T.; Olson, B.L. Phys. Rev. B 1987, 36, 4047.
52. Wang, Z.Z.; Clayhold, J.; Ong, N.P.; Tarascon, J.M.; Greene, L.H.; McKinnon, W.R.; Hull, G.W. Phys. Rev. B 1987, 36, 7222.

RECEIVED July 15, 1988

Chapter 6

Bonds, Bands, Charge-Transfer Excitations, and High-Temperature Superconductivity

A. J. Freeman, S. Massidda, and Jaejun Yu

Department of Physics and Astronomy, Northwestern University, Evanston, IL 60208

Results of highly precise all-electron local density calculations of the electronic structure for the high T_c superconductors ($La_{2-x}M_xCuO_4$, $YBa_2Cu_3O_{7-\delta}$, $Bi_2Sr_2CaCu_2O_8$, $Tl_2Ba_2CaCu_2O_8$ and $Tl_2Ba_2Ca_2Cu_3O_{10}$), as determined with the full potential linearized augmented plane wave (FLAPW) method are presented. In all these high T_c materials, the Cu-O dpσ (anti-bonding) bands arising from the CuO_2 planes cross E_F and are strongly two-dimensional. The Y, Ba, Sr, and Ca atoms are highly ionic with the Ca^{2+} ions serving to insulate the Cu-O planes. In both $Bi_2Sr_2CaCu_2O_8$ and the Tl compounds, the Bi-O and Ti-O planes contribute to the density of states at E_F and hence to the transport properties. The similarity and difference between these systems are described in detail. The results obtained demonstrate the close relation of the band structure to the structural arrangements of the constituent atoms and provide an integrated chemical and physical picture of their interactions and their possible relation to the origin of their high T_c superconductivity. Our calculations indicate the inadequacy of a purely electron-phonon mechanism in explaining the high T_c. An excitonic mechanism of superconductivity is discussed.

As is apparent from the popular press, the discovery of the high T_c superconductors $La_{2-x}M_xCuO_4$ (1) and $YBa_2Cu_3O_{7-\delta}$ (2) has generated excitement among scientists and technologists on an unprecedented scale. The recent discovery (3-5) of superconductivity above 85 K in Bi-Sr-Ca-Cu-O and above 120 K in Tl-Ba-Ca-Cu-O, which do not have a rare-earth element, has added a new dimension to the exciting subject of high T_c. A particularly exciting aspect of having added a third and fourth oxide superconducting material lies in the opportunity for seeking out common features in all four materials which may be

0097–6156/88/0377–0064$06.00/0

relevant to determining the mechanism of their high T_c for which there is now considerable effort underway. One of the starting points is certainly a detailed picture of the electronic structure of the compound, a goal which is achievable by present day supercomputers in combination with highly precise numerical methods to solve the local density functional (LDF) Kohn-Sham equations in a self-consistent way. Even today, the origin of superconductivity in the new metallic oxides remains a challenge despite some intriguing hints obtained from experiment and electronic structure calculations. Still, it is now quite apparent that understanding the electronic structure and properties of the new high T_c superconductors is emerging. This is an important step towards achieving an understanding of the origin of their superconductivity. Detailed high resolution local density band structure results have served to demonstrate what has been our major emphasis, namely the close relation of the physics (band structure) and chemistry (bonds and valences) to the structural arrangements of the constituent atoms; they may also provide insight into the basic mechanism of their superconductivity. The successes of local density functional studies include excellent agreement of their predictions of their anisotropic Fermi surface (6) (see also Smedskjaer, et al., preprint), and transport and thermopower properties (7-8) with experiment. For all these reasons, a highly precise LDF theoretical determination of the electronic structure and properties of the new superconductors is both timely and important.

For the electronic structure calculations, we used the highly precise full-potential linearized augmented plane wave method (FLAPW) (9-10) within the local density approximation and the Hedin-Lundqvist form for the exchange correlation potential. In the FLAPW approach no shape approximations are made to either the charge density or the potential. Results obtained on the systems we have studied - La_2CuO_4, $YBa_2Cu_3O_7$, $YBa_2Cu_3O_6$, $GdBa_2Cu_3O_7$, $Bi_2Sr_2CaCu_2O_8$, $Tl_2Ba_2CaCu_2O_8$ and $Tl_2Ba_2Ca_2Cu_3O_{10}$ - indicate a number of common chemical and physical features, especially the role of intercalated layers such as the CuO chains, Bi_2O_2 and Tl_2O_2 rock-salt type layers. In this paper, we provide a brief summary of the results on the detailed electronic structures of the $La_{2-x}M_xCuO_4$, $LnBa_2Cu_3O_{7-\delta}$, $Bi_2Sr_2CaCu_2O_8$ and $Tl_2Ba_2CaCu_2O_8$ ($Tl_2Ba_2Ca_2Cu_3O_{10}$) systems, compare them, and point out their relations to an excitonic mechanism of high T_c superconductivity.

Electronic Structure and Properties of $La_{2-x}M_xCuO_4$

Early on, the results of our highly precise all-electron local density full potential linearized augmented plane wave (9-10) (FLAPW) calculations of the energy band structure, charge densities, Fermi surface, etc., for $La_{2-x}M_xCuO_4$ (M = Sr, Ba)(11-13) demonstrated: (i) that the material consisted of metallic Cu-O(1) planes separated by insulating (dielectric) La-O(2) planes and (ii) that this 2D character and alternating metal/insulator planes would have, as some of their most important consequences, strongly anisotropic (transport, magnetic, etc.) properties. Thus, the calculated band structure along high symmetry directions in the Brillouin zone shows only flat bands, i.e., almost no dispersion, along the c axis, demonstrating that the interactions between the Cu, O(2) and La atoms are quite weak. However, along the basal plane directions there are very strong interactions between the Cu-O(1) atoms leading to large dispersions and a very wide bandwidth (~ 9 eV).

The band structure near E_F has a number of interesting features (11-12). What is especially striking is that, in contrast to the complexity of its structure, only a single free electron-like band crosses E_F and gives rise to a simple Fermi surface (13). Since this band originates from the Cu $d_{x^2-y^2}$-O(1) $p_{x,y}$ orbitals confined within the Cu-O(1) layer, it exhibits clearly all the characteristics of a two dimensional electron system. Particularly striking is the occurrence of a van Hove saddle point singularity (SPS). Such an SPS is expected, and found to contribute strongly, via a singular feature, to the density of states (DOS). This dominance of the DOS near E_F by the SPS contribution is responsible for many of the striking properties of this material with divalent ion (M_x) additions (including variations of T_c and other properties).

The quasi-2D properties of the electronic structure are also supported by plots of the charge densities of electrons at E_F (summed over the FS) (cf. Figure 1). This charge density consists mainly of Cu $d_{x^2-y^2}$ and O(1) $p_{x,y}$ hybridized orbitals in the plane with some additional contribution of the Cu d_{z^2} and O(2) p_z components. As shown in Figure 4 of Reference 11, the 2D confinement to the Cu-O(1) plane is more pronouned in a plot of the charge density for the state at E_F crossing along the (110) direction, which has only the Cu $d_{x^2-y^2}$-O(1) $p_{x,y}$ (anti-) bonding character due to the symmetry of the (110) line in the BZ. There is essentially no electron density around the La site at E_F. This means that the La atoms do not contribute directly to the dynamical processes involving electrons near E_F. Further, an analysis of the band structure shows that it is a fairly good approximation to consider the La atoms to be described in chemical terms as La^{3+} ions.

In total energy frozen phonon calculations[14] on $La_{2-x}M_xCu_4$, the role and effect of the optical breathing mode turned out to be significant. Since the breathing phonon mode involves the motion of oxygen atoms against the directional bonding of Cu d - O p in the plane, the in-plane Cu $d_{x^2-y^2}$-O(1) $p_{x,y}$ states of the 2D conduction bands are strongly affected by the breathing displacement. On the other hand, the out-of-plane Cu d_{z^2}-O(2) p_z orbitals, which are quite localized in the plane, are not much affected by the same breathing mode. But, because of the relative change of the Cu $d_{x^2-y^2}$-O(1) $p_{x,y}$ and Cu d_{z^2}-O(2) p_z, the charge fluctuations between Cu atoms, which can be as large as 0.3 electrons at the maximum of the O displacement, lead to transitions of the out-of-plane Cu d_{z^2}-O(2) p_z into the in-plane Cu $d_{x^2-y^2}$-O(1) $p_{x,y}$. Since the out-of-plane (anti-bonding) Cu d_{z^2}-O(2) p_z states near E_F are localized, we expect that the localized Cu d_{z^2}-O(2) p_z states, introduced by the charge fluctuation, may couple to the delocalized conduction electrons of the in-plane Cu $d_{x^2-y^2}$-O(1) $p_{x,y}$ orbital and to possibly form an excitonic state. Thus, a key role in possible charge transfer excitations (CTE) is played by excitations between occupied localized Cu d_{z^2}-O(2) p_z and empty itinerant Cu $d_{x^2-y^2}$-O(1) $p_{x,y}$ states. We also emphasized that these could couple resonantly with natural "Cu^{2+}-Cu^{3+}-like" charge fluctuations which exist in the x > 0 compounds, with important consequences for the superconductivity.

Electronic Structure and Properties of $YBa_2Cu_3O_7$

For the 90K superconductor $YBa_2Cu_3O_{7-\delta}$, discovered by Chu et al. (2), we presented (15-16) detailed high resolution results on the electronic band structure and density of states derived properties as obtained from the same highly precise state-of-the-art local density approach. These results demonstrated the close relation of the band structure to the structural arrangements of the constituent atoms and have helped to provide an integrated chemical and physical picture of the interactions.

The important structural features of the $YBa_2Cu_3O_{7-\delta}$ compounds arise from the fact that (2+δ) oxygen atoms are missing from the perfect triple perovskite, $YCuO_3(BaCuO_3)_2$. The O vacancies in the Cu plane (between two BaO planes) give rise to the formation of a linear chain of Cu and O ions (labelled Cu1-O1-Cu1). The total absence of O ions in the Y plane leads to the two Cu ions (called Cu2) in five-coordinated positions - as shown in Figure 2. The double layers of Cu2-O planes in $YBa_2Cu_3O_7$ yield a 2D structure, corresponding to the single CuO_2 plane in $La_{2-x}M_xCuO_4$.

The calculated band structure of stoichiometric $YBa_2Cu_3O_7$ along high symmetry directions in the bottom (k_z = 0) plane of the orthorhombic BZ is shown in Figure 3. The very close similarity in the band structure for the k_z = 0 and $k_z = \pi/c$ planes (15) indicates the highly 2D nature of the band structure. It is seen from Figure 3, that as in the case of La_2CuO_4, a remarkably simple band structure near E_F emerges from this complex set of 36 bands (originating from three Cu (3d) and seven O (2p) atoms). Four bands - two each consisting of Cu2(3d)-O2(p)-O3(p) orbitals and Cu1(d)-O1(p)-O4(p) orbitals - cross E_F. Two strongly dispersed bands C (S_1' and S_4 in Figure 3; the labelling is given by their character at S) consist of $Cu2(d_{x^2-y^2})$-$O2(p_x)$-$O3(p_y)$ combinations and have the 2D character which proved so important for the properties of $La_{2-x}M_xCuO_4$. Significantly, the $Cu1(d_{z^2-y^2})$-$O1(p_y)$-$O4(p_z)$ anti-bonding band A (S_1 in Figure 3) shows the (large) 1D dispersion expected from the Cu1-O1-Cu1 linear chains but is almost entirely unoccupied. This band is in sharp contrast to the π-bonding band B (formed from the Cu1 (d_{zy})-$O1(p_z)$-$O4(p_y)$ orbitals) which is almost entirely occupied in the stoichiometric (δ = 0) compound.

We have predicted the Fermi surfaces (FS) of $YBa_2Cu_3O_7$ determined from our band structure. Two 2D Cu-O dpσ bands yield two rounded square FS's centered around S. These 2D FS show strong nesting features along (100) and (010) directions. In addition, the 1D electronic structure also gives a 1D FS with possible nesting features along the (010) direction. There are two additional hole pockets around Y(T) and S(R) which come from the flat dpπ bands at E_F discussed before. Our predictions of the FS for $YBa_2Cu_3O_7$ have been confirmed recently by positron annihilation experiments (6-7). It is important to note that confirmation of the FS results has significant impact on several theories (e.g., the so-called resonant valence band

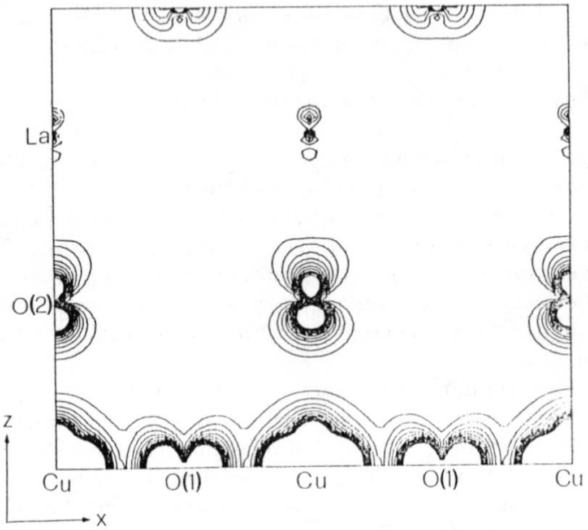

Figure 1 Contour plot in the xz vertical plane of valence charge density at E_F for La_2CuO_4.

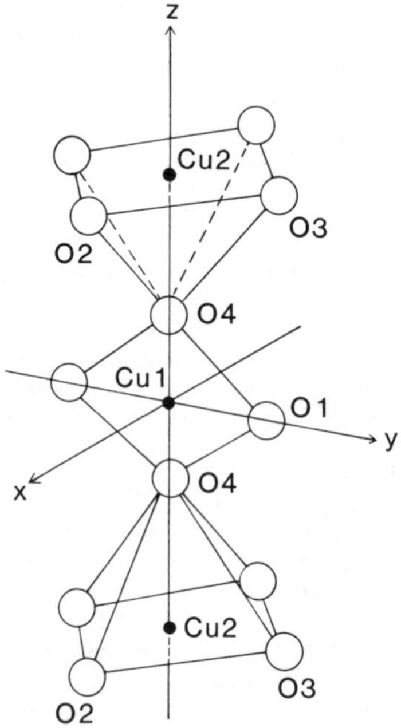

Figure 2 A local environment for the Cu1 and Cu2 atoms in $YBa_2Cu_3O_7$, following the Y-Cu2-Ba-Cu1-Ba-Cu2-Y ordering along z.

or RVB theory) (17), which deny the Fermi liquid nature of the normal ground state in the Cu-oxide superconductors. Significantly, P.W. Andersen has stated (18) that the proven existence of a Fermi surface would necessitate "withdrawal" of his RVB theory.

Here too, charge density calculations (15-16) reflect the structural properties of the material. Charge density plots for the individual states near E_F demonstrate the 2D nature of Cu2-O2-O3 dpσ bands and the 1D nature of the Cu1-O1-O4 dpσ bands. The ionic Y (or R = rare earth) atoms act as electron donors and do not otherwise participate. Also, the partial DOS at E_F for Y give extremely low values for the conduction electrons (the same is true for Gd). These results give an immediate explanation for the observed (19) coexistence of the high T_c superconductivity and magnetic ordering in the $RBa_2Cu_3O_{7-\delta}$ structures. The lack of conduction electron density around the R-site (J. Yu and A. J. Freeman, to be published) means that the unpaired rare-earth f-electrons are decoupled from the Cooper pairs (i.e., magnetic isolation) and so cannot pair-break.

In our calculation for $\delta = 0$, the DOS at E_F, $N(E_F)$, is 1.13 states/eV Cu-atom, which is comparable to the 1.2 and 1.9 states/eV Cu-atom found earlier for $La_{2-x}M_xCuO_4$ at x = 0 and at the peak at x = 0.16, respectively. This means that the $N(E_F)$ per Cu atom values in the high T_c superconductor, $YBa_2Cu_3O_{7-\delta}$, is significantly lower than was found earlier (either experimentally or theoretically) for the (lower) high T_c superconductor, $La_{2-x}M_xCuO_4$. This result has a number of important possible consequences for DOS derived properties and superconductivity, including: reduced screening, an increased role for the polarization of ionic constituents, lowered conductivity (and reduced superconducting current carrying capacity), the important role of charge transfer excitations (CTE), etc.

The Role of Oxygen Vacancies: $YBa_{22}Cu_3O_7$ vs. $YBa_2Cu_3O_6$

Recently, several neutron experiments (20-22) showed that the oxygen vacancies concentrate on the O1 sites and change the composition and symmetry from orthorhombic (in $YBa_2Cu_3O_7$) to tetragonal (in $YBa_2Cu_3O_6$). The absence of oxygens on the O1 site destroys the 1D chain structure in the $YBa_2Cu_3O_6$. The additional oxygen vacancies, therefore, change the local symmetry as well as the electronic configuration around the Cu1 sites. In this geometry, each Cu1 ion would be completely isolated from the other Cu1 ions in the Cu1 plane (having no oxygens lying between the Cu1's) and remain as Cu^+ ions with a completely filled d-shell. Hence, the d-orbital states of Cu1 are expected to be very localized in the Cu1 plane. One notable consequence of the change of structure is that the Cu1-O4 distance in $YBa_2Cu_3O_6$ is even shorter than in $YBa_2Cu_3O_7$.

To examine these expectations and to provide insight into the possible role (16) of the CTE in the 1D chains of $YBa_2Cu_3O_{7-\delta}$, we compare results of calculations (23) for both $YBa_2Cu_3O_7$ and $YBa_2Cu_3O_6$, focussing on the role of chains vs. planes. The calculated band structure (near E_F) of tetragonal $YBa_2Cu_3O_6$ and orthorhombic $YBa_2Cu_3O_7$ both exhibit two strongly dispersed 2D bands crossing E_F consisting of Cu2($d_{x^2-y^2}$)-O2(p_x)-O3(p_y) orbitals in the 2D Cu2-O planes. However, the dominant 1D electronic structure, arising from the linear chains in $YBa_2Cu_3O_{7-\delta}$, is completely absent in $YBa_2Cu_3O_6$. Instead of the 1D structure of dpσ and dpπ states from the Cu1-O1-O4 chains, the O_6 compound has two bands consisting of Cu1 $d_{yz}(d_{zx})$-O4 $p_y(p_x)$ orbitals and are degenerate at the points Γ and M in the Brillouin zone. In addition to the two conduction bands crossing E_F, we find a very small contribution to $N(E_F)$ from the Cu1 complex in the linear chains; therefore, the DOS at E_F of $YBa_2Cu_3O_6$ is reduced to 0.67 states/eV Cu-atom, which is smaller than that of $YBa_2Cu_3O_7$. In addition to the reduction of $N(E_F)$, there is a strong hybridization of the Cu2-O2(3) dpσ bands and the Cu1-O4 dpπ bands along the ΓM direction. However, there are several experimental observations (24-25) of the antiferromagnetic insulators of $YBa_2Cu_3O_6$ as well as La_2CuO_4. These lead to the question whether the antiferromagnetic insulating ground state can be described by a (charge-) spin-density wave state within a band picture. Such a failure of LDA is well known for the case of CoO (26), for example. Later on, there were several reports of a stable magnetic ground state being found (in a band picture), but most of them are not convincing. Although there are still unresolved problems as to the relations of the O vacancies to the anti-ferromagnetic ordering in $La_2CuO_{4-\delta}$ and $YBa_2Cu_3O_{6+x}$, there is now a large effort to overcome this short-coming of the LSDA (Local spin-density approximation). We believe that this part of the phase diagram has no relation to the superconductivity observed for the metallic phases.

Electronic Structure and Properties of $Bi_2Sr_2CaCu_2O_8$

For the new high T_c superconductor Bi-Sr-Ca-Cu-O, we presented (27) results of a highly precise local density determination of the electronic structure (energy bands, densities of states, Fermi surface, and charge densities). As in the case of the other high T_c Cu-O superconductors, we found a relatively simple band structure near E_F and strongly anisotropic highly 2D properties. One of the interesting points in the Bi-Sr-Ca-Cu-O system is that the Bi-O planes contribute substantially to $N(E_F)$ and to the transport properties.

Proposed structures of the Bi-Sr-Ca-Cu-O superconductors include an orthorhombic (quasi-tetragonal) unit subcell with composition $Bi_{2+x}Sr_2Ca_{1-x}Cu_2O_8$ and a superstructure along the b direction suggested by Hazen et al. (28) and Sunshine et al. (preprint); the origin of the superstructure is not yet clear. The subcell structure of $Bi_2Sr_2CaCu_2O_8$ by Sunshine et al. (preprint) shows the presence of two CuO_2 layers (separated by a Ca layer) and of rock-salt type Bi_2O_2 layers; the $(CuO_2)\cdot Ca\cdot (CuO_2)$ layers are separated by single SrO layer from the Bi_2O_2 layers. It is striking that this new system has no rare-earth elements; instead, it has Bi atoms replacing those strongly electro-positive trivalent ions.

The calculated band structure of $Bi_2Sr_2CaCu_2O_8$ shown in Figure 4 has many points in common with those of the other high T_c Cu-oxide compounds (11,15-16): above a set of fully occupied bands (in this case 48) with predominant Cu d-O p character, we find a relatively simple band structure at E_F, which in this compound consists of only three bands crossing E_F. Two (almost degenerate) bands (bands a and

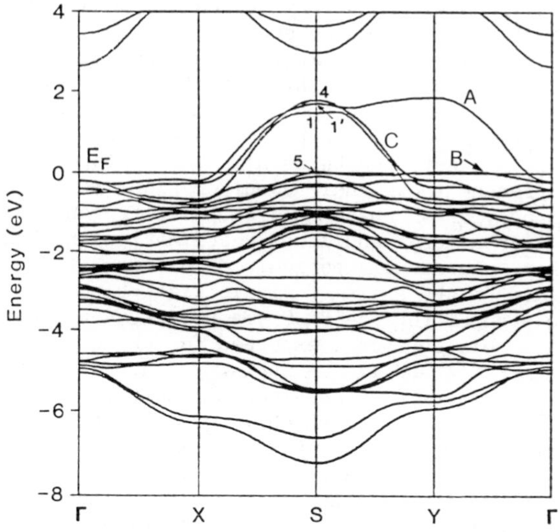

Figure 3 Band structure of $YBa_2Cu_3O_7$ along symmetry directions in the $k_z = 0$ plane of the orthorhombic Brillouin zone.

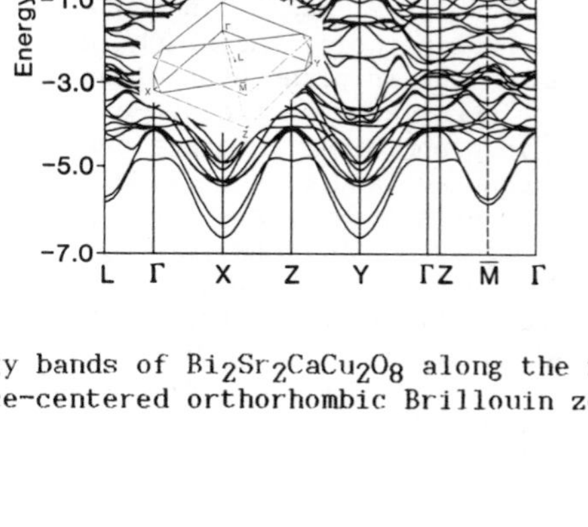

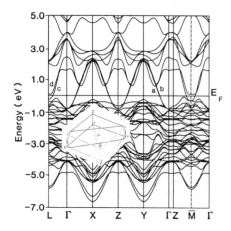

Figure 4 Energy bands of $Bi_2Sr_2CaCu_2O_8$ along the main symmetry lines of the face-centered orthorhombic Brillouin zone shown as an insert.

b in Fig. 4) with strong Cu-O dpσ character cross E_F and have two dimensional character. They do not cross E_F at the midpoint of the Γ-Z direction because of the existence of the Bi-O band which also crosses E_F. Their quasi-degeneracy proves the weakness of the interplane interactions for these states.

At energies (mostly) above E_F we find a set of six bands corresponding to the antibonding hybrids of the p orbitals of the two Bi atoms in the unit cell with the O2 and O3 p states. These bands form electron pockets near the L point and at the midpoint between Γ and Z (which will be referred to as $\bar{M}$). Their dispersion across the BZ is quite different (note their double periodicity) from that of the Cu-O dpσ bands, as a consequence of the different bonding character (ppσ versus dpσ) and local coordination (rock-salt versus perovskite-like). The doubly periodic dispersion of Bi-O ppσ bands can be understood on the basis of simple tight-binding arguments.

The calculated FS for the $Bi_2Sr_2CaCu_2O_8$ system shows the two not quite degenerate hole surfaces, expected for the Cu-O bands, centered at the zone corners (X and Y). A striking feature is the highly 2D nature of these surfaces with considerable nesting. The Bi-O band forms a small electron surface which is close to being a rounded square (dimensions 0.31 ($\frac{2\pi}{a}$) and 0.34 ($\frac{2\pi}{a}$)). Their nesting, however, does not appear to be related to the observed superstructure.

The complicated FS structure near $\bar{M}$ and the non-degeneracy of the Cu-O FS are derived from the band interaction (anticrossing) of the Cu-O bands and Bi-O bands near $\bar{M}$, where, in the absence of band interactions, two Bi-O bands (with upward dispersion) cross three Cu-O bands (with downward dispersion) below E_F in the region near $\bar{M}$, L.

As the band structure of $Bi_2Sr_2CaCu_2O_8$ is complicated by the presence of the Bi-O2 bands near E_F, the Bi p bands also make the total density-of-states (DOS) look different from those of the other Cu-oxide superconductors. The total DOS of $Bi_2Sr_2CaCu_2O_8$ is a superposition of the contributions from the Cu-O1 complex and the Bi-O2 complex with O3 connecting the two. In fact, both Sr and Ca atoms are extremely ionic and give very little contribution to the DOS of the valence bands. The Cu and O1 projected DOS (pDOS) show a wide band-width of ~9 eV, which is typical of dpσ (bonding and anti-bonding) bands from the two-dimensional (2D) CuO_2 layers.

While the pDOS plots of Bi and O2 indicate that the bonding between Bi and O2 is rather ionic, the short in-plane distance between Bi and O2 makes the in-plane hybridization of the Bi $p_{x,y}$ and O2 $p_{x,y}$ orbitals much stronger than the out-of-plane Bi-O2 hybridization. The in-plane Bi-O2 interactions result in a broad- and also partially occupied-band (~ 4 eV wide).

The total density of states at the Fermi level, $N(E_F)$, is 3.03 states/(eV-cell). Large contributions to $N(E_F)$ come from both the Cu-O1 and the Bi-O2 layers. The Bi-O2 contributions are from the pσ bands which create small electron pockets around L (and $\bar{M}$). Therefore (and significantly), both Cu-O1 and Bi-O2 layers provide conduction electrons in this material. This result contrasts with the case of $La_{2-x}M_xCuO_4$ and of $YBa_2Cu_3O_{7-\delta}$, where the cations do not contribute to $N(E_F)$ but give rise to conduction bands which lie 2-3 eV above E_F.

Electronic Structure and Properties of $Tl_2Ba_2CaCu_2O_8$ and $Tl_2Ba_2Ca_2Cu_3O_{10}$

The new high T_c superconductors of the Tl-Ba-Ca-Cu-O system has been discovered (5) and found to have two different but related superconducting phases (29-30), with compositions $Tl_2Ba_2CaCu_2O_8$ (which we refer to as "Tl/2212") and $Tl_2Ba_2Ca_2Cu_3O_{10}$ ("Tl/2223"), with T_c ~ 112 K and ~ 125 K, respectively. For both Tl/2212 and Tl/2223, we presented (31) results of highly precise local density calculations of the electronic structure. A relatively simple band structure is found near E_F and strong 2D properties are predicted - again as in the case of the other high T_c materials.

The crystal structures of Tl/2212 and Tl/2223 determined by Subramanian et al. (30) show essentially the same features as that of $Bi_2Sr_2CaCu_2O_8$. The structure of Tl/2212 consists of two CuO_2 layers (separated by a Ca layer) and of rock-salt type Tl_2O_2 layers, where the (CuO_2)-Ca-(CuO_2) layers are separated by single BaO layers from the Tl_2O_2 layers. Similarly, the Tl/2223 structure is related to the Tl/2212 structure by an addition of extra Ca and CuO_2 layers, where the (CuO_2)-Ca-(CuO_2)-Ca-(CuO_2) layers are separated by single BaO layers from the Tl_2O_2 layers. The notation and local arrangement of atoms in the formula unit of Tl/2212 and Tl/2223 are shown in Figure 5.

The calculated energy bands of Tl/2212 and Tl/2223 (in an extended zone scheme) are shown in Figures 6 and 7. These bands present, as one would expect, strong similarities with those of all the other high T_c Cu-oxide superconductors (11, 13-16, 27). As in $Bi_2Sr_2CaCu_2O_8$, we have in Tl/2212 two Cu-O dpσ bands (one per Cu-O sheet) crossing E_F, while three Cu-O dpσ bands crossing E_F are present in the Tl/2223 compound.

Despite these common features, the Tl systems present some interesting new points. In both the Tl/2212 and Tl/2223 compounds, there exists the presence of electron pockets around the Γ and Z points. A careful analysis of the character of these states, however, reveals important differences with respect to the $Bi_2Sr_2CaCu_2O_8$ case. While the Bi-O bands at E_F in $Bi_2Sr_2CaCu_2O_8$ originate mainly from the in-plane ppσ Bi-O hybrid, the Tl-O bands at E_F in Tl/2212 and Tl/2223 are mostly from oxygen p states hybridized (anti-bonding) with the Tl orbitals. In fact, the major Tl 6s bands are located at about 7 eV below E_F.

The charge density contour map of Figure 8 for the high-lying Γ_1' and Γ_2' antibonding states in a (110) plane of Tl/2212 illustrate the bonding nature of Tl and O in this compound (and, of course, of Tl/2223). These states are both strongly antibonding combinations of the O2 and O3 $2p_z$ orbitals with the Tl 6s and (significantly) $5d_{z^2}$ orbitals. The presence of the d_{z^2} orbitals, which are more localized than the 6s orbitals, may play a role in the stabilization of the very short out-of-plane Tl-O3 (2.03 Å) and Tl-O2 (1.98 Å) distances. On the other hand, since the in-plane Tl-O3 distances (2.46 Å), are larger than the sum of the Tl and O ionic radii (~ 2.2 Å), the in-plane interactions are much weaker; therefore, the O3 $p_{x,y}$ orbitals do not participate significantly in the makeup of these

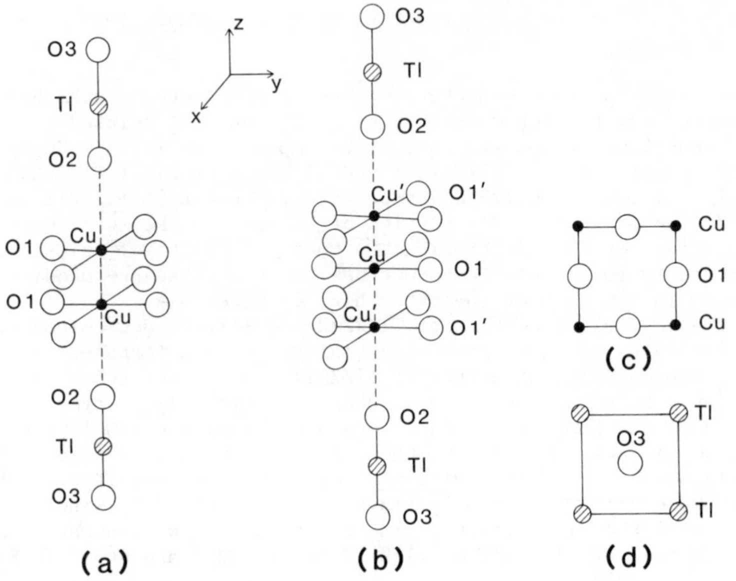

Figure 5 Schematic arrangement of atoms in the formula unit cell of the body-centered tetragonal structure for (a) $Tl_2Ba_2CaCu_2O_8$ and (b) $Tl_2Ba_2Ca_2Cu_3O_{10}$. Not shown are the Ba atoms which lie on the O2-plane and the Ca atoms which lie in planes between the Cu atom planes. The in-plane positions are given in (c) for the CuO_2 plane and in (d) for the TlO plane.

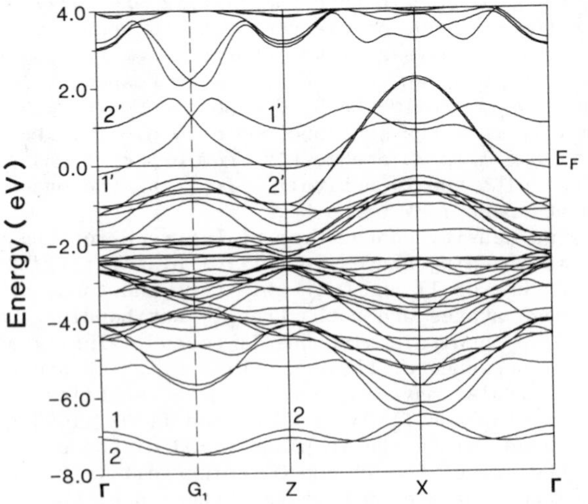

Figure 6 Energy bands of $Tl_2Ba_2CaCu_2O_8$ along the main symmetry lines of the body-centered tetragonal extended Brillouin zone. (Notation from Reference 11.)

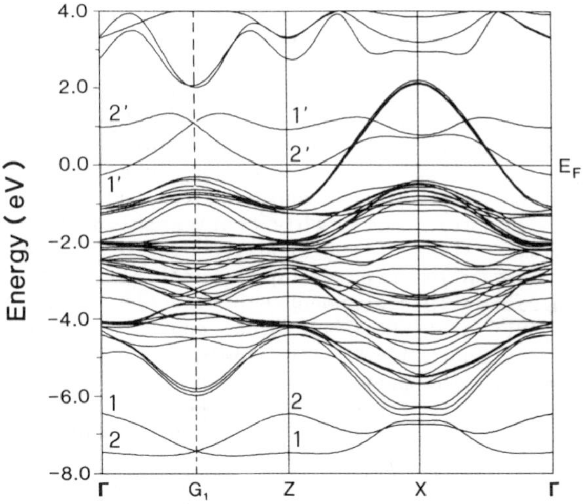

Figure 7 Energy bands of $Tl_2Ba_2Ca_2Cu_3O_{10}$ (shown as in Figure 6).

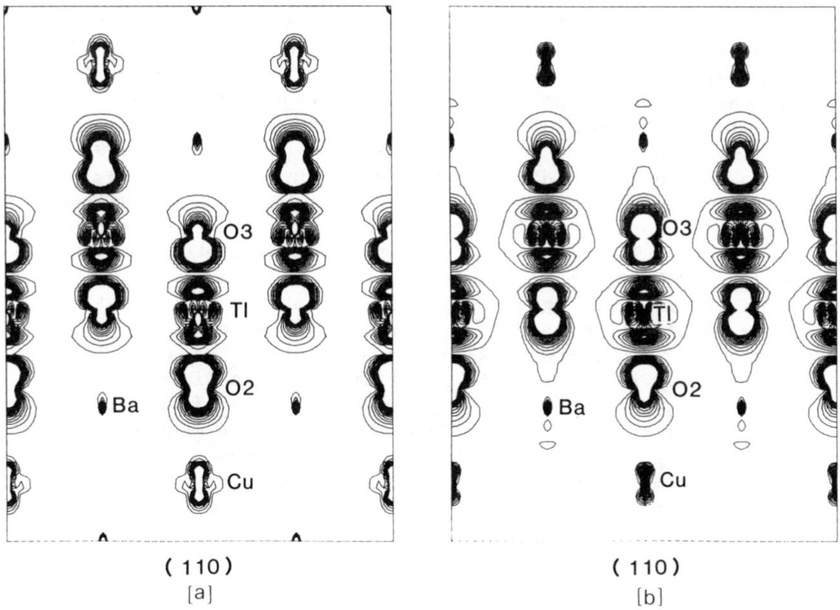

Figure 8 Charge density contour plots for the Γ_1- (a) and Γ_2- (b) states in a (110) plane for $Tl_2Ba_2CaCu_2O_8$. Contours are given on a linear scale and successive contours are separated by 10^{-3} e/a.^{3}.

states, and turn out to be rather non-bonding. The strength of the out-of-plane Tl-O interaction accounts for the observed (30) physical properties of the Tl-O layers: unlike the Bi-O layers with in-plane ppσ bonds in $Bi_2Sr_2CaCu_2O_8$, they do not show mica-like properties.

The O3 $2p_z$ - Tl 6s, $5d_{z^2}$ - O2 $2p_z$ Γ_1- and Γ_2- bands are found to be completely antibonding all over the Brillouin zone (the corresponding bonding states are, of course, the low lying Γ_1 and Γ_2 bands). Since the bonding bands are occupied and the antibonding ones (almost completely) empty, this configuration corresponds to an optimum stability. Also, on the opposite side of the structure, the O2 p_z orbitals see fully occupied Cu d_{z^2} orbitals which exert a Coulombic repulsion. This argument explains the strong elongated tetragonal distortion of the O pyramidal coordination around the Cu sites and the very short Tl-O2 interatomic distances. An analogous situation was found (15) previously in $YBa_2Cu_3O_7$ to stabilize the very short Cu1-O4 and larger Cu2-O4 distances.

We have also found that the 6s electrons of the Tl ions in $Tl_2Ba_2CaCu_2O_8$ are covalently bonded to the out-of-plane oxygens, O2 and O3; similarly, the Bi p orbitals in $Bi_2Sr_2CaCu_2O_8$ form weak covalent bonds with the in-plane O2 oxygens. This result is in contrast to the case of the $La_{2-x}M_xCuO_4$ and $YBa_2Cu_3O_7$ systems where the presence of strongly electro-positive 3+ ions (e.g., La^{3+}, Y^{3+}) is essential. We shall see that this has a significant effect on the electronic structure and may be relevent to understanding the superconducting mechanism.

The DOS of both Tl/2212 and Tl/2223 have the same features as their major contributions arising from both the Cu d - O p and Tl-O2-O3 complexes. As seen in the other high T_c Cu-oxide superconductors, both Ba and Ca remain ionic and give almost no contribution to the DOS of the valence bands. The Cu and O1 projected partial DOS (pDOS) show a wide band width of ~ 9 eV, which is typical of dpσ (bonding and anti-bonding) bands from the two-dimensional CuO_2 layers. While the O1 p orbitals hybridize strongly with the Cu d orbitals, the O2 and O3 p orbitals are decoupled from the Cu-O1 bands and give rise to the major non-bonding peaks located at -3 eV with a narrow band width of about 1 eV. The nature of these non-bonding states is attributed to the O2 and O3 p_x, p_y orbitals.

One of the significant effects of the strong hybridization of the Tl s (d_{z^2}) with O2 and O3 p_z states (discussed above) on the pDOS structure of Tl/2212 and Tl/2223 is the existence of a gap between the non-bonding $p_{x,y}$ bands of O2 and O3 and the anti-bonding Tl s(d_{z^2}) - O2 p_z bands. (This gap, ~ 2.1 eV wide in Tl/2212, is reduced to $\lesssim$ 1 eV in Tl/2223 as a consequence of a Madelung shift of the non-bonding O2 states.) These systems are therefore seen to realize alternating metal/semiconductor superstructures, with the metal Fermi level slightly above the conduction band bottom of the semiconductor, a situation reminiscent of the Allender, Bray, and Bardeen (33) model for excitonic superconductivity (for a critical evaluation of a possible shortcoming of this model, see Reference 34) which we will discuss later.

The calculated DOS at E_F, $N(E_F)$ for Tl/2212 and Tl/2223 are 2.82 states/eV-cell and 3.80 states/eV-cell, respectively. Thus, the additional CuO_2 sheet increases $N(E_F)$ by 1 state/eV-cell while the other components of $N(E_F)$ change by only 10-20%. Consistent with this is the fact that when we subtract the contribution from the Tl-O3-O2 bands, the $N(E_F)$ per Cu-atom is reduced to ~ 1.0 states/(eV-Cu atom), which is about the same as in (27) $Bi_2Sr_2CaCu_2O_8$.

The Fermi surfaces (FS) of $Tl_2Ba_2CaCu_2O_8$ (Tl/2212) are shown in Figure 9 in an extended zone scheme. The electron pockets c and d centered around Γ and Z, respectively, are due to the Tl-O2-O3 bands. The Cu-O dpσ bands produce the two FS indicated by a and b in Figure 9 (there will be a third such surface for Tl/2223 lying between the two shown). These surfaces have a rounded-square shape centered around X. Fermi surface a especially shows striking nesting features along the (100) and (010) directions, with spanning vectors which are not commensurate. This high degree of FS nesting is expected to give rise to singularities in the generalized susceptibility, $\chi(\vec{q})$, of this highly 2D system, and may therefore have important consequences as possible electronically-driven instabilities (e.g., incommensurate charge density waves).

The simple FS of the 2D Cu-O bands in Tl/2212 shown in Figure 9 as a and b should have a simple origin when looked at from the usual tight binding point of view. In a 2D square lattice, the simple tight-binding band is described by:

$$\varepsilon(\vec{k}) = \varepsilon_0 - 2t_1 (\cos k_x a + \cos k_y a) + 4t_2 \cos k_x a \cdot \cos k_y a \qquad (1)$$

where t_1 represents the nearest neighbor interaction and t_2 the next-nearest-neighbor interaction. From a comparison of the tight-binding bands and the dpσ anti-bonding bands of Tl/2212 and Tl/2223, we showed that the Cu-O1 dpσ anti-bonding bands crossing E_F cannot be properly fitted with a nearest-neighbor only tight-binding model (Figure 10(a) for $t_2/t_1 = 0$) while they can be reasonably well described by including the next-nearest-neighbor (most likely to be O-O) interactions (Fig. 10(b) for $t_2/t_1 = 0.45$). We therefore expect that the correct Fermi surface can only be obtained from the fuller tight-binding treatment not from a simple nearest-neighbor tight binding interaction. In fact, the inclusion of the next-nearest-neighbor term in Equation 1 yields a FS which is substantially different from the FS of a simple tight-binding band with only nearest-neighbor interactions. As shown in Figure 11, the square centered at X with perfect nesting along the (110) direction has been transformed dramatically into a rounded square with strong nesting features along the (100) and (010) directions which closely resembles the actual FS of Tl/2212 (and Tl/2223).

Finally, it is important to note that the same result is also true for the $YBa_2Cu_3O_7$ system, (16) where the FS of the 2D dpσ bands at E_F are rounded squares centered at S with nesting along (100) and (010) directions. This result implies that the commonly used tight-binding model Hamiltonian with only nearest-neighbor interactions is not sufficient to describe the anti-bonding bands crossing E_F in Tl/2212 as well as $YBa_2Cu_3O_7$ in that it yields incorrect results. This has important consequences for all such model Hamiltonian descriptions used for explaining the high T_c.

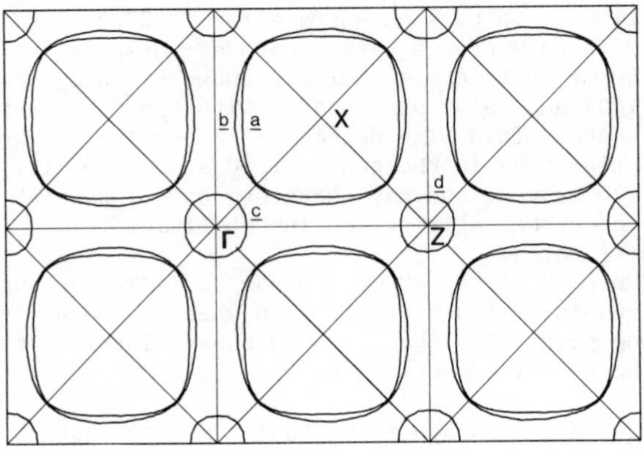

Figure 9 Fermi surfaces of $Tl_2Ba_2CaCu_2O_8$ in an extended zone scheme.

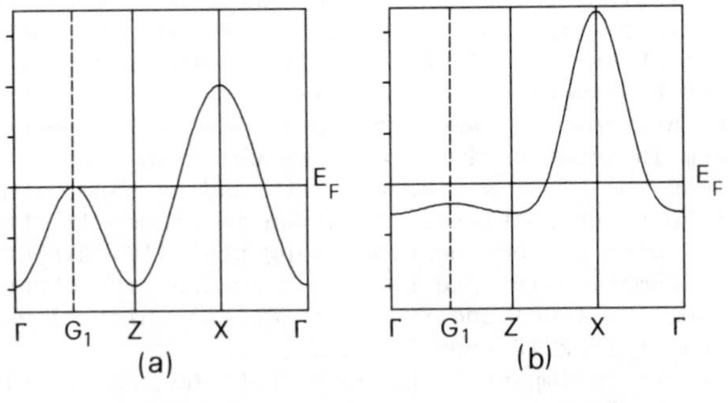

Figure 10 Tight-binding bands drawn in arbitrary units for (a) t_2/t_1 = 0.0 and (b) t_2/t_1 = 0.45, corresponding to the Fermi surfaces in Figure 11.

Origin of High T_c

We have made crude estimates of the electron-phonon interaction in the Cu-oxide superconductors, $La_{2-x}M_xCuO_4$, $YBa_2Cu_3O_7$, $Bi_2Sr_2CaCu_2O_8$ and $Tl_2Ba_2CaCu_2O_8$, using the rigid muffin-tin approximation (RMTA) (34) to calculate the McMillan-Hopfield constant η and the electron-phonon coupling constant, λ. For all the Cu-oxide superconductors, the largest contributions to η come from the Cu and O ions in the CuO_2 planes, indicating the important role played by the "metallized" oxygens. As a crude approximation - and assuming the most favorable conditions, e.g., strong phonon softening $\theta_D \lesssim 100$ K - we estimated the T_c of these systems by using the strong coupling formula of Allen and Dynes (35). The highest calculated T_c is found to be 36 K. Even though the T_c for the $La_{2-x}M_xCuO_4$ is close to the values found in the RMTA calculations, it is unlikely that a purely electron-phonon interaction is responsible for its high T_c because these are most favorable (unrealistic) estimates and the corresponding λ values are much larger than the experimental values (12). For the other systems ($YBa_2Cu_3O_7$, $Bi_2Sr_2CaCu_2O_8$, and $Tl_2Ba_2CaCu_2O_8$), the estimates of T_c are so far off (more than a factor of three) that despite the crudeness of the RMTA approach, they cast doubt on a purely electron-phonon explanation of the observed high T_c. These results suggest the possibility and importance of a non-phonon mechanism of high T_c superconductivity.

Many authors have discussed the excitonic mechanism (36-37) of superconductivity, in which the effective attractive interaction between conduction electrons originates from virtual excitations of excitons rather than phonons. The basic idea of the models proposed is that conduction electrons residing on the conducting filament (or plane) induce electronic transitions on nearby easily polarizable molecules (or complexes), which result in an effective attractive interaction between conduction electrons. As perhaps a striking realization of the excitonic mechanism of superconductivity, $YBa_2Cu_3O_{7-\delta}$ has two 2D conduction bands and additional highly polarizable 1D electronic structure between the two conduction planes.

We have previously discussed (16) the importance of the 1D feature in the electronic structure near E_F, pointing out the possible role played by charge transfer excitations ("excitons") of occupied (localized) Cu1-O dpπ orbitals into their empty (itinerant) Cu1-O dpσ anti-bonding partners. As shown schematically in Figure 12, we can characterize the 1D electronic structure with two types of electronic states in it, one free-electron-like (the well-dispersed dpσ band) and the other localized (the almost flat dpπ state). When the localized hole is created (effectively a "Cu^{4+}-complex") by the excitation, a strong attractive correlation between the hole and excited electron may lead to an electron-hole bound state ("exciton"). Hence, this excitation of the localized dpπ to the extended dpσ with the electron-hole correlation in the 1D electronic structure will give rise to a strong polarization in the 1D chains between two conduction planes and couple to the 2D conduction electrons, which carry most of the superconductivity.

In comparing the electronic states of the four oxide superconductors, a number of common features emerge which supports

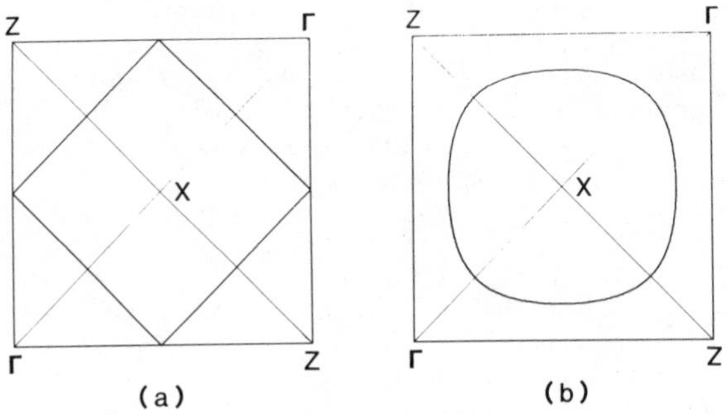

Figure 11 Fermi surfaces of the tight-binding bands for (a) $t_2/t_1 = 0.0$ (with the nearest-neighbor interactions only) and (b) $t_2/t_1 = 0.45$ (with the next-nearest-neighbors interactions included). (See the text for details.)

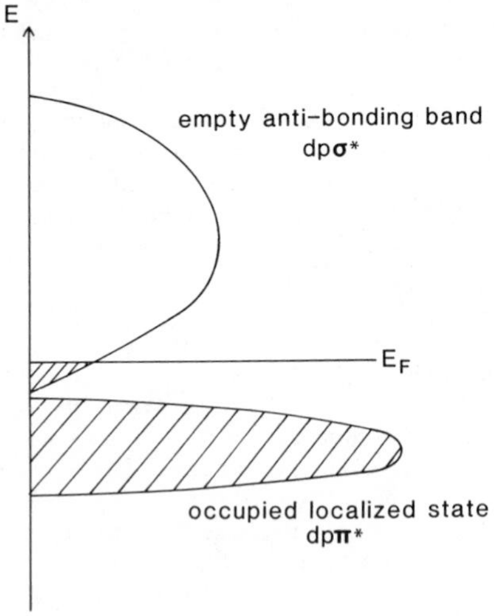

Figure 12 Schematic drawing of the 1D electronic structure in $YBa_2Cu_3O_7$.

the excitonic model of superconductivity: In all materials, the 2D Cu-O dpσ bands dominate the electronic structure near E_F. These bands consist of anti-bonding combinations of Cu $d_{x^2-y^2}$ and in-plane O $p_{x,y}$ orbitals of the CuO_2 planes, which give rise to the strong two-dimensionality of the bands. The remarkable 2D nature of the electronic structure of La_2CuO_4 leads to a simple picture of the conductivity confined essentially to the metallic CuO_2 planes separated by ionic (insulating) planes of the rock-salt type La_2O_2 layers. We note that the slab (LaO)-(CuO_2)-(LaO) has the correct stoichiometry and is charge neutral, where the ionic $La^{3+}O^{2+}$ layers provide residual charge to the CuO_2 layers. Indeed, the (LaO)-(CuO_2)-(LaO) slab becomes a basic building block (with moderate modifications) for the other high T_c Cu-oxide superconductors.

We have seen that in the 90 K superconductor $YBa_2Cu_3O_{7-\delta}$, the building block was modified by introducing the oxygen deficient Y layer between the CuO_2 layers. The new building block for $YBa_2Cu_3O_{7-\delta}$ thus becomes (BaO)-(CuO_2)-(Y)-(CuO_2)-(BaO), where the middle three layers (CuO_2)-(Y)-(CuO_2) correspond to the single CuO_2 layers for La_2CuO_4. Similarly, in $Bi_2Sr_2CaCu_2O_8$, a common building block would be (SrO)-(CuO_2)-(Ca)-(CuO_2)-(SrO) and in $Tl_2Ba_2CaCu_2O_8$, the corresponding one becomes (BaO)-(CuO_2)-(Ca)-(CuO_2)-(BaO). Finally, the one in $Tl_2Ba_2Ca_2Cu_3O_{10}$ is a mere extension of the one in $Tl_2Ba_2CaCu_2O_8$, i.e., (BaO)-(CuO_2)-(Ca)-(CuO_2)-(Ca)-(CuO_2)-(BaO).

For all of these compounds, the La, Ba, Y, Sr, and Ca atoms are purely ionic and supply extra charges to each CuO_2 layer. In contrast to the the strong ionic contribution of the La, Ba, Y, Sr, and Ca atoms, the 2D CuO_2 planes become metallic and give rise to the well dispersed Cu-O dpσ bands at E_F, which are essentially confined within each CuO_2 layer. These 2D Cu-O dpσ bands are essential for all the high T_c Cu-oxide superconductors. In addition, the common structural feature of the layered Cu-oxides superconductor suggests the (intercalated) layer structure as another essential element in the high T_c Cu-oxides. Once we regard the CuO_2 planes as major conduction layers and the other Cu-O chains, Bi_2O_2, and Tl_2O_2 as intercalated semi-metallic or insulating layers, then the role of CuO chains, Bi_2O_2, and Tl_2O_2 layers must be to enhance their superconductivity.

We have discussed some details of the additional electronic structure induced by these intercalated layers. What all of these electronic structures due to the intercalated layers have in common, is almost empty bands having strong covalent (anti-bonding) character. Furthermore, we also find the existence of occupied localized flat bands or non-bonding bands connected to the anti-bonding bands above E_F. This local electronic structure arising from from the intercalated layers can be viewed simply as shown diagrammatically in Figure 12.

As discussed above, in $YBa_2Cu_3O_7$, we have proposed charge transfer excitations of occupied (localized) dpπ states, to empty dpσ bands as a representation of the interband interactions. In $Tl_2Ba_2CaCu_2O_8$ (and similarly in $Bi_2Sr_2CaCu_2O_8$) it becomes clear that the interband interactions between the non-bonding O p states and the almost empty Tl-O sp(dpσ) bands will lead to virtual excitations which couple to the conduction electrons in the CuO_2 planes and may play an important role in their high T_c superconductivity. Indeed, we can consider the role of CuO chains, Bi_2O_2, and Tl_2O_2 layers as providing the low lying charge excitations which couple the conduction electrons in the 2D CuO_2 planes.

We are in the process of quantifying this picture of charge transfer excitations. Such an approach requires detailed calculations of the full dieletric tensor $\varepsilon(\vec{Q}, \vec{Q}')$ including the (important) Umklapp processes using our band structure results as the starting point.

There is an additional element which appears in the case of the Tl superconductors which may be worthy of note and further speculation. We first recall that some years ago, Allender, Bray, and Bardeen (32) (ABB) proposed a specific excitonic mechanism of superconductivity employing a model of semiconductor-metal interfaces. In the ABB model, the metallic electrons at the Fermi energy tunnel into the semiconductor and Cooper pair by exchanging virtual "excitons", i.e., virtual electron-hole pairs. As discussed above, the pDOS structure of Tl/2212 and Tl/2223 shows that both systems can be viewed as a metal-semiconductor (or metal-semimetal) superlattice structure, where the CuO_2 metal layers provide conduction electrons and the Tl-O3-O2 layers serve as low-gap semiconductors. Since the Tl s (d_{z^2}) and O2, O3 p_z states are covalently hybridized, a virtual excitation from the non-bonding $p_{x,y}$ states of O2 and O3 to the anti-bonding Tl s (d_{z^2})-O2,3 p_z state may give rise to a strong polarization of the Tl-O2-O3 bond. We are exploring whether this polarization of virtual excitons can couple to the conduction electrons residing in the CuO_2 planes in a way similar to that described in the ABB model. Here too, calculations of $\varepsilon(\vec{Q}, \vec{Q}')$, including Umklapp processes, will help evaluate the possible validity of this model for high T_c.

Acknowledgments

Work supported by NSF (through the Northwestern University Materials Research Center, Grant No. DMR85-20280) and the Office of Naval Research (Grant No. N00014-81-K-0438). We are grateful to NASA Ames and Kirkland Air Force Base personnel for help with the use of their Cray 2. We thank C.L. Fu, D.D. Koelling, T.J. Watson-Yang and J.H. Xu for collaboration on the early aspects of this work.

References

1. Bednorz, J. G.; Müller, K. A. Z. Phys. (1986) B64, 189.
2. Wu, M. K.; Ashburn, J. R.; Torng, C. J.; Hor, P. H.; Meng, R. L.; Gao, L.; Huang, Z. J.; Wang, Y. Q.; Chu, C. W. Phys. Rev. Lett. (1987) 58, 908.
3. Maeda, H.; Tanaka, Y.; Fukutomi, M.; Asano, T. Jpn. J. Appl. Phys. (1988) 27, in press
4. Chu, C. W.; et. al., Phys. Rev. Lett. (1988) 60, 941.
5. Sheng, Z. Z.; Hermann, A. M.; El Ali, A.; Almason, C.; Estrada, J.; Datta, T.; Matson, R. J. Phys. Rev. Lett. (1988) 60, 937.
6. Manuel, A. A.; Peter, M.; Walker, E. Europhys. Lett. (1987) 6, 61.

7. Allen, P. B.; Pickett, W. E.; Krakauer, H. Phys. Rev. (1987) B36, 3926.
8. Allen, P. B.; Pickett, W. E.; Krakaure, H. Phys. Rev. (1987) B37, 7482.
9. Jansen, H. J. F.; Freeman, A. J. Phys. Rev. (1984) B30, 561.
10. Wimmer, E.; et al., Phys. Rev. (1981) B24, 864.
11. Yu, J.; Freeman, A. J.; Xu, J. -H. Phys. Rev. Lett. (1987) 58, 1035).
12. Freeman, A. J.; Yu, J.; Fu, C. L. Phys. Rev. (1987) B36, 7111.
13. Xu, J. -H.; Watson-Yang, T. J.; Yu, J.; Freeman, A. J. Physics Lett. (1987) A120, 489.
14. Fu, C. L.; Freeman, A. J. Phys. Rev. (1987) B35, 8861.
15. Massidda, S.; Yu, J.; Freeman, A. J.; Koelling, D. D. Physics Lett. (1987) 122, 198.
16. Yu, J.; Massidda, S.; Freeman, A. J.; Koelling, D. D. Physics Lett. (1987) 122, 203.
17. Anderson, P. W. Science (1987) 235, 1196.
18. Anderson, P W. Bull. Am. Phys. Soc. (1988) 33, 459.
19. Willis, J. O.; Fisk, Z.; Thompson, J. D.; Cheong, S. -W.; Aikin, R. M.; Smith, J. L.; Zirngiebl, E.. J. Magn. Matls., (1987) 67, L139.
20. Santoro, A.; Miraglia, S.; Beech, F.; Sunshine, S. A.; Murphy, D. W.; Schneemeyer, L. F.; Waszczak, J. V.. Mat. Res. Bull. (1987) 22, 1007.
21. Jorgensen, J. D.; Veal, B. W.; Kwok, W. K.; Crabtree, G. W.; Umezawa, A.; Nowicki, L. J.; Paulikas, A. P. Phys. Rev. (1987) B36, 5731.
22. Schuller, I. K.; Hinks, D. G.; Beno, M. A.; Capone, II, D. W.; Soderholm, L.; Locquet, J. -P.; Bruynseraede, Y.; Segre, C. U. Zhang, K. Solid State Comm. (1987) 63, 385.
23. Yu, J.; Freeman, A. J.; Massidda, S. Novel Superconductivity; Plenum; New York, 1987; p. 357.
24. Vaknin, D.; et al. Phys. Rev. Lett. (1987) 58, 2802.
25. Brewer, J. A.; et al. Phys. Rev. Lett. (1988) 60, 1073.
26. Terakura, K.; et al. Phys. Rev. (1984) B30, 4734.
27. Massidda, S.; Yu, J.; Freeman, A. J. Physica (1988) C152, 251.
28. Hazen, R. M.; et al. Phys. Rev. Lett. (1988) 60, 1174.
29. Hazen, R. M.; Finger, L. W.; Angel, R. J.; Prewitt, C. T.; Ross, N. L.; Hadidiacos, C. G.; Heaney, P. J.; Veblen, D. R.; Shen, Z. Z.; El Ali, A.; Hermann, A. M. Phys. Rev. Lett. (1988) 60, 1657.
30. Subramanian, M. A.; Torardi, C. C.; Calabrese, J. C.; Gopalakrishnan, J.; Morrissey, K. J.; Askew, T. R.; Flippen, R. B.; Chowdhry, U.; Sleight, A. W. Science (1988) 239, 1015.
31. Yu, J.; Massidda, S.; Freeman, A. J. Physica (1988) C152, 273.
32. Allender, D. W.; Bray, J. M.; Bardeen, J. Phys. Rev. (1973) B7, 1020.
33 Cohen, M. L.; Louie, S. G. In Superconductivity in d- and f-band Metals; Douglass, D. H., Ed.; Plenum; New York, 1976.
34. Gaspari, G. D.; Gyoffry, B. L. Phys. Rev. Lett. (1972) 28, 801.
35. Allen, P. B.; Dynes, R. C. Phys. Rev. (1975) B12, 905.
36. Little, W. A. Phys. Rev. (1964) 134, A1416.
37. Ginzburg, V. L. JETP (1964) 46, 397.

RECEIVED July 12, 1988

Chapter 7

Electron Spectroscopic Data for High-Temperature Superconductors

David E. Ramaker

Department of Chemistry, George Washington University, Washington, DC 20052

An interpretation of spectroscopic data for the high temperature superconductors utilizing a highly correlated CuO_n cluster model shows that a single set of Hubbard parameters predicts all of the state energies. Differences in the data from that for CuO are attributed to an increased Cu-O covalency in the superconductors. The reported temperature effects are attributed to increased metallic screening at lower temperatures.

An abundance of electron spectroscopic data has been reported for the high temperature superconductors. These include the valence band (VB), Cu 2p, and O 1s photoelectron (UPS and XPS) data, the $L_{23}VV$ and $L_{23}M_{23}V$ Auger (AES) data, the O K and Cu L_{23} x-ray emission (XES) data, and the O K and Cu L_{23} electron energy loss (EELS) and x-ray absorption near edge structure (XANES) data. These data reflect 1-, 2-, and 3-valence hole and core-hole density of states (DOS) and therefore can provide direct measures of the Hubbard U and transfer parameters. Unfortunately, this data has proved to be difficult to interpret; not surprisingly since the data for CuO is not even well understood.

Recently we consistently interpreted these data within a cluster model assuming a highly correlated system (1,2). Here, we review the previously reported experimental spectra for polycrystalline and single crystal samples of $La_{2-x}Sr_xCuO_4$ or $YBa_2Cu_3O_{7-x}$ (herein referred to as the La and 123 superconductors (HTSC's)), and compare them with CuO. We also review our interpretation of this data. Previously unassigned features are identified and some are assigned differently from that given by others (3). We also obtain the magnitudes of the U parameters. Knowledge of these magnitudes are important for understanding the possible mechanisms involved; for example, the resonating valence bond (4), exitonic (5), super-exchange (6), and Coulombic (7) pairing mechanisms depend critically on these U's.

The basic VB electronic structure of the HTSC's can be described by a simple LCAO-MO (i.e. tight-binding) or Hubbard model, characterized by the transfer or covalent interaction t, the Cu and O

0097–6156/88/0377–0084$06.00/0

orbital energies ε_d and ε_p, the intra-site Coulomb repulsion energies U_d and U_p, and the intra-site core polarization energies Q_d and Q_p. We also include the inter-site repulsion energies U_{dp} and U_{pp^o} (i.e. between neighboring Cu-O and O-O atoms). An important parameter is the difference between ε_d and ε_p, namely $\Delta = \varepsilon_p - \varepsilon_d$.

All of the spectroscopic data can be understood within a $CuO_n^{(2n-2)-}$ cluster model, which is valid when the U's are large relative to the bandwidths (8,9), i.e. when correlation effects dominate covalent or hybridization effects. Both La and CuO contain CuO_6 groups (10), having 4 short and 2 long Cu-O bonds. The 123 HTSC contains CuO_5 and planar CuO_4 groups (10). The different n may alter the relative intensities of various features as pointed out below, but similar features are present in each case. The different bond lengths may increase the widths of the spectral features, but little else since correlation dominates.

Although relatively isolated CuO_2 planes exists in the HTSC's, the CuO_n clusters are not isolated, for most of the O atoms actually are part of two CuO_n clusters. Consistent with previous work (7), we account for this by defining the effective parameter, $\varepsilon_p = \varepsilon_p' + U_{pe}$, where U_{pe} includes the interaction of a hole in an O p orbital with its environment, i.e. with the neighboring Cu atom or cluster. If for example each of the neighboring Cu atoms contains a hole (we shall see below that this is essentially the ground state), U_{pe} will equal U_{dp}. But in general, U_{pe} will be smaller than U_{dp} due to polarization of the lattice. Throughout the remainder of this work, ε_p is assumed to include this effect.

Since Cu atoms have the electronic configuration $3d^{10}4s^1$ and O the configuration $2s^22p^4$, the $CuO_n^{(2n-2)-}$ cluster has one hole shared between the Cu 3d and O 2p shells in the ground state. The various spectroscopies then reflect either v^2–, v^3–, c–, or cv– hole (c = core hole, v= valence hole) states as indicated in Table 1. We indicate the location of the holes by d (Cu 3d) or p (O 2p). In the case of two holes on the oxygens, we distinguish two holes on the same O (p^2), on ortho neighboring O atoms (pp^o), or on para O atoms (pp^p) of the cluster. Furthermore, neighboring pp^o holes can dimerize (11-13), so we distinguish between two holes in bonded ($pp^o{}_b$) and antibonded ($pp^o{}_a$) O pairs, i.e. the same or different O_2 dimers. Both the doped La and 123 materials contain additional holes which serve as the charge carriers, and some spectroscopies reflect these directly.

Table 1 contains the estimates of the Hubbard parameters which together provide the best agreement with the state energies as reflected in the spectra and the theoretical DOS. The energies relative to the ground state are given in terms of the Hubbard ε and U parameters defined above. The agreement of the estimated energies with the spectral features is also shown in Table 1. We discuss the assignment of the spectral features below.

The Theoretical DOS

The v states are best reflected by the highly accurate theoretical DOS (14), which for the HTSC's can qualitatively be described as having the Cu-O bonding (ψ_b) and antibonding (ψ_a) orbitals centered at 4 and 0 eV and the nonbonding Cu and O orbitals at 2 eV. The O features each have a width $2\Gamma = 4$ eV due to the O-O bonding and

TABLE 1 Summary of hole states revealed in the spectroscopic data, and estimated energies using the following optimal values for the Hubbard parameters in eV[a]:

$\delta_1 = 2$	$\varepsilon_d = 2$	$U_p = 12, 13$	$U_d = 9.5, 10.2$
$\delta_2 = 0.5, 0.8$	$\varepsilon_p = 2, 3$	$U_{pp}^{o} = 4.5, 4$	$U_{dp} = 1$
$\Gamma = 2$	$U_{pp}^{p} = 0.$	$U_{cp}^{o} = 2$	$Q_d = 9$
$\alpha = 1, 0.5$	$\beta = 2$	$\Delta = 0, 1.$	$K = 4$

State[b]	Energy expression	Calc. E. eV[c,d]	Exp. E. eV[c]	Remark
G.S. and IPES, v				
Ψ_a) d	$\varepsilon_d - \delta_1 \mp \Gamma$	0 ∓ 2	-	heavily mixed
Ψ_b) p	$\varepsilon_p + \delta_1 \mp \Gamma$	4 ∓ 2	-	
UPS and XES, v^2				
1)[e] pp^p	$\varepsilon_p + \Delta - \delta_2 + \alpha$	2.5	2.5	heavily mixed
2)[e] dp	$\varepsilon_p + U_{dp} + \delta_2 + \alpha$	4.5	4.2	
3) pp^{o}_{a}	$\varepsilon_p + \Delta + U_{pp}^{o} - \Gamma + \alpha$	5.5	5.	
4) pp^{o}_{b}	$\varepsilon_p + \Delta + U_{pp}^{o} + \Gamma + \alpha$	9.5	9.5	mystery peak
5) d^2	$\varepsilon_d + U_d + \alpha$	12.5	12.5	Cu sat.
6) p^2	$\varepsilon_p + \Delta + U_p + \alpha$	15	16	
Cu 2p XPS, cv				
d → cp	$\varepsilon_c + \Delta + \alpha$	$\varepsilon_c + 1$	E_{2p}	main
cd	$\varepsilon_c + Q_d + \alpha$	$\varepsilon_c + 10$	$E_{2p}+9.2$	sat.
Cu 2p XPS for $NaCuO_2$, pp^p → cv^2				
pp^p→cpp^p	$\varepsilon_c + \delta_2 + \beta$	$\varepsilon_c + 2.5$	$\varepsilon_c+2.2$	main
cpp^{o}_{a}	$\varepsilon_c + U_{pp}^{o} - \Gamma + \delta_2 + \beta$	$\varepsilon_c + 4.5$	ε_c+5	?
cpp^{o}_{b}	$\varepsilon_c + U_{pp}^{o} + \Gamma + \delta_2 + \beta$	$\varepsilon_c + 8.5$	ε_c+9	?
cdp	$\varepsilon_c - \Delta + Q_d + U_{dp} + \delta_2 + \beta$	$\varepsilon_c + 11.5$	ε_c+11	sat.
cp^2	$\varepsilon_c + U_p + \delta_2 + \beta$	$\varepsilon_c + 15.5$	ε_c+14	sat.?
O 1s XPS, cv				
d → cd	$\varepsilon_c + \alpha$	$\varepsilon_c + 1$	E_{1s}	main
cp^p	$\varepsilon_c + \Delta + \alpha$	$\varepsilon_c + 1$	E_{1s}	main
cp^o	$\varepsilon_c + \Delta + U_{cp}^{o} + \alpha$	$\varepsilon_c + 3$	$E_{1s}+2$?	tail
cp	$\varepsilon_c + \Delta + Q_p + \alpha$	?	?	not obs.
pp^p→cdp^p	$\varepsilon_c - \Delta + U_{dp} + \delta_2 + \beta$	$\varepsilon_c+3.5$	$E_{1s}+2$?	tail
Cu L_3VV AES, v^3				
dpp^p	$2\varepsilon_p + 2U_{dp} + \alpha$	7	7	2 cent.
dpp^o	$2\varepsilon_p + U_{pp}^{o} + 2U_{dp} + \alpha$	11.5	-	no mix
d^2p	$\varepsilon_d + \varepsilon_p + U_d + 2U_{dp} - \delta_2 + \alpha$	16	15.5	main
dp^2	$2\varepsilon_p + U_p + 2U_{dp} + \delta_2 + \alpha$	19.5	18-25	sat.

TABLE 1 (cont.)

State[b]	Energy expression	Calc. E. eV[c,d]	Exp. E. eV[c]	Remark
Cu $L_3M_{23}V$ AES, cv^2				
cdp	$\varepsilon_c+\varepsilon_p+Q_d+U_{dp}\mp K+\alpha$	ε_c+9	$E_{3p}+10$	main, ^1L
		ε_c+17	$E_{3p}+18$	main, ^3L
cp^2	$\varepsilon_c+\varepsilon_p+\Delta+U_p+\alpha$	ε_c+15	-	not obs.
cd^2	$\varepsilon_c\infty\varepsilon_d+U_d+2Q_d+\alpha$	$\varepsilon_c+30.5$	-	
Cu L_{23} EELS, c				
d → c	$\varepsilon_c-\varepsilon_d+\delta_1$	$E_{2p}-1$	$E_{2p}-1.4$	edge
cpCB	$\varepsilon_c+\Delta-CB+\alpha$	$E_{2p}-CB$	$E_{2p}+1.2$	upper
pp^p → cp	$\varepsilon_c-\varepsilon_p+\delta_2+\beta$	$E_{2p}-0.5$	E_{2p}	middle
O K EELS, c				
d → c	$\varepsilon_c-\varepsilon_d+\delta_1$	$E_{1s}-1$	E_{1s}	edge
cdCB	$\varepsilon_c-CB+\alpha$	$E_{2s}-CB$	$E_{1s}+1.7$	upper
pp^p → cd	$\varepsilon_c-\Delta-\varepsilon_p+\delta_2+\beta$	$E_{1s}-0.5$	-	not obs.

[a]Parameters for 123 indicated first, those for CuO second.
[b]The dominant character in the hybridized states is given.
[c]The Calc. E and Exp. E columns indicate the results for 123, except for the "Cu 2p XPS, pp^p → cv^2" section, which is for $NaCuO_2$.
[d]The calculated E is defined relative to the ground v^1 (d) state energy = $\varepsilon_d-\alpha$, or to the v^2 (pp^p) ground state energy = $2\varepsilon_p-\delta_2-\beta$. The v^1(d) energy defines the Fermi level relative to the vacuum level at zero.
[e]The dominant character switches as described in the text, and thus the sign in front of δ_2 is the opposite for CuO.

antibonding character and the Cu-O dispersion. The ψ_b and ψ_a wavefunctions can be expressed as (8,9),

$$\psi_a = d\cos\vartheta_1 - p\sin\vartheta_1 \quad (1a)$$
$$\psi_b = d\sin\vartheta_1 + p\cos\vartheta_1 \quad (1b)$$

where $\vartheta_1 = 0.5 \tan^{-1}(2t/\Delta)$. We also define the Cu-O hybridization shift $\delta_1 = 0.5\ \mathrm{sqrt}(\Delta^2+4t^2) - \Delta/2$, which is utilized in Table 1 to give the energies. In this picture, the ground state of an average CuO_n cluster is located at 1 eV having the energy $\varepsilon_d-\delta_1+\Gamma/2 = \varepsilon_d-\alpha$, which we use as a reference energy for all of the higher v^n states. Similarly, the inverse photoemission data (e.g. see Fig. 4c), which reflects the unoccupied v DOS, gives a feature at $\varepsilon_d-\delta_1-\Gamma/2 = -1$ eV, due to the simultaneous Cu-O and O-O antibonding orbitals. Above, α is the average hybridization energy shift in the ground state of a CuO_n cluster. We ignore the O-O hybridization shift for all of the excited states, except for the pp^o states, where it is large.

The DOS for CuO is in fact less well characterized than for the HTSC's theoretically, but generally the hybridization shift Γ is smaller because of the larger O-O distances, and we shall see below that $\Delta=\varepsilon_p-\varepsilon_d$ has increased to 1 eV. This increase can be attributed to an increase in ε_p, or U_{pe}, and reflects a smaller lattice polarization response due to the more ionic character in CuO.

The VB UPS and XPS Data

The photoemission process involves excitation from the v to the v^2 states. The six different v^2 states in Table 1 have a large energy spread. States 1,2,5 & 6 have the same symmetry and mix together; unfortunately, the result cannot be given simply in analytical form as above. The two pp^o states (3 & 4) have different symmetry and mix separately. Only states 1 & 2 and 3 & 4 are heavily mixed so that they are the only ones to experience a significant hybridization shift, δ_2 and Γ, as shown in Table 1. The sudden approximation and the cross-sections for ionization from the O 2p and Cu 3d shells, σ_p and σ_d, can be utilized to give the expected relative photoemission intensities for the six v^2 states. σ_p and σ_d are known to be approximately equal around 40 eV, with σ_d larger above this, and σ_p larger below this energy (i.e. σ_p/σ_d is roughly 2. for 21 eV, 1. for 45, and 0.3 for 100 eV photons (14)). The photon dependence of the UPS data can be understood from the variation in the cross-sections, and from the realization that at low photon energies, the sudden approximation breaks down. The opposite extreme, the adiabatic limit, gives intensity only in the lowest state of each symmetry, 1 and 3, since the system is able to relax before escape of the photoelectron. UPS spectra reflect the case somewhere between these limits.

The valence band features. Examination of the UPS data in Figure 1b shows two features at 3 and 5.5 eV for CuO (15,16) and three for the superconductors (17) at 2.5, 4.2, and 5 eV, which obviously result from the hybridized states 1-3. Photon energy dependent data in Figure 1 shows that the features around 5.5 eV in CuO and 2.5 and 5 eV in 123 arise more from σ_p, and the feature at 3 in CuO and 4.2 eV in 123 from σ_d (17-22). In CuO, we assign the 5.5-eV feature to $pp^o{}_a$ and pp^p and the 3-eV to dp. In 123, we assign the 5-eV to $pp^o{}_a$,

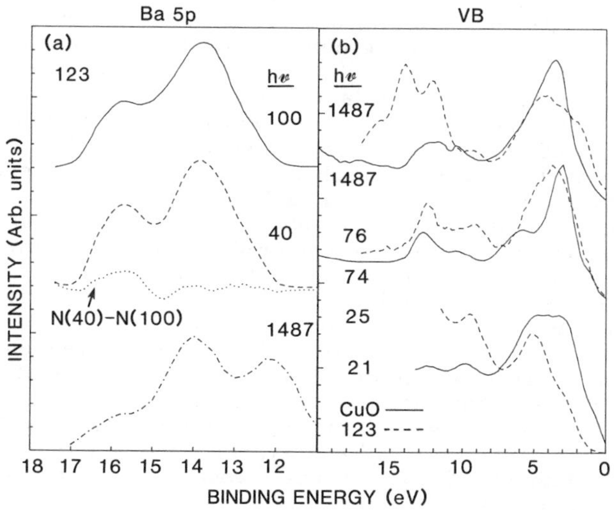

Figure 1a. Comparison of photoelectron spectra in the range 11-18 eV for 123 . Data from refs. 17 (hν = 100 and 40) and 24 (hν = 1487).
1b. Comparison of UPS spectra for CuO and 123 taken with the indicated photon energies in eV. Data for CuO from refs. 23 (hν = 1487), 15 (hν = 74) and 16 (hν = 21). Data for 123 from ref. 17 (hν = 25 and 74) and 24 (hν = 1487).
(Reproduced from Ref. 1, Not subject to copyright.)

the 4.2 to dp and the 2.5 to pp^p, where we indicate the dominant character of each hybridized state. Calculated photoemission intensities and their variation with Δ further confirm these assignments (1).

The character switch of state 1 from mostly dp to pp^p and vice versa for state 2 between CuO and 123 arises because Δ decreases from 1 eV in CuO to 0 eV in 123. The reduction in Δ, due to reduction in ε_p or U_{pe}, is consistent with the Cu 2p XPS data and with the XES data to be discussed below; the latter very dramatically reveals this character switch. States 1 and 2 remain a few eV apart in spite of this switch because of the heavy CI mixing. Since state 1 is primarily of pp^p character in the HTSC's, the additional "charge carrier holes" (present in the La after Sr doping and in the 123 when 7-x is greater than 6.5) are primarily on the oxygens. We use this "pp^p" state as the ground state for those CuO_n clusters containing two holes.

Angle resolved PES data on single crystals of 123 show that the 2.5 eV feature is the only one which shows significant angular dispersion and a photon energy dependence (17). Its angular dispersion of 0.25-0.3 eV is much smaller than the 2 eV expected from band calculations (14). The near lack of dispersion is consistent with our highly correlated cluster model. The small dispersion of the 2.5 eV feature probably comes from inter-CuO_4 cluster interaction, which is expected to be the largest when both holes are on the bordering O atoms. The region within 1 eV of E_F is free of photoyield for $h\nu$ = 30 eV, but has a substantial yield for $h\nu$ = 18 eV; indeed it stretches up to E_F for off normal emission (17). These lower energy states reflect the average v^2 ground state, since only a small fraction of the CuO_n clusters have two holes in the ground state. Thus we use $2\varepsilon_p-\delta_2-\beta$ as the energy of the pp^p ground state relative to E_V, where β represents the 2 eV energy shift between the principal pp^p UPS final state at 2.5 eV and the lowest pp^p states around 0.5 eV from the Fermi level.

The d^2 satellite. The principal multiplet of the d^2 final state for CuO is known to fall at 12.5 with a smaller one around 10 eV (15). The intensity of the d^2 final state can be enhanced by the Cu 2p → 3d resonant excitation process followed by an Auger decay (15). This process is resonant between 72-80 eV. The HTSC's exhibit a similar behavior (18). The satellites in Cu_2O and Cu do not have non-resonant components (15) because the UPS for Cu_2O and Cu reflect the one-hole DOS. However, the VB XPS of CuO and the HTSC's can and do show a significant nonresonant d^2 satellite (see Figure 1) (23); indeed, it should grow as one approaches the sudden limit. This possibility makes it even more difficult to interpret the XPS data for the HTSC's, since the d^2 satellite at 12.5 in the VB XPS falls at or near the same energy as the Ba spin-orbit split 5p features, which have been very controversial.

For the XPS (Figure 1a), Miller et al (24) have indicated that the 12.5 eV feature results from the Ba representative of the bulk, and the 14 and 16 eV features result from Ba bonded to OH^- and $CO_3^=$ on the surface. Steiner et al (25) indicate that the 12.5 eV feature is representative of those Ba atoms surrounded by O atoms, but that the 14 and 16 eV features arise from those Ba atoms with either neighboring O defects or O atoms with holes (i.e. O^- instead of

O^{2-}). Recent data (17) on single crystals cleaved in-situ, when impurities are not expected, reveal the 14 and 16 eV features at glancing emission (i.e. representative of the surface), and two additional features shifted up by about 1 eV at normal emission (i.e. more representative of the bulk). This shift has been interpreted as a surface chemical shift, but it is actually consistent with the Steiner data and interpretation, if one assumes more O defects exist at the surface. Recently Weaver et al (26) reported XPS data for sintered 123, which actually reveal only the features at 12.5 and 14 eV. This indicates either that their surface is free of impurities or that the bulk and surface is totally oxidized (i.e. within the Miller or Steiner interpretations). More experimental data is required here to conclusively decide on these two alternatives and to determine what part if any of the 12.5 eV feature results from the d^2 satellite. In our opinion, the Steiner interpretation appears the more plausible at this time.

The $pp^o{}_b$ feature. The $pp^o{}_b$ state is believed to be responsible for the "mystery" peak found at 9.5 eV in the UPS. Although some earlier reports suggested that this feature might result from impurities such as carbon on the surface (27), more recent single crystal data (17) as well as sintered powder data (22) (Figure 1b) indicate that it is intrinsic to the material . Comparison of the UPS (15,16) in Figure 1 for photon energies of 74 and 21 eV indicates that such a feature also appears for CuO. Thus this feature is not unique to the HTSC's. It does not appear for Cu_2O, as expected since UPS reflects the one-hole DOS in Cu_2O.

UPS data indicate that the 9.5 eV feature has a cross-sectional dependence similar to σ_p (19-22), consistent with the pp^o identification. This feature cannot arise from the p^2 final state because U_p is around 12-13 eV, much too large to cause a feature at 9.5 eV. An upper estimate of the two-center pp^o hole-hole repulsion, U_{pp^o}, can be obtained from the Klopman approximation (28),

$$U_{ij} = e^2/(r_{ij}^2 + (2e^2/\{U_i + U_j\})^2)^{1/2} \qquad (2)$$

where r_{ij} is the interatomic distance and U_i and U_j are the corresponding intra-atomic repulsion energies. Equation 2 gives a value for U_{pp^o} around 4.8 eV, assuming the O-O distance is 2.7 A°. The experimental energies of 9.5 and 5.0 eV for $pp^o{}_b$ and $pp^o{}_a$ in 123 suggests that the average pp^o final state energy is 7.2 eV. This gives an empirical estimate for U_{pp^o} of 4.2 eV, close to the Klopman theoretical result, which does not include the effects of interatomic screening.

The above result shows that metallic screening of two holes, which are spatially separated on neighboring O atoms, is not very significant. This is in contrast to two Cu-O holes, where Table 1 indicates the optimal U_{dp} = 1 eV, while eq. 2 estimates U_{dp} at 6.1 eV, assuming an average Cu-O distance of 1.9 A°. This large reduction in U_{dp} probably results from charge transfer into the Cu 4sp levels to screen the Cu-O holes. Although metallic screening, which results from virtual electron-hole (e-p) pair excitations at the Fermi level, is not expected to be large in an insulator such as CuO, screening effects are expected to be much larger in metals, such as the HTSC's. The above results show that U_{dp} is significantly reduced in both, and U_{pp}^{o} remains large in both. The lack of a significant change in the U's between CuO and the HTSC's indicates that the DOS at the Fermi level in the HTSC's must be very small.

The assignment of the 9.5 eV feature explains some of its interesting characteristics. Comparison of data (19) for $YBa_2Cu_3O_x$ (123_x) with O levels at x = 6.95, 6.5, and 6.05 reveal that the reduced O materials, $123_{6.5}$ and $123_{6.05}$, have two peaks around 9.4 and 11.5 eV. It is known that the oxygen decrease resulting from quenching or heating in vacuum occurs primarily from the CuO_4 chains (29). This may leave distorted CuO_x or even peroxide O_2^- clusters (11-13) which have an O-O distance less than that in the ordered CuO_4 groups, and hence a larger U_{pp}^{o}. A U_{pp}^{o} of 6.5 eV requires an O-O distance of less than 2 A°. Very recent data (30) on the new Bi and Th type HTSC's indicate a single feature at around 10 eV, consistent with the 123 material.

The p^2 feature. Evidence for the existence of the p^2 feature, estimated to appear at 17.5 eV can indeed be found around 17 eV in the XPS for CuO in Figure 1. In the HTSC's, the Ba 5d peaks fall in this region, making it more difficult to identify the d^2 feature. Nevertheless, recent UPS data on single crystal 123 materials may reveal the p^2 feature (17). Figure 1 shows UPS at hν = 100 and 40 eV. An interesting change in the relative intensity of these two peaks is found, when normally one would expect the relative intensity of the $5p_{1/2}$ and $5p_{3/2}$ peaks to remain constant with photon energy. But, the 40 eV spectrum should have a larger σ_p contribution. This suggests that the hν = 40 eV spectrum may have a contribution from the p^2 state, such as that indicated in Figure 1.

The Cu 2p and O 1s XPS Data.

The Cu 2p XPS. The Cu 2p and O 1s XPS data for CuO and the 123 or La materials are shown in Figure 2 (31,32). The primary and satellite features in the Cu spectrum are known to arise from the cp and cd states, respectively (8), having the energies given in Table 1. The relative satellite intensity, I_s/I_m decreases from 0.55 in CuO to 0.37 in 123 as determined from the experimental data (33). The energy separation, $E_{sat} - E_m$ increases from 8.7 eV in CuO to 9.2 in 123 (33). These changes are just that expected for a decrease in ε_p and a possible increase in t_1. These changes reflect an increased covalency in 123.

In the sudden approximation, the satellite intensity increases with change in the hybridization between the v and cv states. In the ground v state, the hole is primarily in the d orbital, in the primary cv state it is mostly in the p orbital. However, some hybridization still occurs among the cv states, since the large width of the primary cp peak is believed to arise from the mixing with the cd state. The cd state has a large width due to the large core-hole, valence-hole interaction, indeed, the satellite actually reveals the cd multiplet structure. Evidence that the primary cp peak width arises from the cd interaction comes from the Cu halide data (8), as well as the different Cu oxide data in Fig. 2a, which show a direct correlation of the primary cp peak width with the satellite cd peak intensity. We do not believe that the primary peak width arises from the O p band width as proposed by others (34).

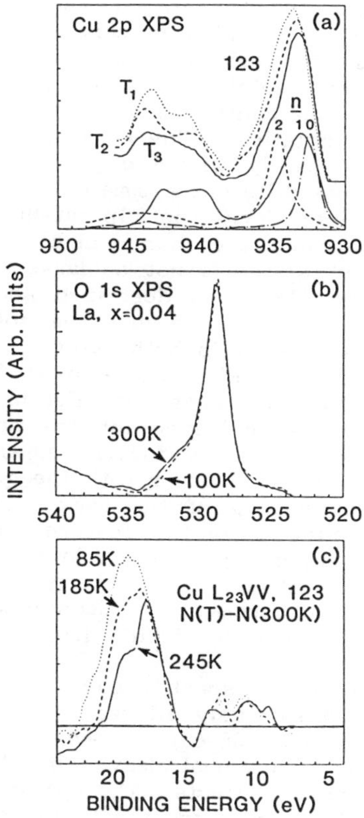

Figure 2a. Cu 2p XPS data for 123 at the temperatures 350, 230K, and 180K and denoted by T_1, T_2, and T_3 in the Figure (from Ref. 32) and for $NaCuO_2$, CuO, and Cu_2O, which reflect the cv^n (n = 2,1, and 0) DOS (from Ref.35). The energy shift seen between 123 and CuO is not real, since data taken by the same authors indicate that they agree to within 0.5 eV (35).
2b. O 1s XPS data from a single crystal La sample at the indicated temperatures (from Ref. 31).
2c. Cu L_{23}VV AES difference spectra (N(T)-N(300K)) for 123 at the indicated temperatures (from Ref. 52).

Figure 2a also contains the Cu 2p XPS spectra for Cu_2O, CuO, $NaCuO_2$ (35). These materials have the formal Cu valence of +1, +2, and +3, but in the current picture they reflect the cv^n DOS, with n=0, 1, and 2. Cu_2O has just a core hole c, with an energy of ε_c, and a negligible satellite. The primary cp feature for CuO has its energy shifted by $\Delta+\alpha$ relative to ε_c, and the cpp^p feature for $NaCuO_2$ by $\delta_2+\beta$ (Table 1), which is consistent with Fig. 2a. The spectrum in Fig. 2a for $NaCuO_2$ also reveals several other cv^2 features at energies consistent with Table 1. Identification of the cpp^o_a and cpp^o_b states as contributing to the features around 5 and 9 eV relative to ε_c is tentative. This is because the pp^o configurations do not mix significantly with the pp^p configuration in the v^2 ground state (they have different symmetry), and hence they should contribute very little intensity. The cdp satellite feature has a lower intensity in $NaCuO_2$, because the change in hybridization between the initial v^2 and final cv^2 states is not as large as for the v and cv states in CuO and the HTSC's as described above. The change in hybridization is small in $NaCuO_2$ because the pp^p and cpp^p configurations have the lowest energy in both cases.

Finally, we note that no evidence exist for the cd^2 feature, even in $NaCuO_2$, so that Cu^{3+} does not exist in either $NaCuO_2$ or in the HTSC's. Nevertheless, evidence for the presence of the cpp^p feature does appear in the spectra for 123, as shown in Fig. 2a. The magnitude of the cpp^p feature varies with the quenching temperature utilized during the processing of the samples (36,37). It has also been shown that the intensity of this feature correlates with T_c; but, the samples are still superconducting at lower T_c, even though this feature appears to be absent (36,37). At still higher quenching temperatures the material becomes a semiconductor, and the c feature appears, indicating the presence of Cu^{+1} (36).

The O 1s XPS. The O 1s spectra have been reported by many authors; however, it is seriously altered by impurities such as OH^- and $CO_3^=$ on the surface (38-40). Recent data (31) from single crystal samples of the La material cleaved in-situ are expected to be reasonably free of impurity effects. The cp^o and cp^p states listed in Table 1 are believed to account for the tailing off of the spectra for the O 1s XPS as seen in Figure 2 (this will be positively identified upon examination of the XES data). This tailing off is much smaller in undoped La samples indicating that the $v^2 \rightarrow cv^2$ transitions also contribute in this region similar to that found in the Cu 2p XPS. Consistent with the sudden approximation, the cp state is not seen in the O 1s XPS because now both the ground v and core hole cv states have similar hybridization, i.e. the valence hole is mostly in the d orbital in both cases.

The Cu $L_{23}VV$, $L_{23}M_{23}V$, and O KVV Auger data.

The Cu $L_{23}VV$ and $L_{23}M_{23}V$ Auger data (9,33,41-43) reflect the v^3 and cv^2 DOS, respectively, and therefore provide further information on the Hubbard parameters. The L_2 and L_3 features and associated satellites are identified in Fig. 3. Table 1 indicates that the energies of these features are predicted accurately by the U and ε parameters established above. We need only discuss the relative intensities.

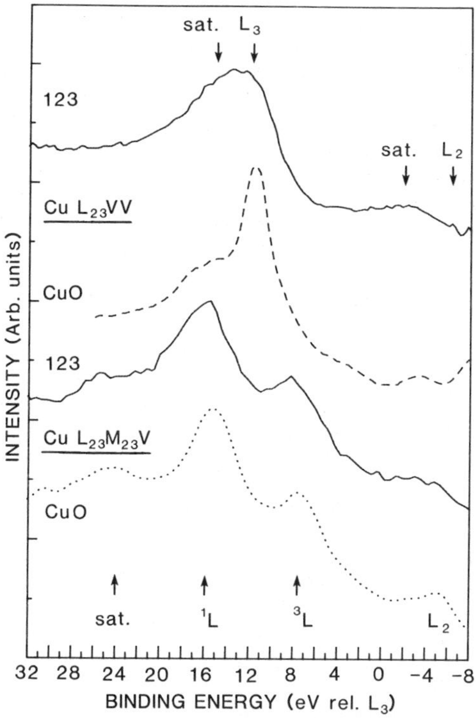

Figure 3. Comparison of Auger data for the materials indicated. $L_{23}VV$ data for CuO and 123 from ref. 49. $L_{23}M_{23}V$ data for CuO from ref. 42 and for 123 from Ref. 2. The $L_{23}VV$ data is on a 2-hole binding energy scale = $E_{L3}-E_{k}$,, and the $L_{23}M_{23}V$ on a 1-hole scale = $E_{L3}-E_{k}-E_{M3}$, where E_{L3} = 933.4 and E_{M3} = 77.3 eV (15,16). (Reproduced from Ref. 2, Not subject to copyright.)

We point out at the outset that the initial state shake-up (SU) process, which is responsible for the Cu 2p XPS satellite discussed above, does not produce a satellite in the Auger lineshape, because the cd final states resulting from this SU process "relax" to cp states before the Auger decay. This is expected because the SU excitation energy is much larger than the core level width. In contrast, the cdp states resulting from the shakeoff (SO) (g.s. + hν $\rightarrow$ L_3dp) and the Coster-Kronig (CK) decay (g.s. + hν $\rightarrow$ L_{12}p $\rightarrow$ L_3dp) processes cannot dissipate before the Auger decay, because an extra hole is present, and it is bound on the CuO_n cluster. Thus the satellites identified in both the $L_{23}VV$ and $L_{23}M_{23}V$ lineshapes arise primarily from the initial state SO and CK processes.

Comparison of the intensities of these satellites reveals a most interesting point however. Although 123 shows an increased satellite in the $L_{23}VV$ relative to CuO, it is not increased in the $L_{23}M_{23}V$. This different behavior between the two satellites suggests strongly that the increased satellite in the L_3VV does not result from the initial state SO and CK processes, since the initial state is the same in both. Thus it must arise from a final state effect. Within our cluster model, a final state effect arises naturally from configuration mixing between the v^3 states listed in Table 1.

Since only the primary cp core-hole state Auger decays, and this process is known to be strictly intra-atomic, the $L_{23}VV$ lineshape reflects the d^2p DOS, as it is distributed among the v^3 final states. The dpp^o state does not mix with d^2p since it does not have the same point group symmetry possessed by all the other v^3 final states and the cv initial state. The d^2p state at 15.5 eV is the main feature, and the dpp^p state at 7 eV is what we have previously referred to as the "two center" feature (33). The dp^2 state falls on top of the satellite feature and obviously accounts for its increased intensity in 123. Its intensity is increased in 123 relative to CuO because the energy separation (before hybridization) between d^2p and dp^2 has decreased from 3.8 eV in CuO to 2.5 eV in 123. The hopping parameter t may also be increased in 123, further increasing the mixing between these two states. We have indicated this mixing in Table 1 by adding the hybridization shifts δ_2 to the energy expressions for these two states.

The $L_{23}M_{23}V$ lineshape reflects the cdp DOS. The mixing of the other states (cd^2, cpp^p, $cpp^o{}_a$, and $cpp^o{}_b$; the latter three are not listed in Table 1) with the cdp state is small because of the large energy separations involved. The cp^2 state is close to cdp; however, it falls in between the ^3L and ^1L multiplets of the cdp state. Although it may have some intensity, it surely does not contribute to the CK + SU satellite around 25 eV in either CuO or 123. The exchange splitting (2K) between the 3p and d holes is known to be very large (8), so we include it explicitly in Table 1 to account for the $^{1,3}L$ multiplets.

The O KVV lineshape is severely altered by impurities in the sintered HTSC's, and no single crystal lineshape data have been reported. The O KVV lineshapes for CuO and Cu_2O have been reported (16), and they have the primary dp^2 or p^2 features, respectively, around 19 eV. A very small satellite appears around 7 eV in Cu_2O which we attribute to the pp^p state. A much larger and broader satellite around 7 to 14 eV in CuO appears, which we attribute to the d^2p state around 14 eV as well as a smaller amount to the dpp^p state around 7 eV. Thus the d^2p and dp^2 states appear in both the Cu $L_{23}VV$ and O Auger lineshapes for Cu^{+2} oxides, except their primary and satellite roles are reversed.

The Cu L_{23} and O K XES data

The Cu L_{23} XES data (44,45) shown in Figure 4 dramatically reveals the switch in character of the 1 and 2 v^2 states between CuO and 123. As in the Auger process, the satellite cd initial state relaxes to the cp state before the decay, and the x-ray emission process is intra-atomic in nature. Therefore, the XES reflects primarily the dp DOS. In CuO the XES spectrum peaks at 3 eV, in the 123 it falls around 4.2 eV, very near where we indicated the dp states fall in the UPS data. The large intensity in the CuO XES extending above the Fermi level is believed to be an experimental artifact (45).

The O K XES data (44) confirms our assignment of the O XPS. The principal XPS peak arises from the cd state, and it decays to the dp state. Therefore the principal O XES peak aligns with the Cu XES and the dp feature in the UPS. The cp^o state does not mix with the primary cd state; therefore, it does not relax before the decay, but decays directly to the pp^o_b (and perhaps a little also to the pp^o_a) state. This accounts for the feature around 6.5 eV in the XES, just 3 eV above the pp^o_b feature in the UPS. The shift of 3 eV matches the energy difference between the cp^o and cd core hole states. The cp^p state does mix with the cd state, therefore it can relax to the cd state, but it does this slowly because of the small excitation energy of 0.5 eV. Therefore, the cp^p state decays either directly to the pp^p state, or relaxes to the cd state, which then decays to the dp state. This explains the photon energy dependence seen in the data. At high photon energy, the sudden approximation is more valid, creating a larger intensity for the cp^p state, and consequently a larger pp^p contribution around 2.5 eV in the XES.

The Cu L_{23} and O K EELS and XANES Data

The EELS and XANES data in Figure 4 (46-48) reflect the contributions from three possible transitions; the dominant d $\rightarrow$ c contribution nearest the Fermi level, the pp^p $\rightarrow$ cv (v = d or p) contribution resulting from the carrier hole states, as well as the cvCB contribution well above the Fermi level (47). Here CB represents an electron present in the higher Cu 4sp or O 3p "conduction band". The latter two contributions are not always resolved, and sometimes have been confused in the literature (46-48). Although very similar, the EELS data appear to have lower resolution than the recent XANES data (48), and the EELS data may emphasize the cvCB contributions, perhaps because of a slight breakdown in the dipole approximation.

The CB DOS must reflect the presence of a core hole as dictated by the final state rule (49). The large peak at -1.7 eV in the O K for CuO and around -1.2 eV in the Cu L_{23} in Figure 4 is believed to arise from these cdCB and cpCB states, respectively (47). Note that an E_{2p}-E_B excitation energy in Table 1 corresponds to a feature at E_B (binding energy) in Fig. 4. The cdCB state in CuO is excitonic-like as the O 3p DOS drops into the 2 eV gap because of the core hole. Cu_2O is a filled band in the ground state, so only a cCB feature appears around -1. eV. This feature in Cu_2O has helped

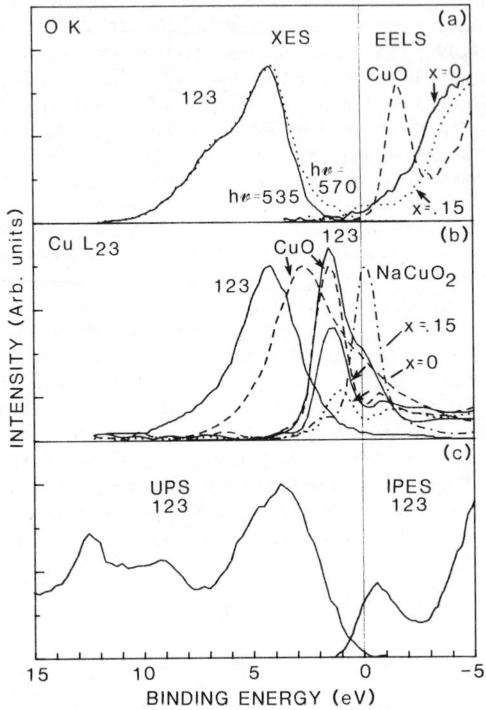

Figure 4a. Comparison of O K XES data for 123 taken at the indicated photon excitation energies (from Ref.44). Also, comparison of O K EELS data for CuO and La at the indicated Sr dopant levels (from Ref. 46)
4b. Comparison of Cu L_{23} XES data for 123 (Ref. 44) and CuO (Ref. 45), of EELS data for CuO and La at the indicated Sr dopant levels (Ref. 46), and of XANES data for 123 and $NaCuO_2$ (Ref. 48).
4c. UPS ($h\nu$ = 74 eV from Ref. 17) and IPES data ($h\nu$ = 18 eV from Ref. 14) for 123.
(Reproduced from Ref. 2, Not subject to copyright.)

to identify the similar features in CuO and the HTSC's. The doped La material, which is metallic, does not exhibit the excitonic feature, but a cdCB contribution is still believed to be present around the same energy in the O K EELS (2).

The dominant d → c features are predicted by our model to be excitonic-like, i.e. bound by $\Delta+\alpha$ at the Cu L_{23} and by α at the O K levels. For La this is predicted to be 1 eV in both cases, however the data in Fig. 4 indicates that this is near zero at the O K. Nevertheless, our model predicts that the c states at the Cu L_{23} level should have a larger binding energy for CuO than for La (1.5 vs. 1 eV), and should be larger at the Cu L_{23} level than at the O K level for CuO (1.5 vs. 0.5 eV). Both of these trends are consistent with experiment.

The difference in the intensities of the d → c features for CuO and La can be understood by invoking the initial state rule (50). Since the c state has no valence holes, the edge features in the EELS data reflect the initial DOS as dictated by the initial state rule. We have previously proposed initial and final state rules which are appropriate for the nearly filled and empty band cases respectively (50). Here, the valence band is nearly filled while the conduction band is nearly empty; hence the c states reflect the initial v DOS, the cvCB reflect the final CB DOS. Thus the edge features in the EELS or XANES reflect directly the unfilled ground DOS, just as the inverse PES (IPES) data does in Figure 4c.

The change in intensities of the edge features between CuO and the undoped La material arises because of the smaller Δ in La. A decrease in Δ from 1 eV for CuO to significantly less than one for La means a larger probability for holes on the oxygens in the ground state of La (see eq. 1). This dictates an increase in the O K, and decrease in the Cu L_{23} d → c features compared with CuO, in agreement with experiment.

The pp^p → cp contribution has been resolved as a separate feature in the recent high resolution Cu L_{23} XANES data of Sarma et al (48), as shown in Fig. 4. As expected this feature dominates for $NaCuO_2$ consistent with the Cu L_{23} XPS data. The separation of the d → p and pp^p → cp features is found to be less than 1 eV in La and has relative intensities of around cp/c = 0.6 (48). This explains why it is not seen in the lower resolution EELS data in Fig. 4. For 123, the separation is 1.4 eV and cp/c is around 0.9 (48). The smaller energy separation in La compared with 123 is consistent with our theoretical model which predicts the separation to be $-\Delta+\delta_2-\delta_1+\beta = -\Delta+0.5$ eV. As Δ decreases from La to 123, the separation should get bigger. The pp^p → cv contribution should increase at both core levels with doping level in La; however, it should increase much faster at the Cu L_{23} level since state 1) has more pp^p character than dp. This is consistent with the EELS data, where the c and cp contributions are seen as one feature.

Temperature effects

Figure 2 summarizes the temperature effects which have been seen in the spectroscopic data for the high T_c materials (51–55). This data has been somewhat controversial, because it is not seen by all investigators. Nevertheless, it has been seen by several groups who have shown that it does not arise from increased impurities on the surface at lower temperatures; e.g. these effects do not appear for the tetragonal 123 material, a non-HTSC, while they do appear for the orthorhombic 123 (54) which is a HTSC. Recent single crystal data (17,31) do not appear to exhibit similar effects, but they may yet be found since single crystals have not been thoroughly studied. If they indeed do not appear in the single crystal data, one might conclude that the grain boundaries in the sintered polycrystalline materials are somehow involved.

All of the temperature effects can be attributed to a single phenomenon, namely a decrease in ε_p due to increased metallic screening (2) or long range polarization, which reduces U_{pe}. Thus the changes seen upon going from CuO to 123 simply continue upon lowering the temperature (2). This is consistent with the decrease in the main cp peak energy in the Cu 2p XPS, while the cd satellite remains unshifted (Figure 2a). The larger energy separation between the cd and cp states decreases the mixing which causes the satellite to decrease in intensity and the main peak to get narrower. Note that the primary cd peak does not shift in the O 1s XPS (Figure 2b), but a slight shift to lower energy is seen in the cp^p and cp^o contributions at lower temperature, as expected with a decrease in ε_p. It is also consistent with the increase in the dp^2 satellite contribution in the Cu $L_{23}VV$ AES as shown in Figure 4c; since the dp^2 and d^2p states become closer in energy and hence mix more. However, no shift in the Auger kinetic energy of the principal peak is seen, this is because both the core hole cp and final d^2p states decrease together with ε_p. Finally, the UPS spectra (not shown, see ref. 55) show a skewing toward the Fermi level at lower temperature, as expected with a decrease in ε_p.

The intensity of the $pp^p \rightarrow cpp^p$ feature around 937 eV in the Cu XPS of Fig. 2b appears to decrease with temperature, although this is not clear. In other data (51,53,54), the breadth of the main peak clearly increases, and we attribute this to an increase in the $pp^p \rightarrow cpp^p$ contribution. Also note that the $d \rightarrow cp$ transition energy decreases with D, but the $pp^p \rightarrow cpp^p$ energy remains fixed, consistent with experiment. As the temperature is lowered, the charge carrier concentration or density of pp^p states apparently increases. This increase is consistent with the decrease in ε_p due to increased metallic screening.

Finally, we should note that no temperature dependence has been seen in the EELS or XANES data (46–48). We note first that the energy of the primary $d \rightarrow c$ contribution does not involve Δ or ε_p, so no shift is expected. We would expect an increase in the intensity of the $pp^p \rightarrow cp$ contribution as well as a small energy shift. Although this is has not been observed (48), it may be difficult to observe because this small contribution is barely resolved from the main feature. Furthermore, the electron excited in the absorption process can screen its own hole, since it generally remains on the same atom, so the long range polarization may be less important. Finally, the EELS and XANES data is more reflective of the bulk than the other spectroscopic data, and perhaps the surface alters the screening process.

The increased metallic like screening which appears to occur at lower temperature may indeed involve the grain boundaries, for it has been shown that the grain boundaries can strongly alter the conductivity and its dependence on temperature in these superconducting materials (56). Since this metallic screening occurs through electron-hole pair excitations at the Fermi level, the extent of this screening is also expected to depend strongly on the DOS at the Fermi level. It should also depend strongly on whether the sample is above or below the bulk T_c, however, insufficient data has been published to determine the exact nature of the temperature dependence around T_c. Evidence has been presented for granular superconductivity in 123 in the range 100 to 160 K, with the islands of coupled HTSC granules increasing in size as the temperature decreases (57). This could possibly contribute to the changes seen well above T_c.

Summary

We have reviewed the spectroscopic data for the high temperature superconductors, and compared them with that for CuO. We have summarized an interpretation of the data utilizing a highly correlated CuO_n cluster model and show that a single set of Hubbard parameters predicts all of the state energies. Changes in the data between CuO and the HTSC's arises primarily from a reduction in ε_p; this reduction continues with decreasing temperature in the HTSC's due to increased metallic screening. Compared with CuO, the HTSC's show an increased covalent interaction between the Cu-O bonds. The large size of U_{pp}^0 and the temperature dependence reveal that metallic screening is incomplete, and hence that the DOS at the Fermi level in the HTSC's is relatively small.

Acknowledgments

This work was supported in part by the Office of Naval Research.

Literature Cited

1. Ramaker, D.E.; Turner, N.H.; and Hutson, F.L. submitted.
2. Ramaker, D.E. submitted.
3. Wendin, G. Proc. 14th Intl. Conf. on X-ray and Inner-Shell Processes, J. Physique (France)(In press).
4. Anderson, P.W. Science 1987, 235, 1196.
5. Varma, C.M.; Schmitt-Rink, S. E.; Abrahams, E. Sol.State. Commun. 1987, 62, 681.
6. Doniach S. et al. In Novel Mechanisms of Superconductors, Wolf S.A. and Fresin V.Z. Ed.; Plenum: NY, 1987, p 395.
7. Hirsch, J.E. et al. Phys. Rev. Letters 1988, 60, 1168.
8. vanderLaan G. et al. Phys. Rev. 1981, 24, 4369.
9. Fuggle J.C. et al. Phys. Rev. 1988, B37, 1123.
10. Greedan J.E. et al. Phys. Rev. 1987, B35, 8770.
11. de Groot, R.A.; Gutfreund, H.; Weger, M. Sol. State Commun. 1987, 63, 451.
12. Folkerts, W. et al. J. Phys. C: Solid State Phys. 1987, 20, 4135.
13. Manthiram, A.; Tang, X.X.; Goodenough, J.B. Phys. Rev. 1988, B37, 3734.
14. Redinger, J. et al. Phys. Lett. 1987, 124, 463 and 469.
15. Thuler, M.R.; Benbow, R.L.; Hurych, Z. Phys. Rev. 1982, B26, 669.

16. Benndorf, C. et al. J. Electron. Spectrosc. Related Phenom. 1980, 19, 77 .
17. Stoffel, N.G. et al. Phys. Rev. 1988, B37, 7952; also preprint.
18. Kurtz, R. et al. Phys. Rev. 1987, B35, 8818.
19. Mueller, D. et al. In ref. 6, p 829.
20. Onellion, M. et al. Phys. Rev. 1987, B36, 819.
21. Tang, M. et al. Phys. Rev. 1988, B37, 1611; also preprint.
22. Samsavar, A. et al. Phys. Rev. 1988, B37, 5164.
23. Rosencwaig, A.; Wertheim, G.K. J. Elect. Spectrosc. Related Phenom. 1972/73, 1, 493.
24. Miller, D.C. et al. In Thin Film Processing and Characterization of High Temperature Superconductors, Harper, J.M.; Colton, J.H.; and Feldman, L.C. Eds., AVS Series No. 3, American Institute of Physics: New York, NY, 1988; p 336.
25. Steiner, P. et al. Appl. Phys. 1987, A44, 75.
26. Weaver, J. et al. preprint.
27. Reihl, B. et al. Phys. Rev. 1987, B35, 8804.
28. Klopman, G. J. Am. Chem. Soc. 1964, 86, 4550.
29. Brewer J.H. et al. Phys. Rev. Lett. 1988, 60, 1073.
30. Chang, Y. et al. preprint; Onellion, M. et al. preprint.
31. Takahashi T. et al. Phys. Rev. 1988, B37, 9788.
32. Kohiki, S.; Hamada, T. Phys. Rev. 1987, B36, 2290.
33. Ramaker, D.E. et al. Phys. Rev. 1987, 36, 5672.
34. Sarma, D.D. Phys. Rev. 1988, B37, 7948.
35. Steiner, P. et al. Z. Phys. B- Condensed Matter 1987, 67, 497.
36. Steiner, P. et al. Z. Phys. B- Condensed Matter 1988, 69, 449.
37. Gourieux, T. et al Phys. Rev. 1988, B37, 7516.
38. Qiu, S.L. et al. Phys. Rev. 1988, B37, 3747.
39. Ford, W.K. et al. Phys. Rev. 1988, B37, 7924.
40. Ramaker, D.E.; Turner, N.H.; Hutson, F.L. In Ref. 24, p 284.
41. van der Marel, D. et al. Phys. Rev. 1988, B37, 5136.
42. Fiermans, L.; Hoogewijs, R.; Vennik, J. Surf. Sci. 1975, 47, 1.
43. Chang, Y. et al. Phys. Rev. B (In press).
44. Tsang, K.L. et al. Phys. Rev. 1988, B37, 2293.
45. Koster, A.S. Mole. Phys. 1973, 26, 625.
46. Nucker, N. et al. Z. Phys. B: Cond. Matter 1987, 67, 9; Phys. Rev. 1988, 37, 5158.
47. Bianconi, A. et al. Solid State Commun. 1987, 63, 1009; Intn. J. Modern Phys. 1987, 131, 853.
48. Sarma, D.D. et al. Phys. Rev. 1988, B37, 9784.
49. Ramaker, D.E. Phys. Rev. 1982, B25, 7341.
50. Erickson, N.E.; Powell, C.J.; Ramaker, D.E. Phys. Rev. Letters 1987, 58, 507.
51. Dauth, B. et al. Z. Phys. B- Condensed Matter 1987, 68, 407.
52. Balzarotti, A. et al. Phys. Rev. 1987, B36, 8285.
53. Sarma, D.D. et al. Phys. Rev. 1987, B36, 2371.
54. Kim, D.H. et al. Phys. Rev. 1988, B37, 9745.
55. Iqbal, Z. et al. J. Materials. Res. 1987, 2, 768.
56. Renker, B. et al. Z. Phys. B- Cond. Matter 1987, 67, 1.
57. Cai, X.; Joynt, R.; Larbalestier, D.C. Phys. Rev. Letters 1987, 58, 2798.

RECEIVED July 16, 1988

Chapter 8

Theory of High-Temperature Superconductivity

Overlap of Wannier Functions and the Role of Dielectric Screening

Daniel C. Mattis

Physics Department, University of Utah, Salt Lake City, UT 84112

Virtual charge fluctuations in one conducting plane can interact with virtual fluctuations in adjacent, parallel, conducting planes and can, under certain conditions, overcompensate the intraplanar Coulomb repulsion. We prove this using the Lindhard dielectric function formalism. Surprisingly, the physical overlap of Wannier functions on neighboring atoms within each plane is highly favorable to High-T_C superconductivity, while analogous overlap between planes is detrimental to it. Thus, small intraplanar m* favors High-T_C superconductivity, and the hard-to-achieve combination of high interplanar resistivity and small interplanar distance, optimizes it.

This Chapter outlines a new concept, and introduces a theory for the High-T_C superconductors (1-3) based on interactions between virtual charge fluctuations on adjacent, parallel metallic planes. The mechanism can be superficially understood by analogy with the well-known van der Waals interaction between neutral atoms and molecules, with one essential difference: the attractive potential in the present instance encourages condensation of the metallic electrons into a BCS (4) superconducting state within each plane.

0097–6156/88/0377–0103$06.00/0

The inclusion of a *small* amount of interplanar hopping does not significantly affect the results, but ensures that superconducting long-range order in the present model (in contradistinction with purely intraplanar schemes) will not violate the important Mermin-Wagner theorem (5). This theorem forbids long-range order in one and two dimensions at finite temperature, permitting only superconducting fluctuations ("para"superconductivity) to survive. With a small amount of interplanar hopping, the correlation length in the vertical direction increases from zero to a finite value. But in many other respects, the present mechanism is inherently low-dimensional. Insofar as the electronic overlap from plane to plane becomes comparable to that within the planes, the attractive mechanism is lost. I therefore do not expect this mechanism to work in isotropic, three-dimensional materials.

But would it help to go to lower dimensions still? Although a similar mechanism *may* conceivably be optimized in certain arrays of polymers, or in other quasi-one-dimensional systems, there are technical reasons, related to a substantial decrease in the relevant phase space, which would prevent T_C from becoming as dramatically large in 1D as in 2D. Thus the discovery of Berdnoz and Müller (1) seems to be singularly related to the two-dimensional, layered structure of the LaSrCuO material they chose to examine, and not to any other unusual properties that it (or the other high-T_C materials (2,3)) might incidentally exhibit, such as Jahn-Teller distortion, superstructures, antiferromagnetism, twinning, or the like.

I present the theory in two stages. In the first, an intuitive picture is drawn and a simple explanation of the phenomenon is given. In the second, a formulation is sketched out which, aside from one or two lengths that can be experimentally obtained, is practically free of adjustable parameters. Theessentialrole played by the *Wannier states' overlap* and by *dielectric screening* is explained. Explicit formulas are exhibited for such properties of the superconductors as T_C.

The Simple Physical Picture

Even in the best of metals, a space- and time-dependent charge distribution $\delta\rho(\mathbf{r},t)$ cannot be efficiently screened out at wavelengths short compared with the screening length nor at frequencies high compared with the plasma frequency. The screening response of an electrical conductor to a test-charge is measured by the dielectric function $\epsilon(\mathbf{q},\omega)$, a complex function of wave-vector and frequency (6) which also depends on such electronic parameters as the Fermi wave-vector k_F. In the present theory, we are only concerned with relatively small wave-vectors $q \approx 2k_F$ and low frequencies $\hbar\omega << E_F = \hbar^2 k_F^2/2m^*$, corresponding to a low-density gas of fermions. Given m^* and lattice parameters $a \approx b$ and d=interplanar separation (the vertical cell dimension c may be a multiple of d,) it becomes possible to estimate quantitatively the optimal parameters for high-T_c superconductivity. At first, let us seek possible sources of electron-electron attractions (a principal requirement, if the BCS theory is to be applicable) within the Coulomb interactions themselves.

The real part of the static dielectric function for a charged gas of fermions, interacting by ordinary Coulomb forces (e^2/r_{ij}), on a single planar conducting sheet, was formulated by F.Stern (7):

$$\epsilon(\mathbf{q},0) = [\,1 + (2/qa_0^*)\,] \quad , \quad q \leq 2k_F \tag{1a}$$

$$= [\,1 + (2/qa_0^*)(1-[1-(2k_F/q)^2]^{1/2}\,)]\,, \quad q > 2k_F \tag{1b}$$

where $\mathbf{q}=(q_x,q_y)$, $a_0^* = \hbar^2/m^*e^2$ being the Bohr radius, the background dielectric constant κ being taken here to be 1. This formula shows that screening is highly effective only for small q. The Hamiltonian for any two parallel conducting sheets is,

$$H = H_1 + H_2 + H'_{1,2} \tag{2}$$

with H_i containing the kinetic (and potential) energy operators of the Fermi fluid over the N cells (each of area a^2) in the i-th plane, and

$$H'_{1,2} = (1/N)\sum_{\mathbf{q}} W_0(\mathbf{q};d)\,\rho_1(\mathbf{q})\rho_2(-\mathbf{q}) \tag{3}$$

is the "bare" interaction connecting charge fluctuations ρ in the two planes, with the operators $\rho_i(\mathbf{q})$ given as:

$$\rho_i(\mathbf{q}) = \sum_{\mathbf{k}\sigma} c^*_{\mathbf{k+q},\sigma,i}\, c_{\mathbf{k},\sigma,i} \tag{4}$$

in the language of fermion anihilation and creation operators, σ ($=\uparrow$ or $\downarrow$) being the spin parameter. $W_0(\mathbf{q};d)$ is obtained as follows:

$$W_0(\mathbf{q};d) = (e^2/a^2)\int d^2r\, e^{i\mathbf{q}\cdot\mathbf{r}}\, [d^2 + r^2]^{-1/2}$$

This integration yields:

$$W_0(\mathbf{q};d) = e^{-dq}\,[2\pi e^2/qa^2] \tag{5a}$$

We note that this is an *over*estimate of the interaction, as both ρ_1 and ρ_2 should be accompanied by their respective *intra*planar screening charges (image charges, in classical electrostatics.) A more conservative estimate is obtained by replacing each $\rho(\mathbf{q})$ in (3) by $\rho(\mathbf{q})/\epsilon(\mathbf{q},0)$. This simulates the net charge fluctuation, after subtraction of the polarization charges. In dividing W_0 by $\epsilon^2(\mathbf{q},0)$ we use the larger value of ϵ^2, Equation 1a, to obtain:

$$W(\mathbf{q};d) = e^{-dq}\,[2\pi e^2/q(1+2/qa_0)^2a^2] \tag{5b}$$

which now *under*estimates the actual interaction. This W vanishes at long wavelengths, $q\to0$, showing the effects of metallic screening; it has a broad maximum (on the order of eV) as a function of q, falling off once $q > 1/d$.

In the homogeneous metal, there are no "real" charge fluctuations, just as in the Helium atom there is no "real" dipole moment. Van der Waals' attraction between two Helium atoms comes about as the result of the correlated quantum fluctuations of a virtual dipole moment induced in each atom by the other. Similarly, an attractive force comes about from correlated, virtual, charge fluctuations in the two planes. I eliminate H' from the Hamiltonian to leading order in W, using the operator generalization of second-order perturbation theory for the energy levels. This yields:

$$H \to H = H_1 + H_2 - (i/2)\int_0^\infty dt\ \ [H'_{1,2}(t),H'_{1,2}] \tag{6}$$

where the time-dependent operators are

$$H'(t) \equiv e^{i(H_1+H_2)t}\ H'\ e^{-i(H_1+H_2)t} \tag{7}$$

as usual. Having previously included the intraplanar screening in $\mathcal{W}$, in evaluating Equation 7 we replace H_1 and H_2 by just their kinetic energy operators (the free Fermions' Hamiltonians,) obtaining:

$$H_{tot.} = \sum_{\mathbf{k},\sigma,i} (\hbar^2/2m^*)(k^2-k_F^2)c^*_{\mathbf{k},\sigma,i}\, c_{\mathbf{k},\sigma,i} + (1/N)\sum_{\mathbf{q},\mathbf{k},\sigma,i} V(\mathbf{k},\mathbf{k}+\mathbf{q})\, c^*_{\mathbf{k}+\mathbf{q},\sigma,i}\, c_{\mathbf{k},\sigma,i}\, \rho_i(-\mathbf{q}) \quad (8)$$

in which $\mathcal{V}$ includes <u>both</u> attractive (interplanar) <u>and</u> repulsive (intraplanar) two-body forces:

$$V(\mathbf{k},\mathbf{k}+\mathbf{q}) = \mathrm{Re}\{- W^2(\mathbf{q};d)(\gamma/N)\sum_{\mathbf{k}''}[f(\mathbf{k}'')-f(\mathbf{k}''+\mathbf{q})]/[D(\mathbf{k},\mathbf{k}'',\mathbf{k}+\mathbf{q})] + W_0(\mathbf{q};0)/\epsilon(\mathbf{q},0)\} \quad (9a)$$

where we defined,

$$D(\mathbf{k},\mathbf{k}'',\mathbf{k}+\mathbf{q}) = (\hbar^2/2m^*)[\,[(\mathbf{k}''+\mathbf{q})^2 - \mathbf{k}''^2] - [(\mathbf{k}+\mathbf{q})^2 - \mathbf{k}^2]\,] \quad (9b)$$

The sum over (i) in $H_{tot.}$ has been extended to all metal planes in the solid. If a given metal layer interacts with 2 neighboring layers, $W^2(\mathbf{q};d)$ is effectively doubled. The factor γ ($1 \le \gamma \le 2$) which we include to take this into account could even exceed 2 (in principle,) if next-nearest or more distant planes contributed additional significant amounts.

Additionally, a screened Coulomb *intra*planar repulsion $W_0(\mathbf{q};0)/\epsilon(\mathbf{q},0)$ is explicitly included in Equation 9a. Under normal circumstances, this repulsion is the dominant feature, and superconductivity prohibited.

Defining the BCS pair operator $b_i^*(\mathbf{k}) \equiv c^*_{\mathbf{k},\uparrow,i}\, c^*_{-\mathbf{k},\downarrow,i}$ in the usual way, the model effective pairing Hamiltonian (2) is now:

$$H_{eff} = \sum_{\mathbf{k},\sigma,i} (\hbar^2/2m^*)(k^2-k_F^2)c^*_{\mathbf{k},\sigma,i}\, c_{\mathbf{k},\sigma,i} + (1/N)\sum_{k,k',i} V\, b_i^*(\mathbf{k}')\, b_i(\mathbf{k}) \quad (10)$$

This pair potential $\mathbf{V}(\mathbf{k},\mathbf{k}')$ is conducive to superconductivity if the attraction exceeds the repulsion in a sufficient amount. With $W(2k_F,0)$ and E_F both typically O(eV), just how negative *is* $\mathbf{V}(\mathbf{k},\mathbf{k}')$?

To answer this question accurately, it is necessary to abandon the perturbation-theoretic approach and treat all the interactions within the solid on the same footing, as is done within the framework of dielectric function formalism in the latter half of this Chapter. But first, the physics merits our attention.

A Question of Overlap. Previous experience with low-temperature superconductors, and with the BCS theory used to analyze them, has made it almost a matter of dogma that large effective mass m* (and consequently, large density-of-states,) are conducive to high-temperature superconductivity. Nevertheless, here the facts dictate otherwise (8,9). The concensusfor the high-T_C materials — based on experiments too numerous and varied to recapitulate here — is that m* is rather smaller than a free electron mass, and that the density-of-states $\mathbf{N}_F$ at the Fermi surface is unusually low. This is but one clue to the origin of the physical mechanism.

A second clue comes from the anisotropy in the conductivity of these layered materials. We noted that the carriers have a rather small m* within the planes, i.e. a greater than usual propensity for motion within each plane. This, however, contrasts sharply with their almost negligible conductivity in the direction normal to the planes, which may be related to an unusually small "hopping" matrix element for motion *from plane to plane*. Both features are reflected in the band structure calculations (10,11), which have specifically remarked upon the absence of dispersion within the conduction bands in the direction normal to the planes, as well as the small density of states at or near the Fermi level. The width of an energy band can be qualitatively identified with 1/m*. It is also related to the overlap of nearest- and next-nearest (and even more distant) Wannier orbitals. A Wannier function $\Psi(\mathbf{r}-\mathbf{R}_i)$ centered on the ith atom or cell, is a compact localized state, similar to an atomic orbital but orthogonalized to all other $\Psi(\mathbf{r}-\mathbf{R}_j)$, and is constructed entirely out of the Bloch states of the conduction bands. It is generally much more extended than an atomic orbital. In constructing the potential energy, and its scattering amplitudes, there might seem to be be an ambiguity in the selection of "best" localized states. In fact, the Wannier functions must be selected, being unique in that they scatter entirely within the given conduction band. One notes that lack of dispersion for motion from plane-to-plane is operationally equivalent to small, or vanishing, overlap of Bloch or Wannier orbitals on distinct planes.

Based on these observations, we can now draw conclusions on the nature and strength of the unscreened Coulomb interactions between two charged carriers in the conduction band, the one centered about an atom or cell located at $\mathbf{R}_i$, and the other centered about $\mathbf{R}_j$:

$$U(\mathbf{R}_i,\mathbf{R}_j) = \int d^3r \int d^3r' \, |\Psi(\mathbf{r}-\mathbf{R}_i)|^2 \; |\Psi(\mathbf{r}'-\mathbf{R}_j)|^2 \,(e^2 / |\mathbf{r}-\mathbf{r}'|) \qquad (11)$$

A Wannier function exists over a reasonably well circumscribed range, out to a distance $a_0 > a$. Thus, if $\mathbf{R}_i$ and $\mathbf{R}_j$ are separated by much more than a_0 , $U \to e^2/|\mathbf{R}_i-\mathbf{R}_j|$. But if $\mathbf{R}_i$ and $\mathbf{R}_j$ are nearby sites, or even equal, the electrostatic repulsion dips below the point charge value $e^2/|\mathbf{R}_i-\mathbf{R}_j|$ owing to the overlap of the functions. This reduction is maximal at $\mathbf{R}_i = \mathbf{R}_j$. For example, on Cu^{+2} in CuO_2 planes, $U(\mathbf{R}_i,\mathbf{R}_i)$ is variously estimated at $U_0 \leq 6$ eV, far less than the corresponding Coulomb energy in the atom, U_{At}=16.6 eV . This reduction by 2/3 cannot be just the result of electronic screening, which is relatively ineffectual at short distance. It must be principally due to the physical size of the Wannier orbitals, here presumably three times more spread out than the parent atomic orbitals.

This observation suggests the following rule: *the smaller m^*, the smaller the so-called "Hubbard parameter" U_0* (all other things being equal.) However self-evident it seems, this rule does not appear to have been previously enunciated.

If two particles within the same plane are nearest- or even next-nearest-neighbors, their interaction also has to be correspondingly reduced from that of point-charges. On the other hand, if the two are in <u>different</u> planes, and if – as we have been assuming -- there is little or no overlap between them, one obtains $U = e^2/|\mathbf{R}_i-\mathbf{R}_j|$ at <u>all</u> distances R_{ij} . (We are neglecting small but possibly important electrostatic multipole corrections relating to the geometrical shape of the orbitals. These will provide additional short-ranged contributions to the interactions, both in-, and out-of-plane.)

We now summarize the conclusions of this Section. On the one hand, the effective repulsion between carriers within the same plane is weakened, as compared to the usual Coulomb value. On the other hand, in the absence of interplanar hopping, the interaction between carriers on different planes is not. The obvious inference, drawn in the following Section, is that interplanar forces can become dominant in a limited region of phase space.

The Two-Body Interactions

The "bare potentials" calculated with the aid of the Wannier functions, scatter Bloch states in the same way an ordinary potential scatters plane waves. To obtain the relevant matrix elements in the Bloch representation, one requires the Fourier series coefficients of $U(\mathbf{R}_i,\mathbf{R}_j)$,

$$U(\mathbf{q},q_z) = \sum_j U(\mathbf{0};\mathbf{R}_j)\, e^{i(q_x X_j + q_y Y_j + q_z Z_j)} \qquad (12)$$

with $\mathbf{q},q_z$ in the first Brillouin Zone. Generally, the $U(\mathbf{q},q_z)$ must be obtained numerically. But if the carrier gas density is not high ($k_F << \pi/a$), such that only long wavelengths ($qa << 1$) are of interest, some sums can be replaced by integrals.

[A justification (based on a Taylor series expansion of F(x) within each cell) follows: if F_n is a symmetric, discrete summand, F(x) being its smooth approximant with $F(xa)=F_n$, then the lattice Fourier sum approaches the Fourier transform to within a palpably negligible error as $q \to 0$, i.e. $\sum e^{iqn} F_n = \{\int dx a^{-1} e^{iqx} F(x)\}\{1 - (qa)^2/24 + O(qa)^4\}$.]

Although this result justifies Fourier transformations at long wavelengths in the x,y plane, one is still required to perform the exact lattice Fourier sums in the perpendicular direction. Let $A=2\pi e^2/a^2$, $B(q) \equiv \exp-(qd)$, $\zeta \equiv \cos(q_z d)$, and let $\Phi(q)$ be the function which incorporates the intraplanar reductions in Coulomb interactions owing to Wannier orbitals' overlap. The end product of our calculation is:

$$U(\mathbf{q},q_z) = (A/q)\{[1- B^2(q)][1-2B(q)\zeta + B^2(q)]^{-1} - \Phi(q)\} \qquad (13)$$

where, specifically,

$$\Phi(q) = q\int_0^\infty dR J_0(qR)[1-(RU(0,R)/e^2)] = qa_0 + O(qa_0)^3 \qquad (14)$$

This defines a_0, the "overlap radius." Later equations simplify considerably if this parameter is henceforth identified with the Bohr radius a_0^* defined in Equation 1. This also accords with the qualitative arguments, that the smaller m^*, the greater Φ becomes.

In the long wavelength limit (q, q_z *both* $\to 0$) the interaction becomes: $U/NN_z \to (4\pi e^2/\Omega)/(q^2+q_z^2)$, with $\Omega=(da^2)(NN_z)$ the volume of the solid. This limit reproduces precisely the three-dimensional Coulomb interaction for "jellium" (6), and is indifferent to Φ.

In the layered materials, however, it is quite a different limit, $q \to 0$, $q_z d \to \pi$, which concerns us. At fixed q_z, the maximum magnitude of $U(\mathbf{q},q_z)$ is at $q=0$, and is given by

$$U(\mathbf{0},q_z) = Aa_0^* \left\{ [(d/a_0^*) -(1-\zeta)] / [1-\zeta] \right\} \equiv Aa_0^* \{ u(\zeta) \} \qquad (15)$$

which serves to define the dimensionless potential, $u(\zeta)$. Interestingly, $(Aa_0^*)^{-1} = \mathbf{N}_0$, the density-of-states at the band edge for carriers of mass m^*. With k_F small, as we have hypothesized, we can reasonably assume that the two-dimensional density-of-states at the Fermi energy remains constant at $\mathbf{N}_F \approx \mathbf{N}_0$. At the long wavelengths of interest for present purposes, Equation 15 is accurate over a sufficient range of q to be an acceptable surrogate for the exact, but more complicated $U(\mathbf{q},q_z)$ of Equation 13.

Note that $u(\zeta)$ is finite at $|q_z|>0$ (i.e. $\zeta>0$), changing sign for $\zeta < 1-d/a_0$. But as $\zeta = \cos q_z d \geq -1$, $u(\zeta)$ can be negative only if $d < 2a_0^*$. Its minimum occurs at $\zeta=-1$, and is $u(-1) = \{(d/2a_0^*) - 1\}$. In the limit $d/a_0^* \to 0$: $u(\zeta) = -1$, is constant, and optimally negative. Although this limit is absurd, it does indicate that the pseudo-attraction has to be sought at small d (or, equivalently, at large a_0^*.)

In phase space, $q_z d = \pi$, $q=0$ indicates where the layered materials are most susceptible to attractive mechanisms of *any* extrinsic origin, such as the electron-phonon interaction. Here it is that the Coulomb repulsion is *least* repulsive, entirely as a consequence of electrostatic *inter*planar interactions of alternating sign summed out to infinity.

The *intrinsic* attractive mechanism under scrutiny here is characterized by Φ. It comes about because of a reduction in the short-range Coulomb repulsion within each plane, caused by the orbitals' overlap. Other mechanisms, such as the electron-phonon interaction, will make additional attractive contributions to the two-body forces, but for lack of space are not considered here.

The reader is cautioned that regardless of the physical correction symbolized by Φ, $U(\mathbf{q},q_Z)$ in Equation 13 is a potential made up entirely of repulsive forces. It is simple to verify that if $d/a_0^* \neq 0$ no low-lying two-body bound state can be formed, either in coordinate space, or in momentum space. *In the presence of many other particles*, however, the situation changes dramatically. *Screening* tempers the repulsions, and may even enhance the attraction. It brings about the possibility of binding of Cooper pairs. Let us now see how and why.

Dielectric Screening. The dielectric function ϵ corrects the "bare potentials" for the current flows and the polarization of the electronic fluid. In the calculation of a screened two-body interaction U_S and of the relaxation time τ, one uses $1/\epsilon$:

$$U_S(\mathbf{q},q_Z\,;\,\omega) \equiv \mathrm{Re}\{1/\epsilon(\mathbf{q},q_Z;\omega)\}U(\mathbf{q},q_Z) \tag{16}$$

$$1/\tau = 2\mathrm{Im}\{1/\epsilon(\mathbf{q},q_Z;\omega)\}U(\mathbf{q},q_Z) \tag{17}$$

Insofar as two-body scattering among similarly charged particles conserves momentum, hence current, τ is unrelated to the electrical conductivity. Nevertheless, τ may figure in connection with other relaxation processes, and is experimentally observable.

The appropriate Lindhard formula for complex ϵ is easily calculated using a standard approach (6,7). With q,k_F both assumed to be small (such that Equation 15 remains a good approximation to the potential,) but k_Z and the ratios ω/E_F and q/k_F having arbitrary values, one finds:

$$\epsilon(\mathbf{q},q_Z;\omega) = 1 + \mathbf{X}(z,\upsilon)\,u(\zeta) \tag{18}$$

where $\zeta \equiv \cos(q_Z d)$ and $u(\zeta)$ have been defined above, $z \equiv q/2k_F$, and $\upsilon \equiv \omega m^*/\hbar q k_F$. At T=0 the complex quantity $\mathbf{X}(z,\upsilon) \equiv R(z,\upsilon) + iI(z,\upsilon)$ is:

$$\mathbf{X}(z,\upsilon)= 1 - (2z)^{-1}C_-[\,|(z-\upsilon)^2-1|\,]^{1/2} - (2z)^{-1}C_+[\,|(z+\upsilon)^2-1|\,]^{1/2} \quad (19)$$

with $C_{\pm}= 1$ if $(z\pm\upsilon)^2 \geq 1$, and $C_{\pm} = i$ if $(z\pm\upsilon)^2 \leq 1$. The sign of the imaginary contribution is chosen to satisfy the dictates of causality. Another consequence of causality (12,13) is a stability condition:

$$\mathrm{Re}\,\{1/\,\epsilon(\mathbf{q},q_Z;0)\,\} \leq 1 \quad (20)$$

Once $u(\zeta)$ becomes attractive (<0) the condition (20) fails to be met for $z \geq 1$. This indicates an incipient instability at $z=1$ ($q\approx 2k_F$) in the x-y planes, which could be against superconductivity and/or against a charge density wave. ("Incommensurate" charge density waves have indeed been seen in the newest high-T_C materials (14). But, having noted this point, we assume henceforth that any deformation of the lattice and/or electronic fluid is either virtual, or if real, sufficiently weak, that we can continue to assume translational invariance in the x-y planes in our study of superconductivity.)

It is then simple to extract the matrix elements for pair scattering out of the screened potential of Equation 16:

$$V(z,\zeta,\upsilon)=Aa_0{}^*u(\zeta)[1+ R(z,\upsilon)u(\zeta)]/([1+R(z,\upsilon)u(\zeta)]^2+ [I(z,\upsilon)u(\zeta)]^2) \quad (21)$$

Here one identifies the momentum transfer $\mathbf{q} = \mathbf{k}-\mathbf{k}' = (q_X, q_y)$ with $z = |\mathbf{k}-\mathbf{k}'|/2k_F$, $\zeta = \cos (k_Z-k_Z')d$ as before , the (dimensionless) energy transfer ($\upsilon=\omega m^*/\hbar qk_F$) to one particle in the pair being therefore $\upsilon=|\mathbf{k}+\mathbf{k}'|/2k_F$. At long wavelengths, $qa<<1$, Equation 21 is essentially an accurate result.

Being independent of q in the long-wavelength limit, this "pair potential" $V(z,\zeta,\upsilon)$ is thus short-ranged in the x-y planes (its spatial "reach" can be larger than the cell parameter a but much less than the typical electronic "size" π/k_F in the low-density limit.) It is, however, a sensitive function of q_Z. Like U, it is attractive in the neighborhood of $q_Zd = \pi$. Unlike U, it does not become infinitely repulsive at long wavelengths, but stays finite (albeit repulsive) even in the limit $q_z \to 0$. If its *average value* is negative, there is a good chance that high-T_c superconductivity will occur.

Applications of BCS Theory of Superconductivity

The model effective pairing Hamiltonian replacing (10) is:

$$H_{eff} = \sum_{\mathbf{k},k_z,\sigma}(E(k)-E_F)c^*(\mathbf{k},k_z,\sigma)c(\mathbf{k},k_z,\sigma) + (1/N)\sum_{k,k_z;\ k',k_z'} V\ b^*(\mathbf{k}',k_z')\,b(\mathbf{k},k_z) \qquad (22)$$

with $E(k)= \hbar^2k^2/2m^*$, $V(z,\zeta,\upsilon)$ discussed above, and the pair operator defined in the usual way, $b^*(\mathbf{k},k_z) \equiv c^*(\mathbf{k},k_z,\uparrow)\ c^*(-\mathbf{k},-k_z,\downarrow)$. To incorporate the effects of a small amount of interplanar hopping on the motional energy, one may replace $E(k)$ by $E(k,k_z)= E(k) + t_z\cos(k_zd)$ and shift E_F correspondingly. (The effect of small t_z on the potential energy can be simulated simply, by shifting d/a_0^* somewhat toward larger values.)

The integral equations of the BCS theory must now be solved numerically, the principal physical parameters being d/a_0^*, k_F, and t_z. It is possible however to anticipate the results of such calculations, and obtain solutions in closed form, simply by fixing ζ,υ in $V(z,\zeta,\upsilon)$ at some appropriately chosen average value. Thus, replacing $R(z,\upsilon)$ by $R(\langle z\rangle,\langle\upsilon\rangle) \equiv \langle R\rangle$, and $I(z,\upsilon)$ by $\langle I\rangle$, one obtains an average $\langle V(\zeta)\rangle$,

$$\langle V(\zeta)\rangle = Aa_0^* \left\{[u(\zeta)][1+ \langle R\rangle u(\zeta)] \ / \ ([1+\langle R\rangle u(\zeta)]^2 + [\langle I\rangle u(\zeta)]^2)\right\}$$
$$\equiv Aa_0^* \{ - \mathcal{K}(\zeta) \} \qquad (23)$$

This serves to define $-\mathcal{K}(\zeta)$ (note the "-"sign.) With $\theta=k_zd$ and $\beta=1/kT$, the BCS equation (4) for the superconducting gap and order parameter $\Delta(T,\theta)$ is:

$$\Delta(T,\theta) = \oint d\theta'/2\pi\ \mathcal{K}(\cos(\theta-\theta'))\Delta(T,\theta')$$
$$\times \int_0^{\epsilon(\theta')} dx\ [x^2 + \Delta^2(T,\theta')]^{-1/2} \tanh\{\beta[x^2 + \Delta^2(T,\theta')]^{1/2}/2\} \quad (24)$$

where $\oint$ indicates an integration from 0 to 2π and $\epsilon(\theta)\equiv E_F - t_z\cos\theta$.

Even at this level, the theory is fully three-dimensional: k_Z enters through θ and **k** through the variable of integration $\mathcal{X} \propto k^2$. At T=0, the nonlinear equation for the gap function $\Delta(\theta)$ ensues:

$$\Delta(\theta) = \oint d\theta'/2\pi\ \mathcal{K}(\cos(\theta-\theta'))\Delta(\theta')\ \mathrm{arcsinh}[\epsilon(\theta')/\Delta(\theta')] \qquad (25)$$

Examine this equation in the limiting case $t_Z=0$ first. If one replaces $\Delta(\theta')$ by a constant Δ_0 , it yields λ_0, the largest eigenvalue of the cyclic kernel $\mathcal{K}$. If $\lambda_0>0$, the equation has a nontrivial ($\Delta\neq0$) solution:

$$\Delta_0 = E_F\ /\sinh(1/\lambda_0) \approx 2E_F\ \exp-(1/\lambda_0) \qquad (26)$$

The correlation distance in the z-direction is zero. For if instead of pairing k_Z with $-k_Z$ we defined a new operator in Equation 22, $b_{qZ}{}^*(\mathbf{k},k_Z) \equiv c^*(\mathbf{k},k_Z+q_Z/2,\uparrow)c^*(-\mathbf{k},-k_Z+q_Z/2,\downarrow)$,absent dispersion in the z-direction we would obtain <u>precisely</u> the same solution for <u>any</u> q_Z . If the pairs can have any momentum in the z-direction, so they can be localized on any plane at no cost in energy. The phase space is then entirely two-dimensional, with fluctuations destroying long-range order at finite T in obedience to the Mermin-Wagner theorem (5).

If $t_Z \neq 0$, however, Equation 25 is no longer so easily solved. It is evident that the canonical pairing ($q_Z=0$) yields the optimum solution. $\Delta(q_Z,\theta)$ is no longer a trivial function of θ, although its coefficients in a Fourier series $\Delta(q_Z,\theta)=\sum C_n(q_Z)\cos n\theta$ can be found by expansion in powers of (t_Z/E_F) . At $q_Z\neq0$, the over-all magnitude of $\Delta(q_Z,\theta)$ decreases, and its profile as function of θ changes. Thus, dispersion in the one-particle spectrum generates dispersion in the pairs, and there will be a finite coherence distance in the z-direction. The phase space will be fully three-dimensional, and long-range order will be able to be sustained. We do not have any quantitative results on this, as yet.

T_C is itself derived from Equation 24 in the limit $\Delta\rightarrow0$. If $t_Z=0$, the onset of "para"superconductivity is at $kT_C = 1.13\ E_F\ \exp-(1/\lambda_0)$, with the ratio $2\Delta_0/kT_C$ having the canonical BCS (4) value, 3.54 . But if $t_Z\neq0$, as it is reasonable to assume, the integral equation for Δ becomes complicated even at very small t_Z . It might be more appropriate to examine $\lambda_0(T)$ than $2\Delta_0/kT_C$. I believe numerical calculations will ultimately reveal the effective eigenvalue λ_0 to be a function of temperature, optimal at T_C and actually decreasing as $T \to 0$.

The Eigenvalue Problem. As already mentioned, nontrivial solutions exist only if the largest eigenvalue of $\mathcal{K}$ is positive, $\lambda_0 > 0$. In the special case $t_Z=0$, λ_0 belongs to a constant eigenfunction, and is:

$$\lambda_0 = \oint d\theta/2\pi \, \mathcal{K}(\cos(\theta)) \tag{27}$$

Even after the simplification brought about in Equation 23, the calculation of λ_0 is nontrivial. To give one concrete example, we obtain at $d/a_0^*=1$:

$$\lambda_0 = (\cos \alpha_0) \, |(1-\mathbf{X})^2 \mathbf{X}(2-\mathbf{X})|^{-1/2} \tag{28}$$,

where $\mathbf{X} = R+iI$ as before, $|\mathbf{X}| \equiv \sqrt{R^2+I^2}$, and
$\alpha_0 \equiv [\tan^{-1}(I/(1-R)) \;\; -(1/2)\tan^{-1}(I/R) \;\; +(1/2)\tan^{-1}(I/(2-R))]$

Picking $z=\upsilon=1$ as representative values for use in Equations 21, 23, we find $R=1-\sqrt{3}/2$, $I=1/2$, and upon evaluating Equation 28, $\lambda_0 = 0.9659$. This corresponds to very high values of T_C and the energy-gap. (With $E_F \approx 1$ eV a value of $\lambda_0 \approx 0.2$ is sufficient to yield $T_C \approx 70$ K while $\lambda_0=1$ implies $T_C \approx 4000$ K (!)) Interestingly, $\lambda_0 = 0.9659$ may be close to the theoretical maximum, for causality can be interpreted (12,13) as limiting λ_0 to be < 1. Certainly, $d/a_0^* \approx 1$ is close to being an optimal value. As d/a_0^* is increased above this value, λ_0 is expected to drop rapidly and vanish strictly beyond $d/a_0^* = 2$ (once $u(-1) > 0$.)

This example demonstrates that our mechanism is quite sufficient to explain the phenomenon. It predicts that with a better choice of parameters, principally a smaller d/a_0^*, one will observe superconductivity at much higher temperatures than have been observed to date.

Conclusion

The reader may consider the present theory somewhat paradoxical. How can purely repulsive forces produce superconductivity? In the course of these pages, I have dwelt on the well-known feature (6) of metal physics, that screening eliminates repulsive interactions between charged carriers at all but the shortest distances. The screening mechanism provides an extra interaction between two carriers, owing to the presence of all the remaining particles. It is basically an attractive correction to the repulsive bare potentials. Given the same bare potentials when there are only 2 particles present, no such attraction is found and no bound state can be formed.

The novelty of the layered-metal-plane materials with small m^* (large a_0^*) lies in the relative weakness of the short-ranged intraplanar repulsion caused by the Wannier orbitals' overlap within the planes, as compared to the unabated vigor of the screening potentials. The paradox is resolved, the mechanism is natural and anchored in common sense. The theory built on it, as outlined in these pages, has only one essential parameter (d/a_0^*) and several auxilliary parameters, all of which are readily measured or understood. It should appeal, if only because of its relative simplicity.

Literature Cited

1. Berdnoz, T. ; Müller, K. Z. Phys. 1986, B64, 189
2. Wu, M.W. *et al* Phys. Rev. Lett. 1987, 58, 908
3. Maeda, H. ; Tanaka, Y. ; Fukutomi, M.; Asano, T. Jpn. J.Appl. Phys. 1988, 27, L209
4. Bardeen, J.;Cooper, L.; Schrieffer,R. Phys. Rev. 1957, 108 , 1175
5. Mermin, N.D.; Wagner, H. Phys. Rev. Lett. 1966, 17 , 1133 & 1307
6. Fetter, A.L.; Walecka, D.W. Quantum Theory of Many-Particle Systems ; McGraw-Hill Book Co.: New York, 1971
7. Stern, F. Phys. Rev. Lett. 1967, 18 , 546
8. Inderhees, S.E.; Salamon, M.B.; Friedmann, T.A.; Ginsberg, D.M. Phys. Rev. 1987, B36, 2401
9. Cheong, S.-W. *et al* Phys. Rev. 1987, B36, 3913
10. Mattheiss, L. Phys. Rev. Lett.1987, 58, 1028
11. Yu, J. ; Freeman, A. ; Xu, J.-H. Phys. Rev. Lett.1987, 58, 1035
12. Ginzburg, V.L.; Kirzhnits, D.A. High-Temperature Superconductivity , Consultants Bureau: New York, 1982; p.47 ff.
13. Dolgov, O.V.; Kirzhnits, D.A. ; Maksimov, E.G. Revs. Mod. Phys. 1981, 53 , 81
14. Subramanian, M.A.; *et al* Science ,1988, 239, 1015

RECEIVED July 12, 1988

Chapter 9

Dynamic Polarization Theory of Superconductivity

F. A. Matsen

Departments of Chemistry and Physics, The University of Texas, Austin, TX 78712

The dynamic-polarization theory employs a negative-U, Hubbard Hamiltonian to compute the spectrum of a one-dimensional array of sites (S). The lowest pair of states are split by a small gap and are predominantly superpositions of a pair of oppositely-polarized, ionic states $|S^-S^+.....S^-S^+>$ and $|S^+S^-.....S^+S^->$. An electric field polarizes the ground state into one or the other of the ionic states producing electron-pair transport with no scattering by the lattice. Excited states are formed by breaking an anion pair and forming a covalent bond with a nearest-neighbor cationic site; e. g., $|S\text{-}S...S^-S^+>$. Standard electrodynamics yields the London equation for the current density and the penetration depth, and the Pippard equation for the coherence length. The two-dimensional, system is discussed briefly and a condensation mechanism suggested.

In this paper we present a non-BCS theory based on a negative U Hubbard Hamiltonian (NUHH) which may serve as an effective Hamiltonian for the several excitonic, pairing mechanisms which have recently been proposed.

The Negative-U, Hubbard Hamiltonian (NUHH) and its States

The dynamic polarization theory subdivides the perovskite crystal into rectangular parallelpipeds called sites (S) each of which contains at its center a metal atom (copper or bismuth) surrounded by an environment of oxygen and counterions. Except for its electroneutrality the electron structure and electron density is unspecified. To the array of sites we apply a Hubbard Hamiltonian with $U<0$ for pairs of conducting electrons on the same site and an effective hopping parameter $t<0$ for conduction electrons on neighboring sites which includes superexchange between metal atoms. A consequence of the negative U is that the site disproportionation reaction, $2S = S^+ + S^-$ is exothermic. We express the NUHH in energy units of $t + |U|$ and in terms of a single (electron correlation) parameter (1),

$$0 \leq x \equiv |U| / (t + |U|) \leq 1$$

0097–6156/88/0377–0118$06.00/0

For x=0 the electrons are delocalized (uncorrelated) and for x=1 the electrons are delocalized (correlated). We decompose NUHH according to $H = H^o + H'$ where the zero order Hamiltonian is

$$H^0 = x((M/2 - \Sigma_r(E_{rr}{}^2 - E_{rr})/2$$

Here M is the number of sites and the zero of energy is the lowest zero order energy. The perturbation Hamiltonian is

$$H' = (x-1)\Sigma_{<rs>}(E_{rs} + E_{sr})$$

The operator E_{rs} is a spin-free generator (1) of the unitary group, U(M) where M is the number of sites. The generators are related to the fermion second-quantized operators by

$$E_{rs} = \Sigma_\sigma a^+{}_{r,\sigma} a_{s,\sigma}$$

and are used to simplify calculation and to emphasize the point that spin-flips do not occur with our NUHH.

The zero order (x = 1) ground states are the pair of completely ionic states, $|A\rangle = |S^-S^+...S^-S^+\rangle$ and $|B\rangle = |S^+S^-...S^+S^-\rangle$ with nearest neighbor anionic (or cationic) sites excluded. Note that |A> and |B> are oppositely polarized; i. e., $D_A = - D_B$ where D_A and D_B are the dipole moments of |A> and |B>, respectively. For point charges and an intersite separation of a, $D_A = - Mea/2$.

The ionic states are highly-correlated, singlet, reference states from which singlet, excited states are constructed by means of the generators. A single excitation breaks one anion electron-pair and forms one covalent bond with its nearest neighbor. The states are classified by the degree of excitation, u, which equals the number of covalent bonds. The uth excited states is written for v odd as follows:

$$|vv'....uA\rangle = E_{v+1,v}........E_{u+1,u}|A\rangle/\sqrt{2^u}$$

and

$$|vv'....uB\rangle = E_{v,v+1}........E_{u,u+1}|B\rangle/\sqrt{2^u}$$

The number of u-excited states $\{|u_jX\rangle, X = A \text{ or } B\}$ is given by the binomial coefficient,

$$W(M/2,u) = (M/2)!/u!(M/2 - u)!$$

From the $|u_jX\rangle$ we form the <u>symmetric state</u>

$$|uX\rangle = \Sigma_j|u_jX\rangle/\sqrt{W(M/2,u)}$$

The remaining W(M/2,u) - 1 u states are made orthogonal to |uA>. The perturbation matrix elements between these states and all other vectors vanish so these states play no dynamic role. The nonzero, perturbation matrix elements are given generally by

$$\langle uX|H'|u+1X\rangle = (x - 1)\sqrt{((2u + 2)(N - u))}$$

where X = A or B. Finally we note that the zero order eigenvalues (x = 1) are given by $E^0(u) = u|U|$.

We illustrate the NUHH calculations with a four-site system:

The Four-Site Zero Order States

u	$E^0(u)$	|uA>	|uB>
2	$2\|U\|$	|2A> = |S-S S-S> =	|2B> = |S-S S-S>
1	$\|U\|$	|1A> = $\|S^-S^+ S\text{-}S> + \|S\text{-}S\ S^-S^+>$	|1B> = $\|S^+S^- S\text{-}S> + \|S\text{-}S\ S^+S^->$
0	0	|0A> = $\|S^-S^+ S^-S^+>$	|0B> = $\|S^+S^- S^+S^->$

The Four-Site NUHH Matrix

|0A>	|1A>	|2A> = |2B>	|1B>	|0B>
0	2(x-1)	0	0	0
2(x-1)	x	2(x-1)	0	0
0	2(1-x)	2x	2(1-x)	0
0	0	2(1-x)	x	2(1-x)
0	0	0	2(1-x)	0

The four-site spectrum is plotted in Figure 1 where the + and the - denote states which are symmetric and antisymmetric with respect to a reflection through a plane at the midpoint and perpendicular to the axis of the fiber. We note that for x < 1, the linear system exhibits a "tunnel" gap, $\gamma = E(0^-) - E(0^+)$, which in linear systems decreases rapidly with M and becomes essentially zero for M>20. As we will see below our expression for the superconducting current depends on γL^2 which decreases more slowly; the two-dimensional gap also decreases more slowly. Further the effect of the Born-Oppenheimer approximation on the gap is unknown. We assume the tunnel gap exists and treat it as an empirical parameter.

The symmetry-adapted zero-order ground states have the form $|0\pm> = |A> \pm |B>$ and since the NUHH spectrum has a (tunnel) gap, the spectrum is analogous to the inversion spectrum of NH_3. (2) Just as for NH_3, the inversion can be interpreted as a classical oscillation between two oppositely polarized ionic states. See Figure 2. The application of an oscillating electric field to the ground state of the fiber dynamically polarizes it by converting the system into one or the other of the oppositely-polarized state so it exhibits electron pair transport with no scattering by the lattice. Single electron transport is excluded because an energy |U| is required to break an electron pair. If the DP fiber is placed in contact with a positive terminal on its left and a negative terminal on its right, the following processes will take place more or less instantaneously: (i) the system polarizes to state |A>; (ii) one pair of electrons transfer from the fiber to the positive terminal and simultaneously, a second pair of electrons tranfers from the negative terminal to the fiber forming the state |B>; (iii) the field of the terminals polarizes state |B> back to state |A>; (iv) the process continuously repeats and produces an electron-pair current with no scattering by the lattice. The current density is proportional to n*, the concentration of the electron pairs.

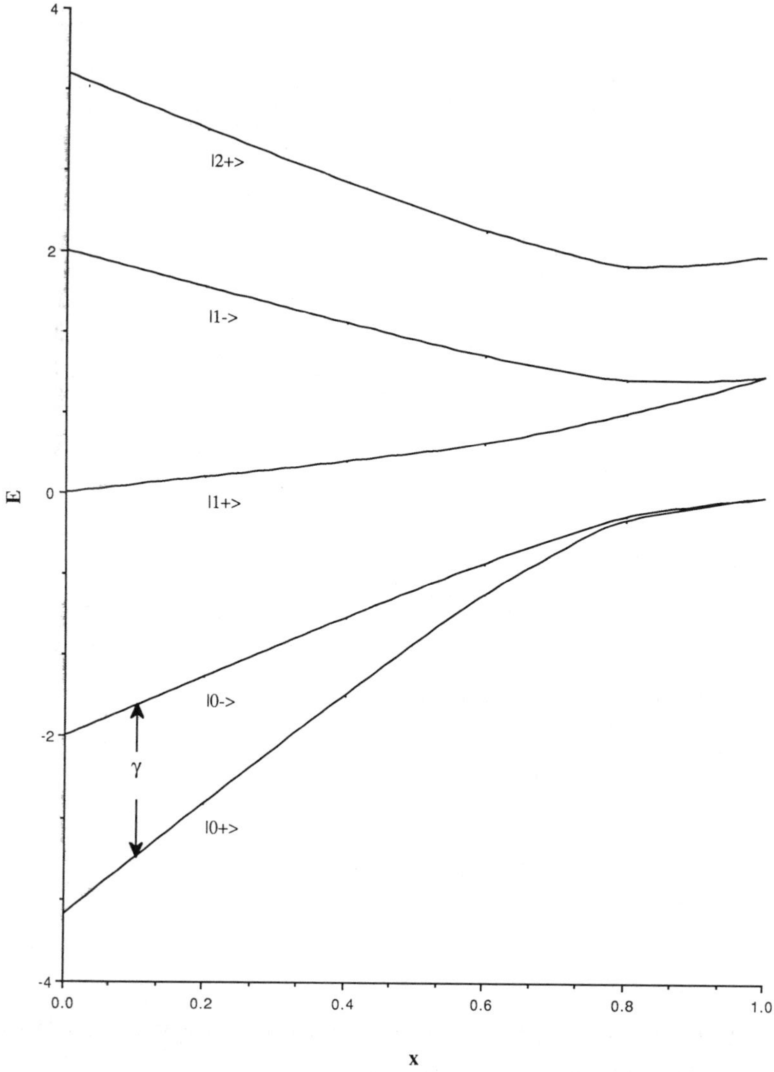

Figure 1. The NUHH spectrum of a linear four site system, x = |U|/(|U|+t).

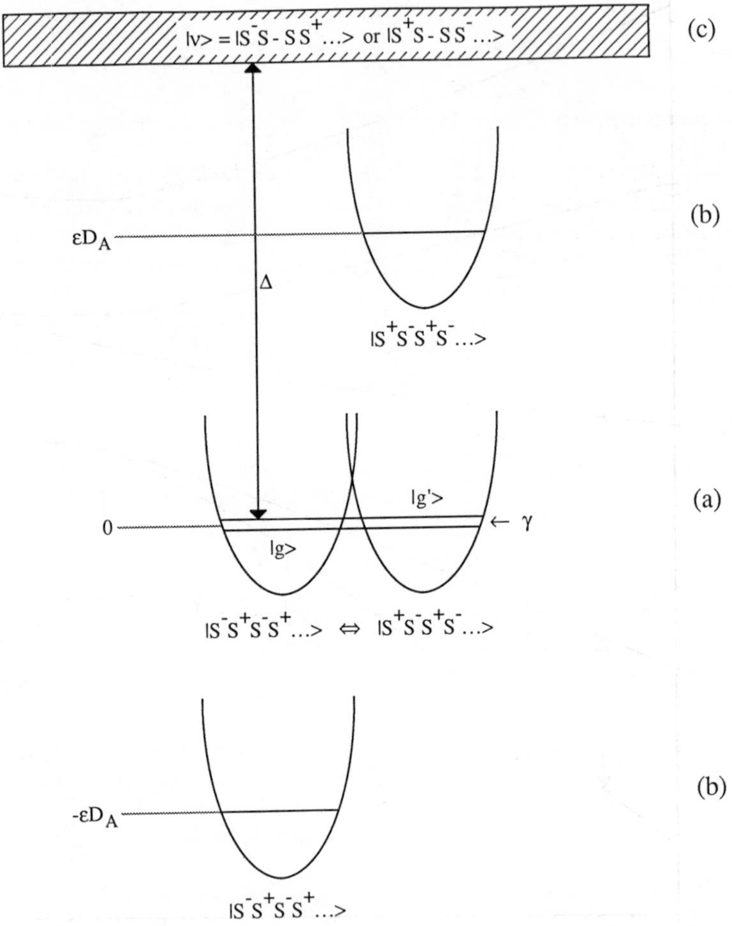

Figure 2. (a) the ground states of a linear system, (b) the polarization by an electric field, (c) the first excited state, $\Delta = |U|$.

The London Theory and the Penetration Depth

London (3) proposed a linear relation between the superconducting current density and the vector potential:

$$J_S(t) = -(c/4\pi\lambda^2)A(t) \tag{1}$$

where

$$\lambda^2 = mc^2/4\pi n_S e^2 \tag{2}$$

is the penetration depth, and where n_S is the concentration of superconducting electrons.

In this paper, we present a novel derivation of the London equation based on DP theory: The application of a time-dependent vector potential, $A(t) = A(0)\cos(2\pi\nu t)$ along a fiber generates a time-dependent ground state vector which for small time and weak field has the form

$$|0+(t)\rangle = |0+\rangle + Q(t)|0-\rangle = (|A\rangle + |B\rangle + Q(t)(|A\rangle - |B\rangle)$$
$$\approx (1 + Q(t))|A\rangle + (1 - Q(t))|B\rangle$$

The time-dependent dipole moment is then

$$D(t) = \langle 0+(t)|\mu|0+(t)\rangle$$
$$= (1 + Q(t))^2\langle A|\mu|A\rangle + 2(1 - Q(t)^2\langle A|\mu|B\rangle + (1 - Q(t))^2\langle B|\mu|B\rangle$$
$$= 4Q(t)D_A$$

We now express the time-derivative of the dipole moment as a current: $dD(t)/dt \cong e^*v^*(t)$ where $e^* = 2e$, is the effective pair charge and $v^*(t)$ is the velocity of the centroid of charge. The current density is then

$$J(t) = n^*e^*v^*(t) = (1/4)D_A dQ(t)/dt$$

where n^* is the concentration of electron pairs. By first order semiclassical radiation theory

$$dQ(t)/dt = -(1/c\hbar^2)\gamma D_A A(t)\exp(2\pi i\gamma t)$$

so that the low-frequency component of the current density, $h\nu \ll \gamma$ is

$$J(t) = -(1/4c\hbar^2)n^*\gamma D_A{}^2 A(t) \tag{3}$$

which has the same form as Equation 1 (the London equation). On comparing Equations 3 and 1, we obtain for the DP penetration depth

$$\lambda^{*2} = 4c^2\hbar^2/n^*D_A{}^2\gamma = 8c^2\hbar^2/n^*e^{*2}L^2\gamma \tag{4}$$

On comparing Equations 4 and 2, we obtain for the effective mass

$$m^* = 2\hbar^2/L^{*2}\gamma \text{ grams} \tag{5}$$

We assume initially that the vector potential exhibits a slow (low frequency) trigonometric growth from zero up to its maximum value. Accompanying this is a slowly increasing ground state polarization, electron-pair transfer and current density up to their maximum values. The maximum value of the current density (which is frequency independent for $h\nu \ll \gamma$) specifies the DC current density and the maximum value of the AC current.

The Pippard Theory and the Coherence Length.

In London theory the vector potential is taken to be space-independent (the rigid approximation). In Pippard theory (4), it is taken to be space-dependent with the following Fourier decomposition: $A(r) = \Sigma_q A(q)\exp(iqr)$. The qth component of the current-density operator is the sum of two terms: $J(q) = J_1(q) + J_L$(London). Here J_L reproduces the London theory. For a weak field the current density is linear in A(q) and can be expressed in terms of a linear response function as follows:

$$J(q) = -(c/4\pi)K(q)A(q)$$

where

$$K(q) = K_1(q) + K_L = K_L(1 + K_1(q)/K_L)$$

is the total response function and where $K_1(q)$ and K_L ($= 1/\lambda^2$) are the response functions for $J_1(q)$ and J_L, respectively. Pippard has shown by a classical argument that for small q the response function has the form

$$K(q) = (1/\lambda^2)[1 - q^2\xi_0^2/5 + ..]$$
$$= (1/\lambda^2) - q^2/5\kappa^2 + ...] \tag{6}$$

where $\xi = \xi_0$ is the coherence length for an infinite mean-free-path and $\kappa = \lambda/\xi_0$ is the Ginszburg-Landau parameter.

We now develop the Pippard formulation from DP theory. Following Tinkham (5) the first order expectation value of $J_1(q)$ is

$$\langle J_1(q),\rangle = -2\mathrm{RE}\Sigma_v \langle A|J_1(q)|v\rangle\langle v|H(q)|A\rangle/|U|$$

The vector potential Hamiltonian is given by

$$H(q) = -(e\hbar/mc)\Sigma_k kA(q)E_{k+q,k}$$

where

$$E_{k+q,k} = \Sigma_\sigma c^*_{k+q,\sigma}c_{k,\sigma}$$

For a finite cyclic system, $k = 2\pi m/M$ where $m = 0, \pm 1, \pm 2, .. \pm M/2$. To convert a molecular-orbtal generator into a site-orbital generator we use the orbital transformation,

$$|k\rangle = (1/M)\Sigma_r|r\rangle \exp(ikr) \quad r = 1 \text{ to } M$$

where r indexes the site orbitals. Then

$$E_{k+q,k} = (1/M)\Sigma_r\Sigma_s(\exp(i((k + q)r - ks))E_{rs}$$

so

$$H(q) = - wA(q)\Sigma_r\Sigma_s\exp(iqr)F_{rs}E_{rs}$$

and similarly,

$$J_1(q) = (wc/V)\Sigma_r\Sigma_s \exp(-iqr)F^*_{rs}E_{rs}$$

where $w = (e\hbar/mc)$, V is the volume and $F_{rs} = (1/M)\exp(ik(r - s)$. Since H(q) and $J_1(q)$ contain only one spin-free generator, E_{rs} only singly excited states occur. Thus,

$$\langle v|H(q)|A\rangle = - (e\hbar/mc)A(q)\Sigma_r\Sigma_s\exp(iqr)F_{rs}\langle A|E_{v,v+1}E_{rs}|A\rangle/\sqrt{2}$$
$$= - (\sqrt{2}wA(q)F\Sigma_v f_{v+1}(q)$$

where $f_{v+1}(q) = \exp(iq(v+1))$ and where

$$F = (1/M)\Sigma_k k\exp ik$$

which we evaluate as follows. We convert to the sine and substitute for $k = 2\pi m/M$

$$F = (4\pi i/M^2)\sum_{m=1}^{M/2} m\sin m\theta$$

where $\theta=2\pi/M$. For M large and by reference 6,

$$F = (4\pi i/M^2)\left(\frac{\sin(\frac{M}{2}\theta)}{4\sin^2(\frac{\theta}{2})} - \frac{\frac{M}{2}\cos(\frac{M}{2}\theta)}{\sin(\frac{\theta}{2})}\right)$$

$$F = 2i$$

Finally,

$$\langle v|H(q)|A\rangle = \sqrt{8}iwA(q)f_{v+1}(q)$$

Similarly,

$$\langle A|J_1(q)|v\rangle = (\sqrt{8}iwc/V)f^*_v(q)$$

so

$$\langle J_1(q)\rangle = -(8w^2cn^*/|U|)\cos(q)A(q) \approx -(8w^2cn^*/|U|)(1 - q^2/2 + ...)A(q)$$

On neglecting unity in comparison to $1/\lambda^2$ and on equating its response function to equation 6, we have

$$K(q) \approx 1/\lambda^2 - q^2/(5/2)\kappa^{*2} + ...$$

where

$$\kappa^{*2} = |U| / 8cw^2n^* \qquad (7)$$

We see that the DP solid satisfies the Pippard conditions for superconductivity; i.e., $K(q) \rightarrow 1/\lambda^2$ as $q \rightarrow 0$. Pippard suggested that the coherence length is the distance over which a perturbing force on a superconductor can act. The DP perturbation breaks an ion pair and then forms a covalent bond with a neighboring site from which it is separated by a distance of ~10^{-7} cm which with the penetration depth of $l = 5\times10^{-6}$ cm yields K* ~10, a Type II superconductor. It would appear that the DP fiber exhibits the properties of a Type II Pippard superconductor. Unfortunately, the substitution of any reasonable set of numbers into 7 yields K*<1 which characterizes a Type I superconductor. Most perovskite superconductors are Type II.

The Effects of High Temperature and Magnetic Field

The London and Pippard derivations employ only the ground state and a few excited states (weak field). Consequently, at T = 0, the weight of the superconducting state, W_S, is essentially one. As the temperature and the applied magnetic field are increased, W_S and superconductivity decreases.

DP Theory in Two Dimensions

The DP theory in two dimensions is considerably more complicated because it involves a larger number of states and more complex connections between them. (See Figure 3). We note that the u = 1 states resemble the states used by Pauling (7) in his theory of metals and the u = 2 states resemble the states used by Anderson (8) in his RVB theory of superconductivity. However in contrast to the RVB theory, DP theory places the RVB states high above the superconducting states so that they cannot participate in the superconductivity at low temperatures.

If NUHH is diagonalized in either the X-space or the Y-space we obtain a spectrum which closely resembles the fiber spectrum shown in Figure 1. This neglect of the Z-space yields the two dimensional DP model composed of intersecting, noninteracting fibers postulated above. The full spectrum is shown in Figure 4.

The Condensation

A DP solid with a nonzero tunnel gap exhibits both London and Pippard superconductivity which decreases with increasing temperature. However, the W_S -

temperature curve does not exhibit a true critical temperature. Further the Hamiltonian and its states maintain reflection symmetry over the entire temperature range and no symmetry-breaking or condensation occurs.

To solve this problem we assume, in addition, that the DP solid is a ferroelectric which undergoes, at some Curie temperature, a ferroelectric condensation into domains with random directions of polarization. We note that macroscopic symmetry-breaking occurs, that condensation occurs in coordinate space rather than momentum space, and that the domain walls are pinned to crystal imperfections. The application of an electric field removes the domain walls and polarizes the crystal in a single direction so that the DP superconducting mechanism can take over. As the temperature and/or the magnetic field is increased, electron pairs are continuously broken by promotion into excited non-superconducting states until the anion concentration is reduced below some critical value at which the domains spontaneously evaporate and symmetry is restored. We further assume that the critical temperature is equal to or lower than the Curie temperature which may or may not be observable due to changes in crystal structure. We note that with this assumption the critical temperature can be quite large because it is not limited by the phonon spectrum. We call this high-temperature superconducting theory the ferroelectric-condensation-dynamic-polarization (FCDP) theory.

The classical example of a ferroelectric is the perovskite, $BaTiO_3$ which has a Curie temperature of 393° K but is not a superconductor. The perovskite $SrTiO_3$ is a superconductor with a critical temperature of 0.5° K and a higher critical temperature which is unobservable as a consequence of crystal melting. The nonobservability of its Curie temperature gave rise initially to the mistaken belief that $SrTiO_3$ is not a ferroelectric. There is no evidence that the high temperature, superconducting perovskites are ferroelectric but the Curie temperature may be nonobservable due to structural changes.

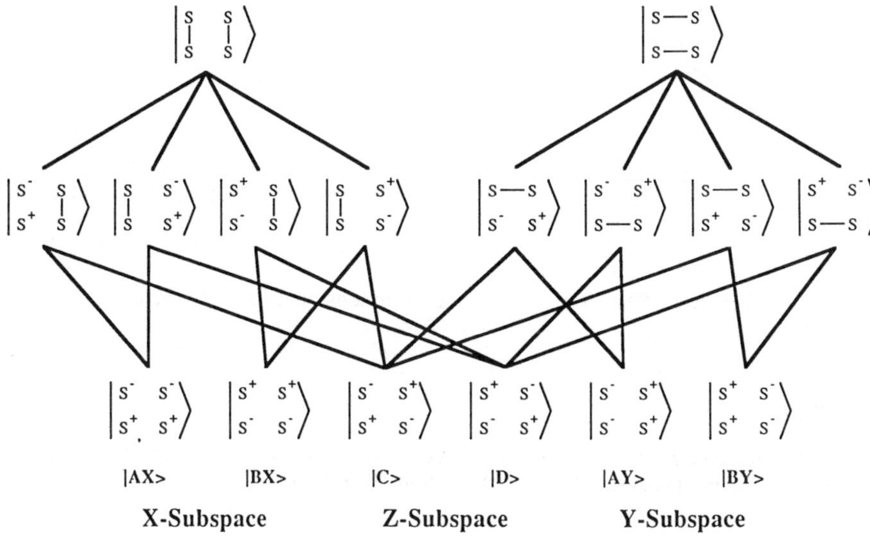

Figure 3. The states of the two-dimensional four-site system.

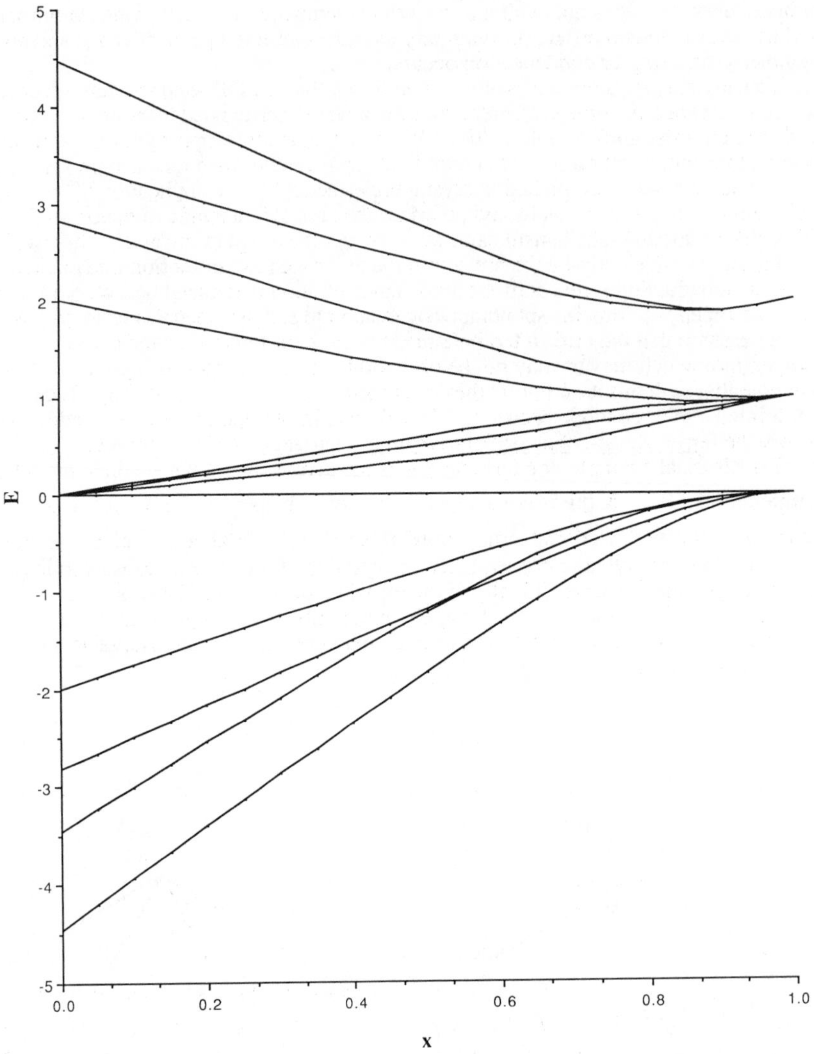

Figure 4. The spectrum of the two-dimensional four-site system.

Conclusion

The ferroelectric-condensation, dynamic-polarization theory of high temperature superconducting perovskites is, at least, a non-BCS theory.

Acknowledgments

It is a pleasure to acknowledge the generous and consistant support of the Robert A. Welch Foundation of Houston, Texas, and the technical assistance of Loudon Campbell and David Feldcamp.

Literature Cited

1. Matsen, F. A.; Pauncz, R. The Unitary Group in Quantum Chemistry; Elsevier: Amsterdam, 1986; Chapter IV, VII.
2. Feyman, R. P.; Leighton, R. B.; Sands, M. The Feymman Lectures on Physics, Vol III; Addison-Wesley: Reading, Mass., 1965; p. 8-11ff.
3. London, F.; London H. Proc. Roy. Soc. (London) 1935, A149, 71.
4. Pippard, A. B., Proc. Roy. Soc. (London) 1953, A216, 547.
5. Tinkham, M. Introduction to Superconductivity: R. E. Krieger Publishing Company; Malabar, Fla; 1975; p. 62.
6. Jolley, L. F. Summation of Series: Dover: New York, 1961; formula 427
7. Pauling, L. Phys. Rev. 1938, 54, 899.
8. Anderson, P. W. Science 1987, 235, 1196.

RECEIVED July 15, 1988

Chapter 10

Polarization-Induced Pairing

An Excitonic Mechanism for High-Temperature Superconductivity

Richard L. Martin, Alan R. Bishop, and Zlatko Tesanovic[1]

Theoretical Division, Los Alamos National Laboratory, Los Alamos, NM 87545

The qualitative aspects of a recently proposed mechanism for superconductivity based on intra and interband Cu<->O charge transfer excitations in the cuprates are reviewed. In this model, the dynamic polarizablility of the environment surrounding the CuO_2 planes plays an important role in enhancing T_c. The "sandwich" structures of $YBa_2Cu_3O_7$, $Bi_2Sr_2CaCu_2O_8$ and $Tl_2Ba_2Cu_3O_{10}$ are ideally suited to this mechanism. We briefly comment on the high T_c design principles suggested by the model.

The structural feature common to all the high temperature cuprate superconductors discovered to date[1] is the presence of two-dimensional sheets[2] of CuO_2. While much has been learned about these materials in the past year, the pairing mechanism remains elusive. Current theoretical debate centers on the importance for the pairing interaction of the spin degrees of freedom vs. the charge degrees of freedom[3]. We have recently proposed a mechanism[4], the polarization induced pairing model, in which the planes are inherently susceptible to a superconducting instability due to the *emergence of low-lying charge transfer excitations when the system is doped* to a mixed-valence stoichiometry (i.e. Cu^{2+}/Cu^{3+}). These low energy charge fluctuations arise essentially because the electron affinity of Cu^{3+} is nearly equal to the ionization potential of O^{2-} in these materials; the configuration $Cu^{3+}O^{2-}$ is nearly degenerate with $Cu^{2+}O^{1-}$. In the same way that the polarizability of the lattice leads to an attractive interaction in conventional

[1]Current address: Department of Physics, The Johns Hopkins University, Baltimore, MD 21218

0097–6156/88/0377–0130$06.00/0

superconductors, the high electronic polarizability associated with this near degeneracy leads to an effective attractive interaction between electrons.

The theory evolved from an examination of *ab initio* configuration-interaction (CI) wave functions for a number of small clusters, as well as from a study of a semi-empirical two-band extended Hubbard Hamiltonian which incorporates the full periodicity of the lattice. The parameters which characterize the latter were chosen to be consistent with the *ab initio* results. For La_2CuO_4, both approaches predict the experimentally observed anti-ferromagnetic spin interaction when the now familiar Cu $d_{x^2-y^2}$ anti-bonding (AB) band is half-full (Cu^{2+}). As the system is doped away from half-filling, the magnetic interactions decrease in importance relative to a low-lying, finite wavevector, collective charge excitation *within* the Cu $d_{x^2-y^2}$ AB band. This intraband collective charge excitation can be likened to a slowly fluctuating charge-density wave without the static deformation of the nuclei, and is effective in mediating a Cooper pairing within the AB band. The expression for the critical temperature in weak coupling is analogous to that in conventional BCS theory,

$$T_c \sim \hbar\omega \; e^{-(1+\lambda)/(\lambda-\mu^*)} \qquad (1)$$

where the effective interaction between two electrons is given by λ-μ^*, with λ the attractive component and μ^* the direct Coulomb repulsion, and $\hbar\omega$ is a characteristic energy[3]. Superconductivity requires a net attractive interaction, $\lambda > \mu^*$. Whereas the characteristic energy $\hbar\omega$ is of the order of the Debye temperature for conventional phonon-mediated superconductivity, it is of the order of the charge transfer energy ("exciton") in the present mechanism, a feature contributing[5,6] to higher T_c. For La_2CuO_4, we estimate it to be of the order of 0.2-0.7eV[4].

While the model suggests that a single CuO_2 plane can support superconductivity, the critical temperature is greatly enhanced if the plane is embedded in a highly polarizable medium. In this regard, the "sandwich" structure[2] of $YBa_2Cu_3O_7$ and the newly discovered TlO and BiO compounds take on special interest. The beauty of the sandwich structures is that the "jelly" (the CuO_3 chains, or the BiO and TlO sheets) can not only enhance the intrinsic attractive interaction within a single plane (λ), but they also offer the possibility of mediating pairing between electrons in two distinct planes, thereby significantly reducing the direct Coulomb repulsion (μ^*).

In what follows, we focus on the manner in which we believe T_c is enhanced in the $YBa_2Cu_3O_7$ material. The details of the model are presented elsewhere[4]. Suffice it to say that it appears to be capable of rationalizing a great deal of the existing experimental evidence. For example, the presence of two super-conducting phases in $YBa_2Cu_3O_{7-\delta}$, one around δ=0 (90K) and another around δ=.2-.3 (60K)[7]; the absence of an isotope effect[8]; the dependence of T_c on the Cu(2)-O(4) bond distance observed by Miceli et al.[9]; the independence of T_c to substitution of other rare earths for Y[10]; and the anisotropy of the gap function[11]. A number of new experimental directions[4] have been suggested which should shed some light on the validity of the model. However, it is our hope that the present communication will encourage the synthesis and characterization of new "sandwiches" with even higher T_c by enhancing the polarizability and coupling of the surrounding environment to the CuO_2 planes.

The structure of the 90K superconductor $YBa_2Cu_3O_7$ consists of two sheets of CuO_2 in the ab plane (the "bread"), separated by chains of CuO_3 which run along the b-axis[2](the "jelly"). The planes are weakly coupled to the chain in the c-direction by bridging oxygen atoms (O(4), Figure 1). What is the nature of the charge transfer which mediates the pairing? In addition to the intraband collective charge excitation occuring in the planes, our calculations point to two additional excitations of importance involving the CuO_3 chain; an intraband exciton analogous to that in the plane, and an interband excitation. The interband(transverse) excitation involves the transfer of an electron from the non-bonding p_π band into the anti-bonding Cu $d_{y^2-z^2}$ band(Figure 1). The transition dipole is directed perpendicular to the chain. *Ab initio* CI studies on the Cu_3O_{12} cluster shown in Figure 1 suggest that the transverse excitation carries a large oscillator strength and requires an energy of the order of 0.1eV-0.5eV. This excitation enhances intraplane pairing. In order to understand this qualitatively, imagine that an electron passes through the Cu in the upper plane, and excites the charge transfer from O(4) -> Cu(1).

The matrix element governing this process should be essentially Coulombic, i.e. $W \sim 1 / r_{Cu(2)\text{-}O(4)}$. This creates a partial positive charge on O(4) which can then attract another electron in the plane(Figure 2a). The net enhancement[4] is given by

$$\lambda(\mathbf{k},\mathbf{k}') = -2n_c N_2(0)\, W^2 \int d\omega\, F(\mathbf{k},\mathbf{k}';\omega) \,/\, \omega \qquad (2)$$

where n_c is the density of chains, $N_2(0)$ is the density of states at the Fermi energy in the planes, and $F(\mathbf{k},\mathbf{k}';\omega)$ is essentially the

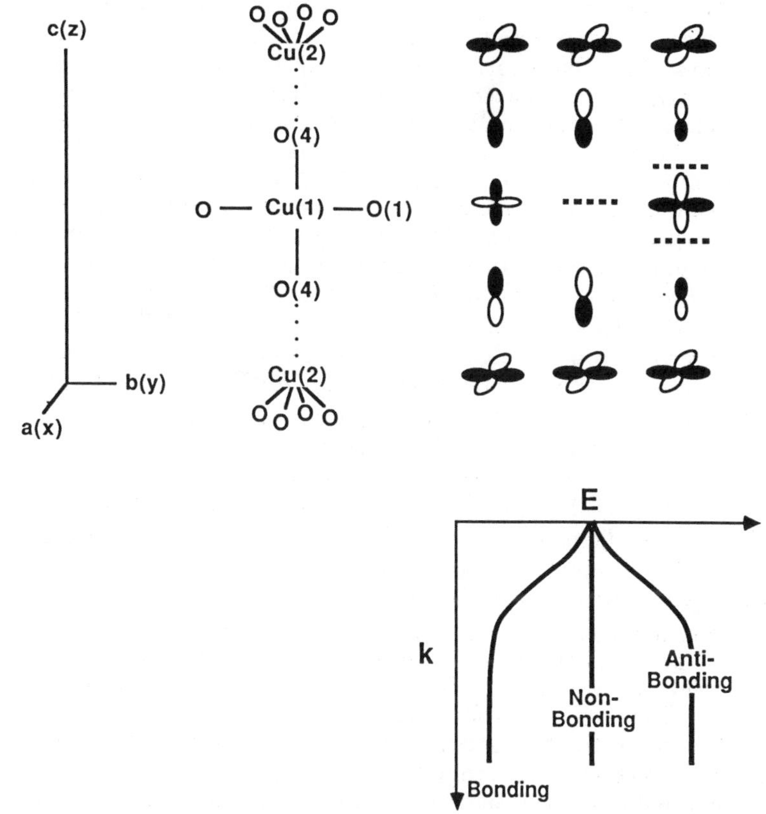

Figure 1. A Cu_3O_{12} cluster model of the sandwich structure of $YBa_2Cu_3O_7$ is shown on the left. It consists of three Cu atoms and their nearest-neighbor oxygens. The upper and lower Cu(2) have the coordination of the planes, while the central Cu(1) is an element of the chain which runs along the b-axis. Two important bond distances are labeled r_1 and r_2. To the right of the cluster are shown the three molecular orbitals which evolve from the $d_{y^2-z^2}$ orbital on Cu(1) and the p_z atomic orbitals on the two O(4) sites. They are classified by the number of nodes as bonding(B), non-bonding(NB), and anti-bonding(AB), and contribute to the B,NB, and AB bands of the chain shown below. The phase of the orbitals is denoted by color; the $d_{x^2-y^2}$ orbitals in the planes are shown for the reader's orientation. In the limit that the chain was purely Cu^{3+}, the B and NB bands would be full, and the AB band empty. The transverse excitation described in the text corresponds to an interband transition in which an electron is transferred from NB ->AB. The longitudinal(intraband) excitation at $2k_f^c$ corresponds to a transition within the AB band.

wavevector and frequency dependent polarizability. The quantity $F(\mathbf{k},\mathbf{k}';w)$ can be determined experimentally, or computed from the parameters which characterize the extended Hubbard Hamiltonian[4]. Very roughly, it is maximized when the oscillator strength is large (as is typically the case for charge transfer transitions) and the excitation energy is small (which is accomplished here through the near degeneracy of $Cu^{3+}O^{2-}$ and $Cu^{2+}O^{-}$). Dimensionality effects can also play a role, with one-dimensional systems exhibiting anomalies in the polarizability not present in otherwise identical systems in two and three dimensions.[12] It is also important to note that as the distance between the plane and O(4) decreases, the coupling ($W \sim 1/r_{Cu(2)-O(4)}$) between the conduction electrons and the charge transfer excitation in the chain increases. The model thus suggests that this distance should correlate with T_c. Finally, note in Figure 2a that the orientation of the transition dipole for the transverse excitation is repulsive for interplane pairing.

The intraband(longitudinal) excitation occurs within the $d_{y^2-z^2}$ AB band of the chain(Figure 1). In this case, the transition dipole is oriented along the chain. Because of the extended nature of the chain, this is manifested as a collective excitation which can move down the chain with wavevector $2k_f^c$, where k_f^c is the Fermi wavevector for the chain. This precursor of a charge density wave is not necessarily commensurate with the lattice, but it can be pictured qualitatively as in Figure 2b. It is strictly analogous to the collective excitation discussed earlier for a single plane, but because of the perfect Fermi nesting associated with one-dimensional systems is even more effective in mediating Cooper pairing[12].

While the transverse exciton favors intraplane pairing and is repulsive for interplane interactions, the longitudinal excitation has the proper symmetry to mediate both inter- and intraplane pairing. These two excitations enter into the expression for λ additively, so that the net attraction is greater for intraplane as opposed to interplane pairing. We suspect, however, that the reduced direct Coulomb repulsion associated with interplane pairing will dominate, and that it is responsible[13] for the 90K transition in $YBa_2Cu_3O_7$. As oxygen is removed, the vacancies disrupt the chains, and the effectiveness of the longitudinal excitation decreases. At some point, presumably near $YBa_2Cu_3O_{6.7}$, the more localized transverse excitation becomes the dominant mode and we associate it with intraplane pairing and the 60K plateau. As further oxygen is removed, the magnetic interactions become more important. At $YBa_2Cu_3O_{6.5}$, we reach a formal Cu^{2+} valence(the 1/2-filled band), and the Cu^{3+} character necessary for the low-lying charge transfer excitation vanishes.

If this model proves to be correct, it suggests that other high T_c materials should have the following design. The "bread" should be a good conductor (metallic) with a large density of states at the Fermi

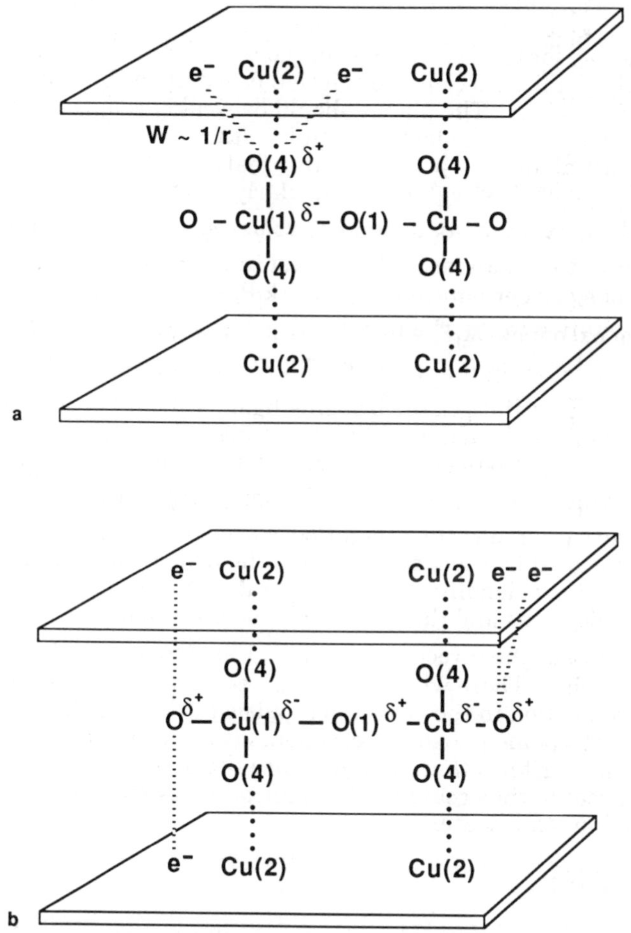

Figure 2. a) A schematic of the way in which the transverse (interband) excitation enhances the intraplane pairing. Excitation of an electron from the NB band to the AB band results in charge transfer from the bridging oxygen to the central Cu. This leads to a partial positive charge on O(4) which can mediate intraplane pairing. The enhancement is dependent on the Cu(2)-O(4) distance as well as the energy and oscillator strength of the charge transfer transition. Note that the symmetry of the transition dipole is repulsive for interplane pairing. b) A schematic of the collective excitation within the anti-bonding band, the longitudinal exciton. This excitation has the proper symmetry to mediate both inter- and intraplane pairing.

energy. It should be obvious that CuO plays no special role in the theory, except for the important feature that the electron affinity of Cu^{3+} is similar to the ionization potential of O^{2-} in this lattice, thereby paving the way for the low-energy charge transfer modes[14]. In this sense, likely candidates could be selected based upon their proximity to a CDW instability. The planes should be weakly coupled to a highly polarizable medium. The latter might be either one- or two-dimensional, but the perfect nesting features at any band-filling associated with chains lead to a particularly enhanced polarizability at $2k_f^c$. The maximum attraction λ is achieved when the Fermi wavevector in the chain is similar to that in the plane, since the formation of a Cooper pair in the plane $(k_f^p,-k_f^p)$ requires a momentum exchange $2k_f^p$, which is precisely what is available in the collective excitation in the chain if $k_f^c=k_f^p$. In order to maximize the coupling to this excitation, the bridging ligands should be as close to the planes as possible and weakly interacting. The latter is ensured in $YBa_2Cu_3O_7$ by having the conduction band composed of $d_{x^2-y^2}$ character, which is orthogonal to the bridging $O(p_z)$ orbital. If this were not the case, the system might be expected take on three-dimensional metallic behavior and the benefits associated with the reduced dimensionality and the separation between the conduction electrons and the polarizable medium would be lost. In order to minimize the direct repulsion μ^*, the planes should be as far apart as possible. There will be, however, some optimal separation where the reduction in the Coulomb repulsion is balanced against the loss in the interplane pairing interaction. It is, of course, easier to stipulate these criteria than it is to realize them, but there seems to be no *a priori* reason they could not be artificially designed and synthesized in other materials.

Literature Cited

1a) J.G.Bednorz and K. A. Mueller, Z. fur Physik **B64**, 189(1986);
 b) C.W. Chu et al., Phys. Rev. Lett. **58**, 405(1987);
 c) C. Michel et al., Z. Phys. B **68**, 421(1987);
 d) H. Maeda et al., Jpn. J. Appl. Phys. **27**, L209(1988);
 e) Z.Z. Sheng and A. M. Hermann, Nature **332**, 55,158(1988).
2a) J.D. Jorgensen et al., Phys. Rev. Lett. **58**,1024(1987);
 b) M.A. Subramanian et al., Science **239**,1015(1988);
 c) M.A.Subramanian et al., Nature **332**, 420(1988);
 d) C.C. Torardi et al., Science **240**, 633(1988).
3. "Novel Superconductivity", S.A. Wolf and V.Z.Kresin, Eds. (Plenum Press, New York, N.Y.), 1987.
4. Z. Tesanovic, A.R.Bishop, and R.L. Martin, LAUR-88-405, submitted to Phys. Rev. Lett., February 5,1988.
5. W.A.Little, Phys. Rev. **A134**, 1416(1964).

6. V.L. Ginzburg, Sov. Phys. JETP, **20**, 1549(1965); Contemp. Phys. **9**,355(1968).
7. B. Batlogg et al., ref.3, p. 653.
8. L.C.Bourne et al., Phys. Rev. Lett., **58**, 2337(1987).
9. Miceli, et al., to appear in Phys. Rev. B and private communication; R. Cava et al., AT&T preprint.
10. Z. Fisk, J.D.Thompson, E. Zirngiebl, J.L.Smith, and S.W.Cheong, Solid State Comm. **62**,743(1987).
11. P. Horn et al., Phys. Rev. Lett. **59**, 2772(1987).
12. V.J. Emery, in "Highly Conducting One-Dimensional Solids", Devreese, Evrard, and vanDoren, Eds., (Plenum Press, New York, N.Y.) 1979.
13. Another possibility for the two critical temperatures involves intraplane pairing only. The effectiveness of the longitudinal excitation is maximized with well-ordered chains, i.e. $YBa_2Cu_3O_7$. In the 90K phase, both the longitudinal and transverse excitations contribute to intraplane pairing. Near $YBa_2Cu_3O_{6.7}$, the oxygen vacancies have disrupted the chain, and only the transverse excitation remains to enhance T_c.
14. The extended Hubbard model discussed in reference 4 exhibits a superconducting instability when $\epsilon_d - \epsilon_p \leq t_{pd}$ and $V > U/4$, where ϵ_d, ϵ_p are the bare site energies for a Cu3d, O2p electron, t_{pd} is the hopping integral, and V and U are nearest-neighbor and on-site Coulomb repulsions projected onto the AB band. These quantities are defined more precisely in reference 4.

RECEIVED July 13, 1988

NEW MATERIALS

Chapter 11

Determination of the Homogeneity Range of La_2CuO_4

Joseph DiCarlo, Chun-Ming Niu, Kirby Dwight, and Aaron Wold

Department of Chemistry, Brown University, Providence, RI 02912

Pure La_2CuO_4 was prepared by decomposition of the nitrates followed by an oxygen anneal at 500°C and 600 psi of oxygen. The extent of anion vacancies was studied by thermogravimetric analysis, x-ray diffraction and magnetic susceptibility. The magnitude of this deficiency is less than can be unambiguously ascertained by direct thermogravimetric analysis which has a limit in accuracy in x of .01 for the composition La_2CuO_{4-x}. However, significant shifts in the Néel temperatures confirmed the variation in anion vacancy concentrations. Attempts to modify the cation ratio of La:Cu = 2:1 failed to yield single phase material.

La_2CuO_4 was reported by Longo and Raccah (1) to show an orthorhombic distortion of the K_2NiF_4 structure with a = 5.363Å, b = 5.409Å and c = 13.17Å. It was also reported (2,3) that La_2CuO_4 has a variable concentration of anion vacancies which may be represented as La_2CuO_{4-x}. The value of x is dependent on the method of preparation.

La_2CuO_{4-x} undergoes a second order tetragonal-orthorhombic phase transformation and the temperature at which this transformation occurs is dependent on the value of x (2). A plot of the transformation temperature vs anion defect concentration in La_2CuO_{4-x} indicates that for a change of .03 in the value of x, the transition temperature is increased by ∿ 75 K. A magnetic susceptibility anomaly also occurs which is due to long-range antiferromagnetic ordering in the sample (3, 4). It was also shown that the Néel temperature is sensitive to the oxygen vacancy concentration and increases as the anion vacancies are increased.

There appear still to be some uncertainties concerning the magnetic properties of La_2CuO_{4-x} as x approaches zero. Whereas there is agreement concerning the observed increase in T_N as the anion vacancy concentration is increased, the disappearance of observable antiferromagnetic behavior in some of the samples prepared (2,5,6) has been difficult to reproduce.

0097–6156/88/0377–0140$06.00/0

Experimental

In this study samples were prepared which contained controlled concentrations of both anion and cation vacancies. All samples were prepared by dissolving copper (5-9's Aesar Chemical Co.) and La_2O_3 (4-9's Aesar Chemical Co.) in 1:1 dilute nitric acid. The solution was evaporated on a hot plate to dryness, then placed in a furnace, and heated to 500°C for 24 hours in order to decompose the nitrates. The product was ground and reheated to 800°C for 24 hours. After a final grinding the product was heated for 24 hours at 950°C and allowed to slow cool in air to room temperature. For the sample corresponding to $La_2CuO_{4.00(1)}$ the product was annealed for 12 hours at 500°C and 600 psi of oxygen. In order to prepare samples deficient in oxygen the products were heated in a predried, pure argon atmosphere (flow rate = 50 sccm/min) from 500-800°C. All samples were analyzed by reduction in a H_2/Ar (15/85) atmosphere to copper metal and lanthanum oxide.

Characterization of Products

Powder diffraction patterns were obtained with a Philips-Norelco diffractometer using monochromatic high-intensity CuKα_1 radiation (λ = 1.5405Å). For qualitative identification of the phases present, the patterns were taken from $12° < 2\theta < 72°$ with a scan rate of 1° 2θ/min and a chart speed of 30 in/hr. The scan rate used to obtain x-ray patterns for precision cell constant determination was 0.25° 2θ/min with a chart speed of 30 in/hr. Cell parameters were determined by a least-squares refinement of the reflections using a computer program which corrects for the systematic errors of the measurement.

Temperature programmed reductions of all samples were carried out using a Cahn System 113 thermal balance. Each sample was purged at room temperature in a stream of H_2/Ar (15/85) for 2 hours. Then the temperature was increased to 1000°C at a rate of 60°/hr. The flow rate of the gas mixture was 60 sccm/min.

Magnetic susceptibility measurements were carried out using a Faraday Balance from 77 to 300 K with a field strength of 10.4 kOe. Honda-Owen (field dependency) measurements were carried out at 77 and 300 K.

Results and Discussion

X-ray diffraction data of the various phases prepared and analyzed are compared in Table I with those reported by previous investigators (1,5,6). It can be seen that the best agreement is with the parameters reported by Longo and Raccah (1) and the greatest discrepancy is with those of Saez Puche et al. (6). Thermogravimetric analysis of the pure samples resulted in a composition determined by weight loss on reductions to La_2O_3 + Cu to be $La_2CuO_{4.00(1)}$. In order to establish the range of homogeneity of the phase La_2CuO_4, varying ratios of lanthanum to copper were reacted where La:Cu = 2.05:1 to 1.90:1. The samples were prepared precisely according to the method used for La_2CuO_4 and the results are given in Table II. It can readily be seen that the pure

Table I. Cell Parameters Reported for "Pure" La_2CuO_4

Composition	T_N	a	b	c	Ref.
		Cell Constants (Abma orientation)			
La_2CuO_4	--	5.363	5.409	13.17	(1)
La_2CuO_4	--	5.366	5.402	13.15	(5)
La_2CuO_4	--	5.342	5.434	13.16	(6)
$La2CuO_{4.00(1)}$	240	5.358	5.400	13.16	This work

compound exists over an extremely narrow range of composition with respect to the La:Cu ratios. For the sample with nominal composition, La:Cu = 1.95:1, x-ray diffraction patterns indicated the presence of a single phase even at the maximum sensitivity of the Philips instrument used. However, counts for 400 seconds were made over the range $38.5 \leqq 2\theta \leqq 39.5°$. The (111) peak of Cu occurred at 38.9°. From these measurements the presence of CuO could barely be detected (Figure 1).

Table II. Product Phases Observed for Various Cation Ratios

Nominal Composition La : Cu	Phases Identified by X-ray Analysis
2.05 : 1	$La_2CuO_4 + La_2O_3$
2:00 : 1	La_2CuO_4
1.95 : 1	La_2CuO_4 (step counting showed trace of CuO)
1.90 : 1	La_2CuO_4 + CuO

The samples prepared by double decomposition of the nitrates followed by an anneal at 500°C and 600 psi oxygen resulted in maximum oxygen content as determined by thermogravimetric analysis. Table III compares the above compound with those annealed in an argon or argon/air atmosphere at 500°C and 800°C. Samples heated in argon between 500 and 800°C were single phase as indicated by x-ray analysis and showed no measurable difference in cell constants.

Table III. Variation of Cell Parameters and Magnetic Properties with Type of Anneal for La_2CuO_4

Analyzed Composition or Preparative Condition	T_N	χ(emu/mol)	Cell Constants a	b	c
$La_2CuO_{4.00(1)}$ 500°C, 600 psi O_2	240	9×10^{-5}	5.358(2)	5.400(2)	13.16(1)
500°C Ar/Air 80/20	260	13×10^{-5}	5.358(2)	5.405(2)	13.15(1)
500°C Ar	300	13×10^{-5}	5.355(2)	5.406(2)	13.14(1)
800°C Ar	300	13×10^{-5}	5.356(2)	5.408(2)	13.15(1)

However, at 900°C there was decomposition of the products as indicated by the appearance of a strong lanthanum oxide peak in the x-ray diffraction pattern. As can be seen from Table III and Figure 2, the Néel point (which is chosen as the maximum in the plot of X vs T) shifts from 240 K for $La_2CuO_{4.00(1)}$ to 300 K for a composition containing a small oxygen deficiency. The magnitude of this deficiency is less than can be unambiguously ascertained by direct thermogravimetric analysis which has a limit in accuracy of x of .01. Furthermore, changes in cell parameters are capable only of differentiating between pure $La_2CuO_{4.00(1)}$ with T_N = 240 K and the oxygen deficient composition with a T_N of 300 K.

It was shown by Mitsuda et al (3) that in the system La_2CuO_{4-x} the Néel temperature is exceedingly sensitive to the oxygen vacancy concentration. Samples of pure La_2CuO_4 have been reported to show no antiferromagnetic ordering down to 4.2 K (2,5,7). However, despite all efforts to reproduce the procedures indicated by these authors, no sample could be synthesized in this laboratory that did not show a T_N. It can be seen from Figure 2 that the removal of a small amount of oxygen shifts T_N from ∿ 240 to 300 K. This agrees with the work of Uemura et al. (8) in which two samples were prepared at 950°C in air and one of them was heated at 260 psi of O_2 at 500°C. Samples of $La_2CuO_{4.00(1)}$ prepared in this laboratory at 500°C up to 2000 psi of O_2 still showed a Neel temperature of 240 K. The shift in the Néel temperature of a sample slightly deficient in oxygen is consistent with the observation by Kasowski et al. (9) that oxygen vacancies occur in the Cu-O plane. Such vacancies could result in the strengthening of the antiferromagnetic interactions between Cu-O-Cu planes along the c-direction in the structure.

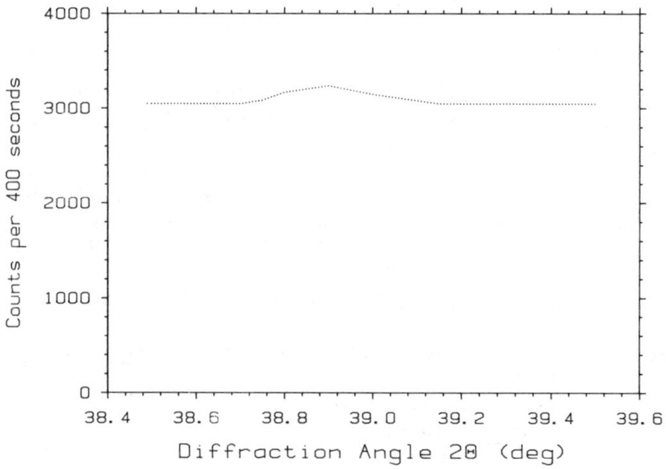

Figure 1. X-ray step counts for the nominal sample of $La_{1.95}CuO_{3.925}$, showing the strongest peak of CuO.

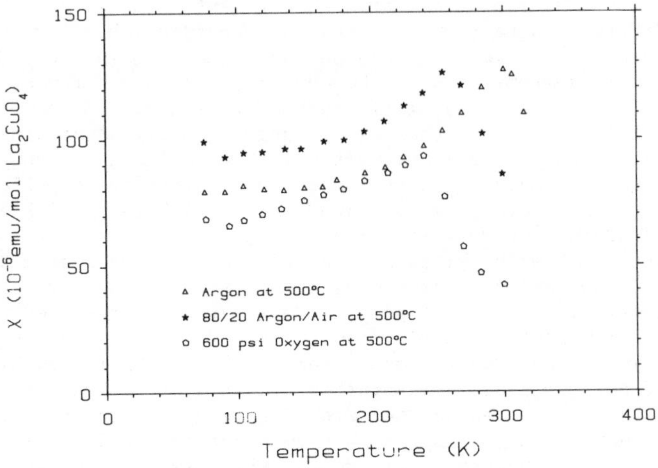

Figure 2. Magnetic susceptibility as a function of temperature for samples of La_2CuO_4 annealed at 500°C in 600 psi of oxygen, in an 80/20 argon/air mixture, and in pure argon.

Acknowledgments

This work was supported in part by the Office of Naval Research. The authors also wish to express their appreciation to Dr. John Longo of Exxon Production Research for helpful discussions.

Literature Cited

1. Longo, J.M.; Raccah, P.M. J. Sol. State Chem. 1973, 6, 526.
2. Johnston, D.C.; Stokes, J.P.; Goshorn, D.P.; Lewandowski, J.T. Phys. Rev. B 1987, 36, 4007.
3. Mitsuda, S.; Shirane, G; Sinha, S.K.; Johnston, D.C.; Alvarez, M.S.; Vaknin, D.; Moncton, D.E. Phys. Rev. B 1987, 36, 822.
4. Freltoft, T.; Fische, J.E.; Shirane, G.; Moncton, D.E.; Sinha, S.K.; Vaknin, D.; Remeika, J.P.; Cooper, A.S.; Harshman, D. Phys. Rev. B 1987, 36, 826.
5. Nguyen, N.; Studer, F.; Raveau, B.; J. Phys Chem. Solids 1983, 44, 389.
6. Saez Puche, R.; Norton, M.; Glaunsinger, W.S. Mat. Res. Bull. 1982, 17, 1429.
7. Freltoft, T.; Shirane, G.; Mitsuda, S.; Remeika, J.P.; Cooper, A.S. Phys. Rev. B 1988, 37, 137.
8. Uemura, Y.J.; Kossler, W.J.; Yu, X.-H.; Kempton, J.R.; Schone, H.E.; Opie, D.; Stronach, C.E.; Johnston, D.C.; Alvarez, M.S.; Goshorn, D.P.; Phys. Rev. Letts. 1987, 59, 1045.
9. Kasowski, R.V.; Hsu, W.Y.; Herman, F.; Phys. Rev. B 1987, 36, 7248.

RECEIVED July 13, 1988

Chapter 12

Physical Properties and Phase Identification in Yttrium–Alkaline Earth–Bismuth–Copper Oxide Systems

Nicholas D. Spencer and A. Lawrence Roe

W. R. Grace and Company–Connecticut, Research Division, 7379 Route 32, Columbia, MD 21044

$Bi_2Sr_2CaCu_2O_{8+\gamma}$ reacts with $YBa_2Cu_3O_{7-\delta}$ at 950°C to produce a new, face-centered cubic phase with a=8.55Å containing Y, Bi and Ba. This phase appears to be isomorphous with the high temperature form of $Cd_3Bi_{10}O_{18}$. Reacted samples are superconducting at values of x as high as 0.2 in the compositional formula $(YBa_2Cu_3)_{(1-x)}(Bi_2Sr_2CaCu_2)_xO_y$ with a T_c of ~83K. x=0.2 corresponds to a material with ~60 wt% cubic phase and 40 wt% residual $YBa_2Cu_3O_{7-\delta}$. The cubic phase is not superconducting above 77K, but when prepared via a solution-phase route, shows a semiconductor-to-metal transition at ~120K. Critical temperatures for the $0 < x \leq 0.2$ compositions appear to be independent of x.

$YBa_2Cu_3O_{7-\delta}$(Y-123) and $Bi_2Sr_2CaCu_2O_{8+\gamma}$ (Bi-2212) are both superconductors with $T_c < 77K$, but they have quite distinct crystal structures (1-2). There is already a considerable literature on substitution of transition metal ions into the Y-123 structure (3-6), but not, so far as we are aware, on the reaction of the two superconducting materials together.

In this paper we examine the product of the reaction between Y-123 and Bi-2212 at a temperature that lies between the melting points of the two materials (950°C). We also compare these products with materials of similar composition, synthesized by a solution-phase process, starting with the salts of the individual component metals.

Experimental

Two sets of materials of general formula $(YBa_2Cu_3)_{(1-x)}(Bi_2Sr_2CaCu_2)_xO_y$ were prepared, and are listed in Table I.

0097–6156/88/0377–0145$06.00/0

Table I. Preparation Method, Overall Composition and T_c for $(YBa_2Cu_3)_{(1-x)}(Bi_2Sr_2CaCu_2)_xO_y$ Samples

Synthetic Method	x	T_c (zero resistance)/K
Composite	0.053	83
Composite	0.077	83
Composite	0.158	83
Composite	0.200	83
Composite	0.429	semiconductor
Coprecipitated	0.053	83 (non-zero)
Coprecipitated	0.077	83
Coprecipitated	0.158	83 (non-zero)
Coprecipitated	0.200	85 (non-zero)
Coprecipitated	0.429	semiconductor to 120K, metal below

In one set, composites of (Y-123) and (Bi-2212) were synthesized by grinding the individual superconductors together in the correct ratios in an agate mortar, pressing them into pellets at 20,000 psi and firing. The starting superconductors (W. R. Grace and Co., Super T_c-Y123 and Super T_c-Bi2212) had a particle size of 1-10μm and a chemical purity of 99.9%. The firing schedule involved heating at 950°C for six hours, cooling to 600°C over two hours and then cooling to 400°C over eight hours. All firing was carried out in pure oxygen. In the other set, the samples were prepared by coprecipitation as carbonates from a solution containing the mixed metal nitrates in the correct stoichiometric ratios. The precipitated carbonates were dried at 110°C and then heated at 540°C in air for 2 hours. The mixed oxides and carbonates produced in this way were then heated in oxygen at 800°C for 12 hours, cooled to 600°C over two hours and to 400°C over eight hours. The gray-black powders were then pressed into pellets and fired, using the same conditions as described for the composite pellets. The metal nitrates (Aldrich Chemical Co.) were of ≥99.9% purity. In both composite and coprecipitated samples, the value of x was deliberately restricted to <0.5 in order to avoid melting of the (bismuth-rich) samples at the particular processing temperature chosen.

X-ray powder diffraction patterns were obtained with a Philips APD 3600 diffractometer (graphite monochrometer and theta-compensating slits) using Cu-Kα radiation and a scan rate of 2 deg 2θ min^{-1}. The reported line positions were corrected using an internal silicon standard (SRM-640b). Y-123 concentrations were calculated by assuming that the intensity of the lines from the Y-123 phase in the reacted samples was the same as from Y-123 itself, and by constructing a calibration curve based on the x-ray mass absorption coefficients for a binary mixture of Y-123 and Bi-2212.

Scanning electron micrographs were taken on a Hitachi model 5570 with a Kevex 7000 energy dispersive x-ray fluorescence (EDX) unit attached. All images were recorded with a 20kV electron beam energy.

A.C. magnetic susceptibility data were taken using a Quantum Technology Corp. Meissner Probe operating at a frequency of 20kHz and a maximum a.c. field of 1 Oe. Resistivity was measured with a four-point resistance meter (Keithley 580) using indium contacts to the sample pellet.

Results and Discussion

In Figure 1, x-ray powder diffraction patterns for the Y-123 starting material and all composite samples are shown. As might be expected, lines characteristic of the Y-123 become less intense as x increases. It is more surprising that by x=0.429 these lines have totally disappeared, having become replaced by a majority phase displaying a cubic pattern, together with traces (i.e. relative intensities <10%) of CuO and an unidentified phase. This cubic phase is already clearly visible at x=0.053. In none of the samples were lines characteristic of the Bi-2212 superconductor to be seen. The cubic pattern indexes as a face-centered cubic structure with a=8.55Å (see Table II).

Table II. Powder X-ray Diffraction Lines for the Cubic Phase in $(YBa_2Cu_3)_{(1-x)}(Bi_2Sr_2CaCu_2)_xO_y$, obtained using Cu-Kα radiation

h	k	l	d(obs.)	d(calc.)	I/I_0
1	1	1	4.934	4.936	8
2	0	0	4.276	4.275	2
2	2	0	3.027	3.023	100
3	1	1	2.578	2.578	8
4	0	0	2.139	2.138	33
3	3	1	1.963	1.962	5
4	2	2	1.746	1.745	51
4	4	0	1.510	1.511	22
5	3	1	1.447	1.445	4
6	0	0	1.424	1.425	2
6	2	0	1.351	1.352	18
4	4	4	1.234	1.234	5
6	4	2	1.141	1.143	23

This phase appears to be isomorphous (Figure 2) with the high-temperature form of $Cd_3Bi_{10}O_{18}$ reported by Kutvitskii et al (7). However, the cadmium compound is reported to have a unit cell dimension of a/2, 4.24Å.

The corresponding x-ray patterns from the coprecipitated set of materials are shown in Figure 3. The very strong similarity between the sets suggests that the formation of the cubic phase is a thermo-dynamically controlled phenomenon rather than a diffusion-controlled process. In the coprecipitated set, small peaks associated with traces of unreacted barium carbonate at 2θ ~ 24° can be seen. Also, Y-123 peaks, which persist to x=0.158 in the composite set, appear to collapse to single peaks, even at the lowest level of Bi-2212 incorporation (x=0.053). This suggests that, in addition to the

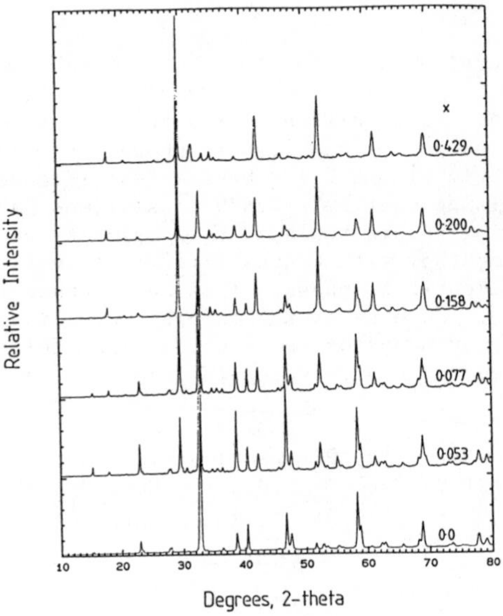

Figure 1. X-ray powder diffraction patterns for $(YBa_2Cu_3)_{(1-x)}(Bi_2Sr_2CaCu_2)_xO_y$ produced by "composite" method.

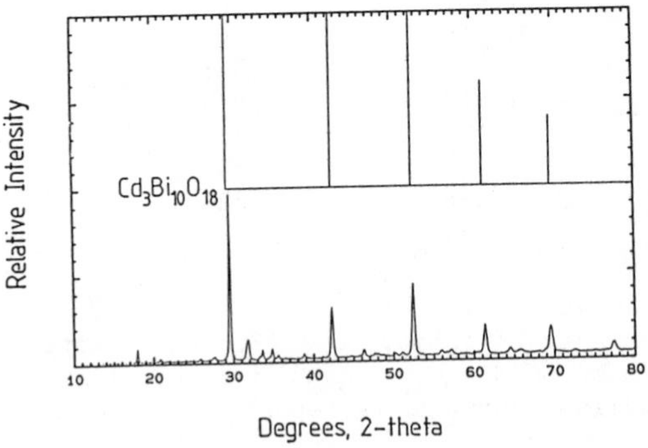

Figure 2. X-ray powder diffraction pattern for x=0.429 "composite" sample, compared to pattern for $Cd_3Bi_{10}O_{18}$ reported by Kutvitskii et al (7).

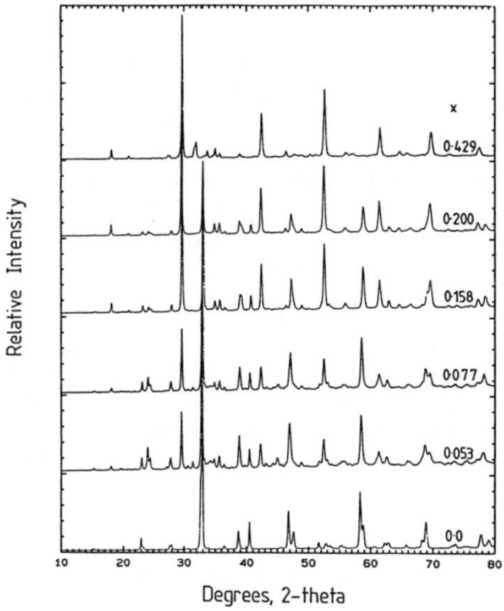

Figure 3. X-ray powder diffraction patterns for $(YBa_2Cu_3)_{(1-x)}(Bi_2Sr_2CaCu_2)_xO_y$ produced by "coprecipitated" method.

formation of the cubic structure, the Y-123 phase is being transformed to a tetragonal structure in the presence of the Bi-2212 components.

From the x-ray patterns, it was calculated that the composite sample with x=0.2 has a Y-123 concentration of 40(±10) wt%. This implies that approximately 1/2 of the Y-123 has reacted with the Bi to form the cubic phase.

Figure 4A shows a scanning electron micrograph of a fracture surface produced by breaking a pellet made from the x=0.158 composite material. Two different morphologies are clearly visible in this micrograph: a smooth, dark, sintered region (Region A), and areas consisting of ~1μm-sized discrete, white particles (Region B). Bismuth and copper x-ray fluorescence maps of this same region are shown in Figures 4B and 4C, respectively. The copper appears to be associated mainly with the smooth regions and the bismuth with the discrete particles. EDX spectra of these two regions show that Region A contains Y, Ba, Cu, with only traces of Ca and Sr, while Region B contains Y, Bi and Ba, with traces of Ca and Cu. These two compositions are presumably associated with the Y-123 and cubic phases, respectively.

Figure 4D shows the equivalent x=0.158 material, prepared by coprecipitation. In this sample, the large, smooth, dark areas are replaced by 5-10 μm chunks. These contain Y, Ba, Cu, and traces of Ca, and Sr, as in the composite sample, but they also contain a trace of bismuth. The presence of Bi may induce the Y-123 phase to adopt the tetragonal structure seen in the x-ray pattern of this material. This behavior has been observed in Y-123 that has been doped with iron (3-6): as little as a 2% substitution of iron for copper in Y-123 can lead to an orthorhombic-to-tetragonal transformation, without loss of superconductivity. The Region B particles are also present in the coprecipitated sample and have a similar Y-Bi-Ba composition (with traces of Ca and Cu) to those seen in the composite material.

At x=0.429, the Region B particles corresponding to the cubic phase dominate the structures as seen by S.E.M. As before, for both composite and coprecipitated samples, this phase appears (by EDX) to consist primarily of Y, Bi and Ba, with traces of Ca and Cu. The remainder of the copper is mostly present as small, smooth chunks, which, in the composite material, contain only traces of the other elements, and in the coprecipitated sample, contain large concentrations of both strontium and calcium. It seems likely that both phases in these complex systems exist as solid solutions, and that the exact partition of the elements between the phases is a kinetically controlled phenomenon, determined by the starting materials from which they were synthesized.

T_cs derived from resistivity vs. temperature measurements are summarized in Table I. The starting Y-123 shows zero resistance at 92K. All samples except those with the highest bismuth concentration (x=0.429) show a sudden drop in resistance with an onset temperature of ~88K. All the composite samples that show this drop reach a measured zero resistance at 83K (see, for example, Figure 5A). This is strong evidence for the persistence of an interconnected superconducting phase. Three of the coprecipitated samples show a resistance vs. temperature curve with a similar shape to the superconducting composite samples. However, in these materials, the

Figure 4. (A) Electron micrograph of fracture surface from an x = 0.158 pellet produced by "composite" method. The distance between the black dots on the white bar corresponds to 10 μm. (B) X-ray fluorescence map of bismuth for the same region as in (A). White points indicate th presence of detected element. Continued on next page.

Figure 4. (C) X-ray fluorescence map of copper for the same region as in (A). White points indicate the presence of detected element.
(D) Electron micrograph of fracture surface from an x = 0.158 pellet produced by "coprecipitated" method. The distance between the black dots on the white bar corresponds to 10 μm.

sudden resistance drop terminates in a small, finite value at ~83K. This small residual resistance is probably due to a thin coating of $BaCO_3$ (the presence of which was indicated by XRD) between superconducting grains. Both of the two x=0.429 samples show an increasing resistance with temperature down to 120K. In the composite sample, this trend continues down to the lowest temperatures investigated (81K), while in the coprecipitated sample, the temperature coefficient of resistance appears to change sign at ~120K (see Figure 5b). Resistance drops in all other samples correlate with a large diamagnetic signal detected in A.C. susceptibility measurements. However, in the x=0.429 coprecipitated material, no such signal was detected, suggesting that the resistance drop is due to a semiconductor-to-metal transition. Although the corresponding composite sample appears to be semiconducting at all temperatures

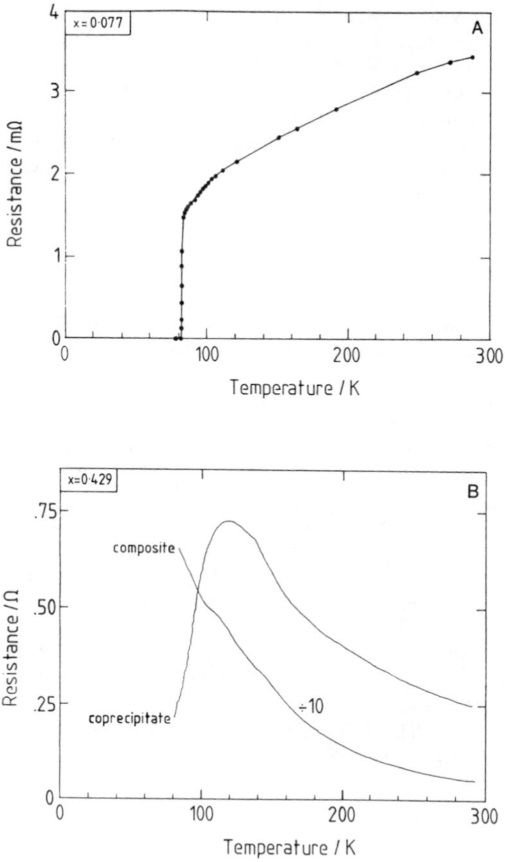

Figure 5. (A) Resistance vs. Temperature curve for x=0.077 sample produced by "composite" method.
(B) Resistance vs. Temperature curves for x=0.429 samples.

investigated, careful scrutiny of Figure 5B shows that a resistance fluctuation is present at 120K, indicating that a small fraction of this sample may also be undergoing a similar transition. The differences between the samples are probably due to minor variations in the cubic phase solid solution composition.

A.C. susceptibility measurements on the composite samples at 77K showed a monotonic decrease to zero in diamagnetic signal, as x was increased from 0 to 0.429. This, together with the x-ray data, suggests that the superconductivity of these samples was simply due to the remaining Y-123 material.

Summary and Conclusions

Bi-2212 reacts completely with Y-123 at 950°C to produce a new, face-centered cubic phase of side 8.55Å, containing yttrium, bismuth and barium, with traces of copper and calcium in solid solution.

XRD analysis of the x=0.2 composite sample showed only the cubic and Y-123 phases to be present, the latter making up ~40 wt% of the total. In other words, this sample was superconducting at 83K, despite the presence of ~60 wt% of non-superconducting material, as shown by both resistivity and magnetic susceptibility measurements. The T_c of both composite and coprecipitated superconducting materials is quite constant at ~83K, irrespective of Bi content. This is surprising, given the extensive solid solution formation and cation substitution by Ca and Bi, that in the composite case is sufficient to change the Y-123 from an orthorhombic into a tetragonal structure.

Acknowledgments

We would like to thank Jeff DeBoy, Carole Carey, Tim Boyer, and Eric Taylor for their skillful experimental work.

Literature Cited

1. Siegrist, T.; Sunshine, S.; Murphy, D. W.; Cava, R. J.; Zahurak, S. M. Phys. Rev. B 1987, 35, 7137-9.
2. Subramanian, M. A.; Torardi, C. C.; Calabrese, J. C.; Gopalakrishan, J.; Morrissey, K. J.; Askew, T. R.; Flippen, R.B.; Chowdhry, U.; Sleight, A. W. Science 1988, 239, 1015-7.
3. Maeno, Y.; Tomita, T.; Kyogoku, M.; Awaji, S.; Aoki, Y.; Hoshino, K.; Minami, A.; Fujita, T. Nature 1987, 328, 512-4.
4. Oda, Y.; Fujita, H.; Toyoda, H.; Kaneko, T.; Kohara, T.; Nakada, I.; Asayama, K. Japan J. Appl. Phys. 1987, 26, L1660-3.
5. Maeno, Y.; Kato, M.; Aoki, Y.; Fujita, T. Japan J. Appl. Phys. 1987, 26, L1982-4.
6. Takayama-Muromachi, E.; Uchida, Y.; Kato, K. Japan J. Appl. Phys. 1987, 26, L2087-90.
7. Kutvitskii, V. A.; Kosov, A. V.; Skorikov, V. M.; Koryagina, T. I. Inorg. Mater. (USSR) 1975, 11, 1880-3.

RECEIVED July 15, 1988

Chapter 13

Valence, Charge Transfer, and Carrier Type for $Bi_2Sr_2Ca_{n-1}Cu_nO_{2n+4+\delta}$ and Related High-Temperature Ceramic Superconductors

T. E. Jones, W. C. McGinnis, R. D. Boss, E. W. Jacobs, J. W. Schindler, and C. D. Rees

Naval Ocean Systems Center, Code 633, San Diego, CA 92152–5000

Samples of the bismuth-based high temperature superconducting family $Bi_2Sr_2Ca_{n-1}Cu_nO_{2n+4+\delta}$ have been prepared and characterized by x-ray diffraction, and temperature-dependent resistivity and thermoelectric power measurements. Both of the high temperature superconducting phases reported in the literature, with transition temperatures near 80K and 110K, have been observed. Evidence from thermoelectric power measurements is presented which shows that this family of ceramic superconductors has contributions to the electrical transport that is both electron-like and hole-like. However, all of the superconducting transitions observed involve hole-like states.

With the recent discovery of the new high temperature ceramic superconductors based on both bismuth (1,2) and thallium (3,4), there is renewed interest in the electronic states of these materials and, in particular, the charge transfer to and from the Cu-O planes vis-a-vis the $YBa_2Cu_3O_7$ family. With no Cu-O chains in the bismuth (5) and thallium (4) materials, the role of the Bi-O and Tl-O planes is being investigated via thermoelectric measurements. The absolute thermoelectric power (thermopower) of a material generally yields the sign of the dominant charge carrier and, as in Hall effect measurements, distinguishes between electron-like (n-type carriers) and hole-like (p-type carriers) conduction. In the previous high temperature superconductors, $(La,Sr)_2CuO_4$ and $YBa_2Cu_3O_7$, both the normal state conductivity and the superconductivity have been shown to be due to hole-like states (6,7). In the lanthanum material, doping with a divalent element, barium or strontium, in place of the trivalent lanthanum, effectively removes electrons from the Cu-O planes leaving conducting hole-like states. Also, x-ray photoemission spectroscopy (8,9) shows that doping creates holes in the Cu-O planes. However, in the heavily doped lanthanum material (25% strontium replacement), evidence from thermoelectric power measurements extended above room temperature have been reported which suggest that there is also a contribution to the normal conductivity from electrons (6). In the yttrium material, empirical atom-atom potential calculations (10,11), extended Hückel molecular orbital calculations (12) and bond-valence calculations (13) imply electron transfer from the Cu-O planes to the Cu-O chains. In addition, the results of substitutional studies with zinc and gallium by Xiau *et al.* (14), show that it is the Cu-O planes which are important for superconductivity, rather than the Cu-O chains, in $YBa_2Cu_3O_7$.

In the bismuth family of materials, one could argue that electrons may be transferred from the Bi-O planes to the Cu-O planes. Because bismuth presumably goes into the material in the valence +3 state, partial conversion to bismuth +5 would require hole states on the bismuth planes and electron states on the Cu-O planes. If such electron transfer occurs and the charge carriers in these materials reside on the Cu-O planes (as in the other copper oxide superconductors), then conduction would be due to n-type electronic states. This would be quite different than in the other copper oxide high temperature superconducting ceramics, where the superconductivity comes from paired hole states. On the other hand, band structure calculations of Hybertsen and Mattheiss indicate that the Bi-O planes in effect dope the Cu-O planes with additional holes, that is, that electrons are transferred from the Cu-O planes to the Bi-O planes (15). Assuming these calculations are correct, there may be an n-type contribution to the conductivity from the Bi-O planes, but the superconductivity would be very similar to that observed in all other copper oxide superconducting ceramics to date, that is, due to pairing of holes in the Cu-O planes.

These considerations have provided the motivation for the experiments described in this paper. Specifically, we would like to answer two questions: Is the conduction in the bismuth family of superconductors due to electrons or holes, and which carriers condense into the superconducting state at each of the observed superconducting transitions in the bismuth family?

Sample Preparation and Characterization

Samples were prepared from Y_2O_3 (Aesar, 99.99%), $BaCO_3$ (Aesar, 99.997%), CuO (Aesar, 99.999%), Bi_2O_3 (Aesar, 99.9998%), $SrCO_3$ (Aesar, 99.999%), and $CaCO_3$ (Aesar, 99.9995%), all of which were used as received. The $YBa_2Cu_3O_7$, sample 1, was prepared from the constituent oxides and $BaCO_3$. The powders were mixed and calcined at 950°C for 16 hours in air, ball-milled to a fine powder, pressed into pellets and sintered at approximately 955°C for 12 hours in oxygen. It was then cooled at 2°C/min. to approximately 550°C in oxygen, annealed at 550°C for 6 hours in oxygen, cooled to 350°C at 2°C/min. in oxygen, and furnace cooled to 200°C in oxygen.

Samples 2-5 were prepared with nominal compositions of Bi:Sr:Ca:Cu in the molar ratios of 2:2:1:2, 2:2:2:3, 4:3:3:6, and 4:3:3:6, respectively. Samples 2 and 3 were calcined in porcelain crucibles for 5 hours at 860°C, then ground and pressed into pellets which were then baked on alumina disks at 860°C for 86 hours. Sample 3 was then baked for an additional 72 hrs at 875°C.

Samples 4 and 5 were calcined for 12 hours at 860°C, ground, then recalcined at 860°C for 12 more hours. After regrinding and pressing into pellets, the pellets were baked on an alumina disk at 865°C for 65 hours in air, concluding with a slow cool in oxygen. This material was then baked further in air for 41 hours at 880°C. During this 880°C bake the sample material reacted to produce a bronze colored coating on the exposed surfaces and reacted with the alumina substrate to produce a green colored material. This green material formed small stalagmites which lifted the pellets about 1 mm above the alumina disk. The crystalline-appearing pellet material supported by the stalagmites constitutes sample 4, with properties reported in this paper. The green stalagmite material is being chemically analyzed. Sample 5 was processed as number 4, but with 70 hours of additional furnace time at 880°C. This sample resembled sample 4 but it also had a glossy black crystalline appearing surface.

All samples were characterized by powder x-ray diffraction, using Cu K-α radiation on material taken from each pellet after all processing. The diffraction pattern for sample 1 agrees with literature data for the yttrium 1:2:3 layered perovskite structure. The diffraction patterns for samples 2-4 are shown in Figure 1. The pattern for sample 3 is in excellent agreement with those of Tarascon _et al_. (_5_) and Takayama-Muromachi _et al_. (_2_). Therefore, we believe sample 3 is mostly $Bi_2Sr_2Ca_{n-1}Cu_nO_{2n+4+\delta}$ with n=2. While all three patterns are similar there are some differences, which may be reflective of the basic differences in the electrical properties of the samples. It should also be noted that a diffraction pattern from sample 4 taken prior to the final 880°C bake was essentially equivalent to that of sample 3, with the exception of a large peak at an angle 2Θ=32°.

An examination of sample 2 with an optical microscope shows no apparent crystallinity on that scale. Similarly, sample 3 shows little bulk crystallinity on an optical scale, but several small domains of microcrystallinity were dispersed throughout the material. The appearance of both samples 2 and 3 were independent of whether or not the surface had been in contact with the alumina substrate. Conversely, samples 4 and 5 had different appearances for the surface in contact with the alumina and the other surfaces. The surface which was in contact with the alumina was black but essentially microcrystalline everywhere. Microscopy of the other surfaces revealed small golden colored bars, with the remaining area being black in coloration. It is also of note that the surface in contact with the alumina had, prior to the final 880°C bake, grown flat plate-like whiskers which were not present after the 880°C bake. The stalagmite-like formations were not very crystalline visually, and contained regions of various colors.

Electronic Properties

Each of the five samples described above were shaped with hand tools into rectangular bars approximately 2 mm x 2 mm x 4 mm. They were mounted in a closed-cycle He-4 refrigerator in such a way that the thermopower and four-probe resistance of all samples could be measured during a single cooling/warming cycle. This arrangement is shown schematically in Figure 2.

The thermopower data was obtained using a slow a.c. differential technique (16) in which a small, slowly oscillating thermal gradient is produced across the sample, and the resultant thermoelectric voltage measured at the sample ends. The temperature gradient was produced by alternately heating two parallel quartz blocks (each wrapped with Manganin wire heaters) which were bridged by the samples. The blocks were bonded (using G.E. 7031 varnish) to a brass sheet, which was in turn attached (with a thin layer of Apiezon N grease) to the copper stage on the cold-head of the refrigerator. The temperature T_{stage} of the refrigerator stage was monitored with a calibrated silicon diode. The samples were soldered, along with the thermoelectric voltage leads (#34 copper; Belden 8057), to the edges of the quartz blocks with indium. The temperature difference ΔT between the two quartz blocks (that is, the temperature drop across the sample) was measured using a copper/constantan/copper differential thermocouple whose junctions were indium soldered to the quartz. The time-averaged temperature difference δT between one of the quartz blocks and the refrigerator stage was also monitored with this type of thermocouple. The average sample temperature T is then given by $T_{stage} + \delta T$.

At a given temperture, the thermopower of each sample was measured as follows. The voltage of the block-to-block thermocouple was plotted on an X-Y recorder versus the thermoelectric voltage measured across the sample, ΔV, as the blocks were alternately heated at a frequency which varied with temperature, but which was in the range of 0.02-0.05 Hz. The thermocouple voltages were converted to a temperature difference using tabulated values (17) of $dV/dT = S_{constantan} - S_{Cu}$. The resultant trace, apart from transient behavior as the heater current was switched from one quartz block to the other, was a straight line whose temperature-converted slope, $\Delta V/\Delta T$, equals the sample thermopower minus the thermopower of the copper voltage leads. After the heating current is switched to the other block, the trace is completed to form a closed loop. The block-to-block temperature difference, ΔT, was usually less than 1K.

Normally, the thermopower measurements would be performed at a constant T_{stage}. It was more convenient, however, to take mea-

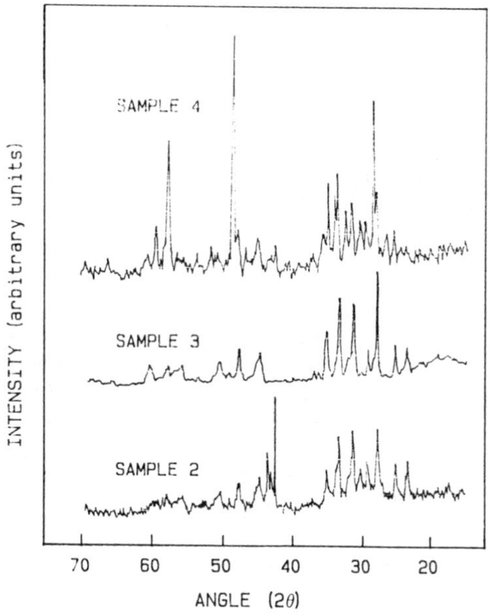

Figure 1. X-ray scans for samples 2-4.

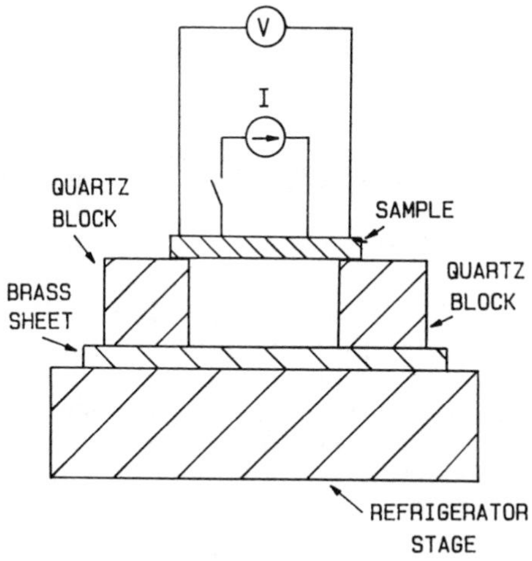

Figure 2. The apparatus for measuring the thermopower.

surements while the refrigerator, which was continually pumped by a liquid nitrogen trapped diffusion pump, very slowly warmed at a rate of about 0.5 K/min. As a typical trace required about 30 seconds to complete, this mode of operation introduced very little error and greatly sped data acquisition. Before each set of thermopower traces at a given temperature, readings of the four-probe resistance were taken by closing the current source switch shown in Figure 2. For convenience, the current was sent through the center connections on the samples (using the same type of wire as the voltage leads) and the voltage measured at the end connections. Reversing the current and voltage leads gave the same resistance values. Within experimental accuracy, the thermopower values measured with and without the current wires attached to the samples were the same.

The temperature dependent thermopower of the copper voltage leads is required in order to obtain the samples' absolute thermopower. For temperatures below 91K, sample 1 is superconducting and therefore has zero thermopower, which allowed a direct determination of the absolute thermopower of the copper voltage leads. The copper thermopower values from this self-calibration agree with the results of Crisp et al. (18), as corrected by the new thermopower scale of Roberts (19), throughout this temperature range. The values of S_{Cu} used to analyze the thermopower data of the samples were a composite of a fit to the voltage lead data taken on sample 1, for $T < 91K$, and the high temperature data of Crisp et al., for $T > 91K$.

Errors in thermopower measurements can arise from several sources. The main source of error in this experimental configuration was an approximately 10% uncertainty in the temperature difference ΔT across the samples due to uneven heating of the quartz blocks. For temperatures above 91K, values of the samples' absolute thermopower may be off by as much as 0.25 μV/K, depending on the values of S_{Cu} used. The error in the sample temperature T was relatively small, approximately ±0.3K. The resistivity data contain a fixed error of about 10% due to uncertainty in the measured sample dimensions.

Results of the thermopower and resistance measurements for samples 1-4 are displayed in Figures 3-6, respectively. In Figure 3, the $YBa_2Cu_3O_7$ resistivity data show the usual superconducting transition at 94K. The thermopower for this sample is positive above this temperature, indicating hole-like carriers, and drops to zero at the transition. The data are very similar to that obtained by Uher et al. (7) As shown in Figure 4, the thermopower and resistivity of sample 2 both vanish at approximately 80K. The thermopower reaches its maximum positive value just above the superconducting transition, and decreases from there with rising temperature. The thermopower changes sign from positive to negative at about 200K, indicating that electrons become the dominant carriers above this temperature. Figure 5 shows that the thermopower and resistivity of sample 3 both go to zero at about 90K. In contrast to sample 2, holes appear to be the dominant carrier throughout the temperature range investigated.

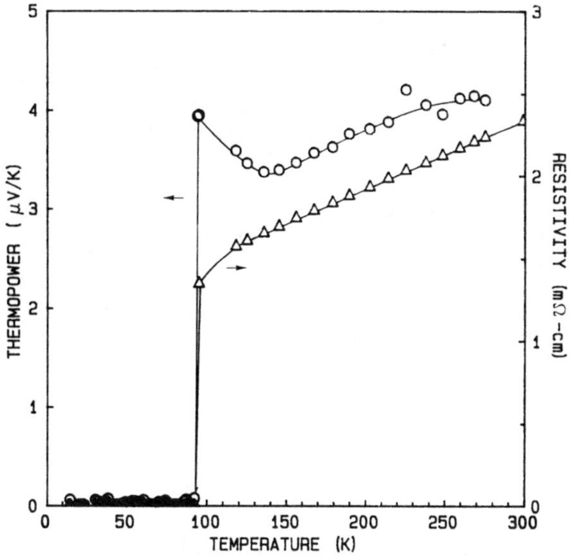

Figure 3. The resistivity (triangles) and absolute thermoelectric power (circles) for sample 1, $YBa_2Cu_3O_7$, as a function of temperature.

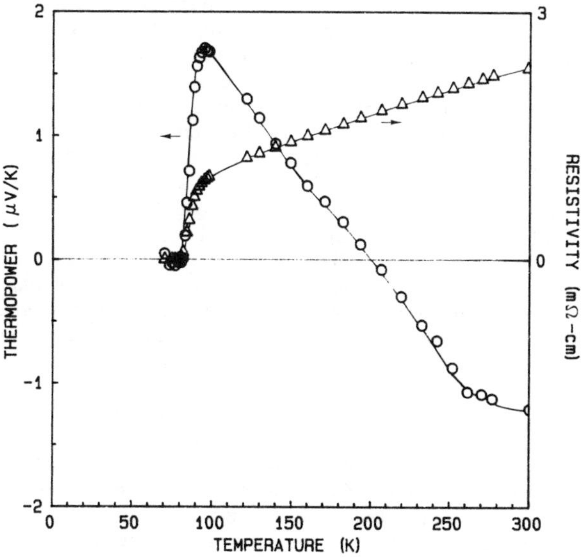

Figure 4. The resistivity (triangles) and absolute thermoelectric power (circles) for sample 2, nominal composition $Bi_2Sr_2Ca_1Cu_2O_{8+\delta}$, as a function of temperature.

The data in Figure 6 show that sample 4 is a mixed-phase sample with superconducting transitions near 110K and 80K. This may indicate that sample 4 has the composition $Bi_2Sr_2Ca_{n-1}Cu_nO_{2n+4+\delta}$ with a mixture of the n=3 and n=2 phases, respectively. The thermopower data suggest that both phases are dominated by hole-like carriers.

The resistivity versus temperature for sample 5 is illustrated in Figure 7. Sample 5 was not made in time to be included in the thermoelectric measurements. This sample has enough of the higher temperature phase that no lower transition is apparent in this measurement. The onset transition temperature is about 115K and the midpoint of the transition is 106K. Magnetization and thermoelectric power measurements on this sample are in progress and will be reported in a subsequent publication.

Discussion

The bismuth family of layered superconducting structures presents a formidable challenge to understand. The set of crystal structures represented formally as $Bi_2Sr_2Ca_{n-1}Cu_nO_{2n+4+\delta}$ may be an idealized oversimplification of the possible structures. Unlike the $YBa_2Cu_3O_7$ structure, where the correct structure is obtained by reacting the constituent materials in the desired final amounts, these bismuth materials apparently require initial compositions which are off stoichiometry. This makes it more difficult to achieve a desired composition and results in extraneous x-ray peaks due to non-reacted materials and other by-products of the synthesis. Three of the four bismuth samples described in this paper were made from different nominal compositions, they all showed distinct x-ray spectra, and they all had different transport properties. However, there are several observations which can be made. Three of the four bismuth samples showed a lower temperature superconducting transition with a midpoint in the range of 80-90K. Many samples similar to those described here show partial higher temperature superconducting transitions near 110K. As illustrated in Figure 6, sample 4 has a significant fraction of this higher temperature transition showing in both the resistivity and thermopower. Also, as shown in Figure 7, sample 5 appears to be all high temperature phase. However, the resistivity is not the best measurement to study multiphase samples. The magnetization measurements currently in progress should reveal what fraction of sample 5 is in the higher temperature superconducting phase, and how much remains, if any, in the lower temperature phase.

In all of the superconducting transitions observed, the superconductivity is due to the pairing of holes, as evidenced by the abrupt decrease in the thermopower from a positive value. The thermopower decreases to zero at the lower transition as expected. However, there are numerous known cases where the thermopower has the opposite sign of the carrier charge. The noble metals are a classic example. The measured thermopower has both positive and negative contributions, from holes and electrons, respectively, each of which can be enhanced by such effects as phonon drag and

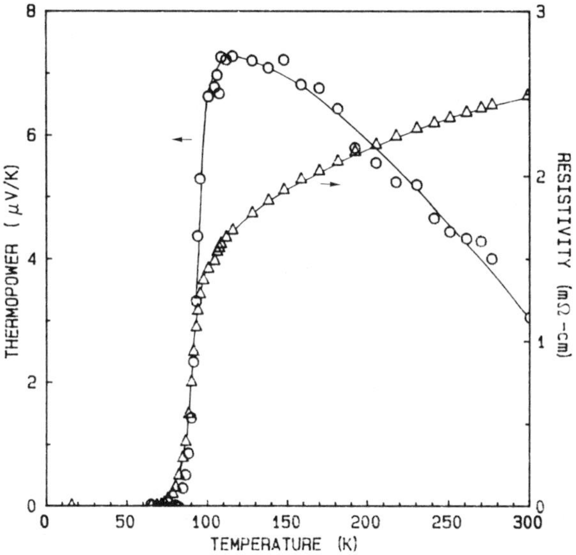

Figure 5. The resistivity (triangles) and absolute thermoelectric power (circles) for sample 3, nominal composition $Bi_2Sr_2Ca_2Cu_3O_{10+\delta}$, as a function of temperature.

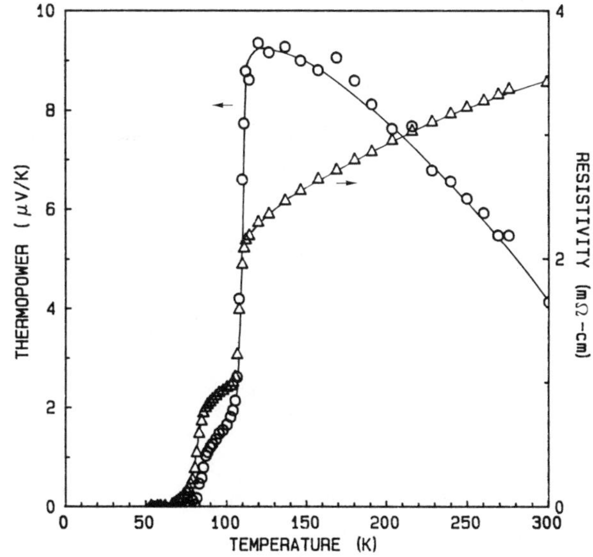

Figure 6. The resistivity (triangles) and absolute thermoelectric power (circles) for sample 4, nominal composition $Bi_4Sr_3Ca_3Cu_6O_x$, as a function of temperature.

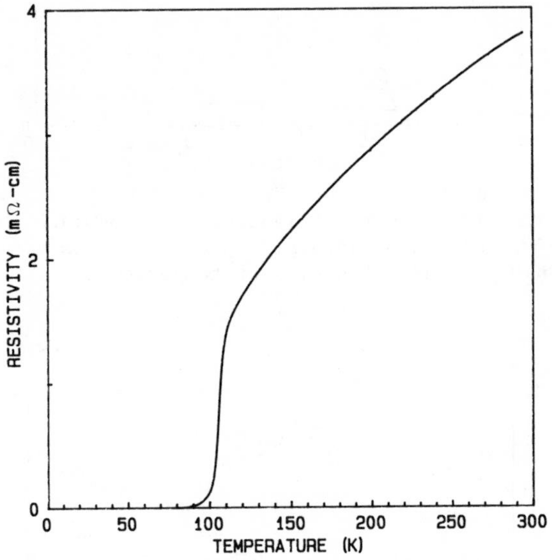

Figure 7. The resistivity for sample 5, nominal composition $Bi_4Sr_3Ca_3Cu_6O_x$, as a function of temperature.

inelastic scattering. Significant electronic contributions to the thermopower are therefore possible even though the net thermopower is positive. However, at the transition, those carriers which pair into the condensed superconducting state lose their ability to generate a diffusion-driven thermoelectric voltage. Thus, the abrupt drop in the thermopower for the mixed phase sample 4, more than the fact that the thermopower is positive, implies pairing of holes. That is, the thermopower abruptly becomes less positive or more negative in an algebraic sense, precisely at the superconducting transition because the pairing of holes diminishes their positive contribution to the diffusion thermopower. This is particularly incisive for sample 4, the mixed phase sample, because this abrupt drop takes place at the higher temperature transition, coincident with the resistance drop, even though there is no percolation path shorting the sample, which would mask the effect. Once a superconducting percolation path shorts the sample, the thermopower will go to zero independent of whether it is due to electrons or holes.

This picture is consistent with the band structure calculation mentioned in the introduction (15), where electrons are removed from the Cu-O planes leaving p-type carriers. It is these p-type carriers which give the superconductivity in this class of bismuth superconductors at both the lower and higher temperature transitions. The question to be answered then, is what acts as the sink for the transferred electrons? The bismuth cations are unlikely sites for the transferred electrons (when $\delta=0$) because the bismuth is already in its lowest valence state. However, it is possible that the transferred electrons are due to extra oxygen atoms between adjacent Bi-O layers. In this case, δ becomes finite and some Bi ions in the Bi-O planes can now be considered formally valence +5 and the Bi-O planes can then serve as a sink for electrons from the Cu-O planes. Evidence that this might be the case can be inferred from x-ray data of Tarascon et al. (5), showing a partial extra oxygen occupancy near 0.06, implying that $\delta \simeq 0.06$.

Thus, in this picture, the Bi-O planes function in an analogous way to the chains of Cu-O in the $YBa_2Cu_3O_7$. That is, electrons are transferred from the Cu-O planes due to the partially occupied oxygen sites in the Bi-O interplanar regions in the former material, and to the Cu-O chains in the latter. The details of the charge transfer are different, but the results are the same.

The thallium material, believed to form in a similar set of layered structures specified as, $Tl_2Ba_2Ca_{n-1}Cu_nO_{2n+4+\delta}$, is not discussed experimentally in this paper. However, one can easily conjecture that a similar transfer of electrons may occur in that material. This transfer may be greatly facilitated in the thallium compound because the thallium has a lower valence state available. Hence, the thallium cations themselves could serve as sinks for the transferred charge without the ancillary requirements of Cu-O chains, as in the yttrium material, or of extra oxygen atoms in the structure, as may be the case for the bismuth superconductors.

The x-ray data show that the intraplanar Bi-O distance is rather large, approximately 2.71 A in these materials (5). Thus, it is not clear that, in the stoichiometric structure, the electrons in the intraplanar bismuth regions can contribute to the electrical conductivity. However, the addition of oxygen ions to the interplanar regions alters the position of the bismuth ions resulting in a new intraplanar Bi-O distance of approximately 1.99 A. The reduction of the Bi-O distance from 2.71 A to the value of 1.99 A was calculated from the data presented in reference 5, due to the oxygen site being fractionally occupied at 0.065. Thus, these bismuth electrons might contribute at least to the normal state conductivity. The thermopower results presented here definitely show that such electrons do not contribute to the superconductivity. However, the thermopower data for sample 2, which are negative above 200K, provide some evidence that electrons contribute to the normal state conductivity in these materials. The thermopower of the material is a function of the densities of electrons and holes, their respective mobilities, and details involving their scattering. The mobilities are temperature dependent, as are the mean free paths. Clearly, the thermopower data show that holes dominate for all compositions at low temperatures, and at room temperature as well except for sample 2. Since the contribution to the positive thermopower increases as the samples are processed to yield more of the higher T_c phase, it is likely that electrons contribute some negative thermopower for all compositions, but the thermopower is dominated by the holes below 200K.

Conclusions

In summary, we have presented conclusive evidence that the two high temperature superconductive transitions, near 80K and 110K, in the $Bi_2Sr_2Ca_{n-1}Cu_nO_{2n+4+\delta}$ family of high temperature superconductors are due to hole-like states as is the case with all other high temperature copper oxide superconductors discovered to date. The p-type carriers most likely arise from electron charge transfer from the Cu-O planes to the Bi-O interplanar regions, perhaps due to extra oxygen sites which are partially occupied. Further, even though holes dominate the electrical transport, we have presented evidence that electrons as well as holes contribute to the normal state conductivity in this new class of layered materials.

Acknowledgments

This work was supported by the Independent Research program at the Naval Ocean Systems Center. The authors are indebted to Drs. Eugene Cooper and Alan Gordon. J.W.S. is partially supported by the Office of Naval Technology through the American Society for Engineering Education.

Literature Cited

1. Chu, C.W.; Bechtold, J.; Gao, L.; Hor, P.H.; Huang, Z.J.; Meng, R.L.; Sun, Y.Y.; Wang, Y.Q.; Xue, Y.Y. Phys. Rev. Lett. **1988**, 60 (10), 941-943.
2. Takayama-Muromachi, Eiji; Uchida, Yoshishige; Ono, Adira; Izusi, Fujio; Onoda, Mitsuki; Matsui, Yoshio; Kosuda, Kosuke; Takakawa, Shunji; Kato, Katsuo preprint submitted to Jpn. J. Appl. Phys.
3. Sheng, Z.Z.; Hermann, A.M.; El Ali, A.; Almasan, C.; Estrada, J.; Datta, T.; Matson, R.J. Phys. Rev. Lett. **1988**, 60 (10), 937-940.
4. Hazen, R.M.; Finger, L.W.; Angel, R.J.; Prewitt, C.T.; Ross, N.L.; Hadidiacos, C.G.; Heaney, P.J.; Veblen, D.R.; Sheng, Z.Z.; El Ali, A.; Hermann, A.M. Phys. Rev. Lett. **1988**, 60 (16), 1657-1660.
5. Tarascon, J.M.; LePage, Y.; Barboux, P.; Bagley, B.G.; Greene, L.H.; McKinnen, W.R.; Hull, G.W.; Giroud, M; Hwang, D.M. submitted to Phys. Rev.
6. Uher, C.; Kaiser, A.B.; Gmelin, K.E.; Walz, L. Phys. Rev. B **1987**, 36 (10), 5676-5679.
7. Uher, C.; Kaiser, A.B. Phys. Rev. B **1987**, 36 (10), 5680-5683.
8. Tranquada, J.M.; et al., Phys. Rev. B **1987**, 36, 5263.
9. Shen Z.X.; et al., Phys. Rev. B **1987**, 36, 8414.
10. Evain, Michel; Whangbo, Myung-Hwan; Beno, Mark A.; Williams, Jack M. J. Am. Chem. Soc. **1988** 110, 614-616.
11. Whangbo, Myung-Hwan; Evain, Michel; Beno, Mark A.; Geiser, Urs; Williams, Jack M. Inorg. Chem. **1988**, 27, 467-474.
12. Curtiss, L.A.; Brun, T.O.; Gruen, D.M. Inorg. Chem. **1988**, 27, 1421-1425.
13. O'Keeffe, Michael; Hensen, Staffan J. Am. Chem. Soc. **1988**, 110, 1506-1510.
14. Xiao, Gang; Cieplak, M.Z.; Gavrin, A.; Streitz, F.H.; Bakhshai, A.; Chien, C.L. Phys. Rev. Lett. **1988**, 60 (14), 1446-1449.
15. Hybertsen, Mark S.; Mattheiss, L.F. Phys. Rev. Lett. **1988**, 60 (16), 1661-1664.
16. Chaikin, P.M.; Kwak, J.F. Rev. Sci. Instrum. **1975**, 46 (2), 218-220.
17. Manual on the Use of Thermocouples in Temperature Measurement, ASTM Special Technical Publication 470, American Society for Testing and Materials, 1916 Race Street, Philadelphia, Pa. 19103, **1970**.
18. Crisp, R.S.; Henry, W.G.; Schroeder, P.A. Philos. Mag. **1964**, 10, 553-577.
19. Roberts, R.B.; Philos. Mag. **1977**, 36 (1), 91-107.

RECEIVED July 15, 1988

Chapter 14

Preparation of the High-Temperature Superconductor $YBa_2Cu_3O_{7-x}$

Leonard V. Interrante, Zhiping Jiang, and David J. Larkin

Department of Chemistry, Rensselaer Polytechnic Institute, Troy, NY 12180–3590

Organic-soluble precursors to the high-Tc "1-2-3" superconductor, $YBa_2Cu_3O_{7-x}$, have been prepared using polyesters derived from the interaction of citric acid (CA) or ethylenediaminetetraacetic acid (EDTA) with ethylene glycol. Mixtures of the nitrate, carbonate and/or acetate salts of the three metal ions with CA or EDTA and ethylene glycol were heated to obtain homogeneous green solids containing the desired 1:2:3 ratios of Y^{3+}, Ba^{2+}, and Cu^{2+} ions. Pyrolysis at 850 ^{0}C - 960 ^{0}C in oxygen gave the superconducting, orthorhombic, $YBa_2Cu_3O_{7-x}$ phase. Films of the superconductor were prepared by dropping ethylene glycol solutions of these precursors onto zirconia and alumina substrates, followed by pyrolysis to 950 ^{0}C in O_2. The results of TGA/DSC and FTIR studies of these precursors, as well as XRD, FTIR, pyrolysis/GC, electrical and magnetic property measurements on the pyrolysis products are presented and discussed in the context of the probable precursor structures and pyrolysis chemistry. Preliminary studies were also carried out on the chemical vapor deposition(CVD) of Y_2O_3 using the tris-(2,2,6,6-tetramethyl-3,5-heptanedionato)- yttrium complex. The pyrolysis products are identified by g.c./FTIR and Auger studies of the Y_2O_3 films are described.

The recent discovery(1,2) of the high-Tc superconductor, $YBa_2Cu_3O_{7-x}$, has stimulated intensive investigation into its synthesis, structure and potential applications. A key requirement for the practical application of this and related high-Tc oxide superconductors is a better method for the fabrication of these materials in useful final forms (*e.g.*, fibers and thin films)

0097–6156/88/0377–0168$06.00/0

The methods currently in use for the preparation of $YBa_2Cu_3O_{7-x}$ typically involve either repeated grinding and sintering of the parent oxides(3) or co-precipitation(4-6) from aqueous solutions with carbonate, oxalate or organic carboxylates followed by calcination. These methods generally require the use of high calcining temperatures (>900 ^0C) and further annealing in oxygen up to 950 ^0C to produce the desired orthorhombic $YBa_2Cu_3O_{7-x}$ phase. Moreover the powders so obtained are typically inhomogeneous and contain other oxide phases. Use of this powder in the fabrication of pressed pellets has resulted in unacceptably low critical currents, owing, in part, to the presence of impurity phases, voids and grain boundary defects.(7) Morever it is extremely difficult to prepare thin films or continuous fibers by means of the above approaches.

A number of groups have successfully prepared $YBa_2Cu_3O_{7-x}$ films by means of vacuum deposition techniques.(8-10) However, these depositions are expensive to perform and are not easily adaptable to large scale production or to fiber preparation. A potential solution to this problem is to employ metal-organic polymers which could be used to make thin films by spin- or dip-coating followed by pyrolysis. Furthermore, such precursors or their appropriately modified relatives could also be useful in the fabrication of continuous fibers by dry or melt spinning methods. Recently $YBa_2Cu_3O_{7-x}$ thin films have been made using acetate\acetic acid(11) and higher carboxylate salt(12) systems.

Another potential source of processible precursors is the citric acid/ethylene glycol system which has been employed previously in the preparation of highly dispersed perovskite, spinel and related complex oxides. This method provides soluble, metal-organic, polymer precursors which have been used for the fabrication of oxide thin films as well as for the production of oxide powders with excellent homogeneity, good stoichiometry control and uniform sizes at relatively low temperatures(13,14).

Recently, the preparation of thin films of the $YBa_2Cu_3O_{7-x}$ superconductor has been reported using this system(15); however, the structure and pyrolysis reactions for these precursors have apparently not been investigated. In this paper we report the preparation of $YBa_2Cu_3O_{7-x}$ superconducting powders and thin films using both citric acid/ethylene glycol- and ethylenediaminetetraacetic acid [EDTA]/ethylene glycol-derived polyesters as well as the results of studies of the structure and pyrolysis chemistry of the precursors.

Experimental Section

Syntheses of Precursors

Citric acid/ethylene Glycol Precursor to $YBa_2Cu_3O_{7-x}$. $Y(NO_3)_3 \cdot 6H_2O$, $BaCO_3$ and $Cu(OAc)_2 \cdot H_2O$ or $Cu(NO_3)_2 \cdot 3H_2O$ (molar ratio 1:2:3) were mixed with excess citric acid and 40 mL ethylene glycol in a small beaker. The mixture was heated and stirred at about 90 ^0C until all the solids had dissolved. The resultant solution was maintained at

120 °C with stirring for *ca*. 2 h until *ca*. 10 mL of green viscous solution remained; the evolution of NO_x along with the excess ethylene glycol was noted during this period. This solution was heated *in vacuo* at 140 °C for several hours, leaving a green glassy solid. This solid could be redissolved in ethylene glycol and was also readily soluble in 2-methoxyethanol and tetrahydrofuran.

EDTA/ethylene Glycol Precursor to $YBa_2Cu_3O_{7-x}$. The same procedure was followed as above, only using EDTA in place of the citric acid. The precursor after heating and evaporation of excess ethylene glycol was also a green glassy solid.

Citric acid/ethylene Glycol/Cu^{2+} System. $CuSO_4 \cdot 5H_2O$ (1g) and citric acid (0.77g)(molar ratio 1:1) were mixed with 25 mL ethylene glycol, heated at 100 °C for 5h, then the resulting solution was heated *in vacuo* at 150 °C, leaving a green glassy solid.

Citric acid/ethylene Glycol/Ba^{2+} System. $BaCO_3$ (1g) and citric acid (4.87g) (molar ratio 1:5) were mixed in 40 mL ethylene glycol and heated at 140 °C. The $BaCO_3$ gradually dissolved forming a brown solution. After all the solid had dissolved, the solution was heated *in vacuo* at 150 °C, yielding a brown glassy solid.

Polyester from Citric acid and Ethylene Glycol. Citric acid was mixed with excess ethylene glycol in a small beaker. The clear, colorless solution was maintained at 100 °C for 8h, acquiring a pale yellow color. After heating *in vacuo* at 140 °C, a yellow glassy solid was obtained.

Tris-(2,2,6,6-tetramethyl-3,5-heptanedionato)-yttrium This complex was obtained as a white solid by the method described by Sievers, et. al.(16) It was recrystallized from hexane and sublimed *in vacuo* before use (m.p. = 170-2 °C).

Chemical Vapor Deposition Procedure. A horizontal hot-wall quartz-tube reactor was employed. The precursor was heated using an oil bath at 120-140 °C using air or nitrogen as a carrier gas. The pressure in the reactor was maintained at *ca*. 0.5-2.0 torr by continuous pumping. The substrates were Si wafer pieces or SiC fibers which were maintained at *ca*. 600 °C using an external tube furnace. Deposition rates varied from *ca*. 0.1 - 1 μm/hr depending on the precursor temperature and the flow rate into the reactor.

Pyrolyses of Polymeric Precursors. Pyrolyses were carried out in a quartz tube under a flow of oxygen with the precursor contained in an alumina boat. The temperature was increased using a programable controller from room temperature to 960 °C over 8 h, held at this temperature for several hours and then slowly cooled to room temperature, all under flowing oxygen. The resultant black product gave an XRD powder diffraction spectrum consistent with expectations for the superconducting orthorhombic phase of $YBa_2Cu_3O_{7-x}$; no other crystalline phases were evidenced.

Fabrication of $YBa_2Cu_3O_{7-x}$ Films Using the Polyester Complexes. Solutions of the $YBa_2Cu_3O_{7-x}$ precursors (10% by weight) in ethylene glycol were dropped onto zirconia and alumina plates and dried slowly on a hot plate at 200 °C. The resultant samples were placed in a furnace and heated at 500 °C in air for 1h. This coating, drying and heating process was repeated 5 times; after which the coated samples were heated at 950 °C for 1h in O_2. Scanning electron micrographs of a fracture cross-section of this film indicated a relatively fine grained, dense microstructure and a film thickness of *ca.* 70μm.

Characterization of Precursor and Pyrolysis Products. X-ray powder diffraction patterns were measured using a Philips diffractometer employing Cu Kα radiation and a Ni filter. DSC and TGA measurements of the precursors were carried out using a Perkin-Elmer DSC7/TGA7 Thermal Analysis System with heating rates of 10 °C/min. Infrared spectra were determined using a Perkin-Elmer FT-1800 Infrared Spectrophotometer in KBr-pellet samples. Pyrolysis/GC experiments were performed using a SHIMADZU GC-9A Gas Chromatograph with a silicone-packed column by heating the precursor in an atmosphere of oxygen and collecting the volatile liquid fraction in a dry-ice cooled trap. The gaseous fraction was separately analyzed using a molecular sieve column. G.C./FTIR measurements were carried out on the products of the $Y(THD)_3$ pyrolysis using the Perkin-Elmer FT-1800 Infrared Spectrophotometer. In this case the compound was sublimed under N_2 at *ca.* 0.6 torr through a quartz tube held at 600 °C and the volatile pyrolysis products collected in a cold trap.

Electrical Conductivity and Magnetic Susceptibility Measurements. The electrical conductivity of a pressed pellet sample was measured as a function of temperature using a four-probe ac technique. A *ca.* 2 x 3 x 1 mm sample was cut from a cold-pressed pellet of the $YBa_2Cu_3O_{7-x}$ powder obtained from the citric acid/ethylene glycol precursor. After annealing in oxygen at 950 °C for 5 h the sample was wired with Au leads using Ag paint contacts. Measurements were carried out in a specially designed cryostat from room temperature down to 77 °K using a 0.1 μA ac current. The magnetic susceptibility measurements were carried out with an ac inductance bridge from room temperature to 80 °K.

Results and Discussion

The $YBa_2Cu_3O_{7-x}$ precursors derived from the citric acid/ethylene glycol and EDTA/ethylene glycol mixtures were green, homogeneous, clear, glassy solids. These solids were soluble in ethylene glycol and other organic solvents. On heating to 350 - 400 °C for a half hour on the hot plate both precursors darkened and shrank in volume without melting or foaming; the products were black, apparently still homogeneous, glassy solids.

Figure 1 shows the results of TGA studies of these two precursors. The precursors derived from citric acid and EDTA display a similar curve shape, suggesting that a similar pyrolysis process occurs for these solids in the presence of air. Both precursors show weight losses in three distinct regions. A gradual weight loss is first observed above *ca*. 150 ^{0}C. This is followed by a sharp weight loss at 360 - 450 ^{0}C. The third weight loss occurs in the range 750 - 810 ^{0}C.

The volatile by-products of the pyrolysis of the citric acid/ethylene glycol system in air were identified using gas chromatography. In addition to a liquid, identified as mainly ethylene glycol, CO_2 and CO were observed as gaseous products between 200 and 350 ^{0}C. Above 350 ^{0}C, CH_4 and C_2H_6 were found in addition to CO_2 and CO. The proportion of CO_2 to CO in each case was at least 5:1 and the relative amounts of the two hydrocarbons was substantially less.

Differential scanning calorimetry measurements are also consistent with the loss of ethylene glycol in the early stages of the conversion of this precursor to the 1-2-3 superconductor. As is shown in Figure 2, a small endotherm is observed at around 200 ^{0}C which is attributed to the evaporation of ethylene glycol. This is followed in the DSC by a broad exotherm from 220 to 400 ^{0}C, which associated with the thermal decomposition reaction.

Comparing the TGA curves for the citric acid/ethylene glycol precursor in air and nitrogen, almost identical behaviour is observed up to *ca*. 350 ^{0}C, whereupon the curves deviate sharply, with appreciably less weight loss occuring under N_2, presumbly due to conversion of some of the organic components of the precursor to carbon. The close similarity in these curves up to 350 ^{0}C and the observation of CO_2 and CO as gaseous products, suggest that the thermolysis of citric acid along with the loss of ethylene glycol is responsible for the weight loss in this region. The rapid drop in weight beyond about 350 ^{0}C in air suggests that oxidation of the organic components is occurring in this region, leading to CO_2 and CO and presumbly H_2O. The CH_4 and C_2H_6 which are observed in this region presumbly arise from thermolysis of the ethylene groups.

Figure 3 shows the XRD patterns of the solid pyrolysis products of the citric acid/ethylene glycol precursor after extended heating in oxygen at different temperatures from 500 - 960 ^{0}C. From these results, it is apparent that a mixture of CuO, Y_2O_3 and $BaCO_3$ is present by at least 500 ^{0}C and the orthorhombic phase of $YBa_2Cu_3O_{7-x}$ starts to appear at 800 ^{0}C. This phase was fully developed after heating at 960 ^{0}C for 5h in oxygen. It was also found that the EDTA/ethylene glycol precursor forms the same intermediates and the orthorhombic 1-2-3 phase at similar temperatures. The formation of $BaCO_3$ as an intermediate in both systems and its subsequent thermal decomposition to BaO (or to the 1-2-3 phase directly) also explains the third weight-loss at 750 - 810 ^{0}C observed by TGA.

The magnetic susceptibility measurements of the 1-2-3 powder obtained from the pyrolysis of the polymer precursor at 960 ^{0}C for

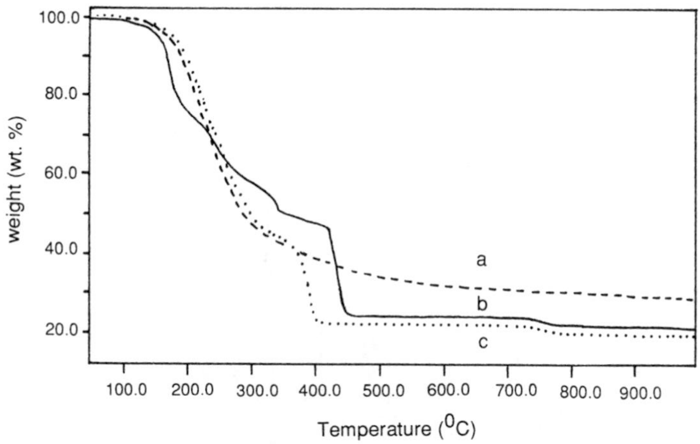

Figure 1. TGA data for polymeric precursors to $YBa_2Cu_3O_{7-x}$.
a : Precursor from citric acid/ethylene glycol in N_2.
b : Precursor from EDTA/ethylene glycol in air.
c : Precursor from citric acid/ethylene glycol in air

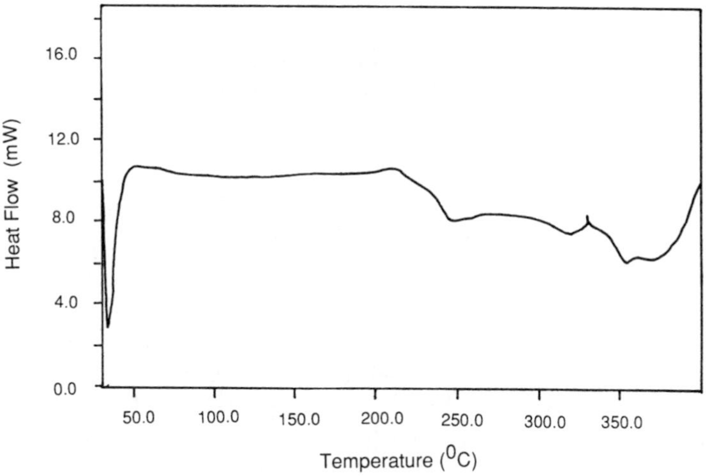

Figure 2. DSC curve for the citric acid/ethylene glycol 1-2-3 precursor system.

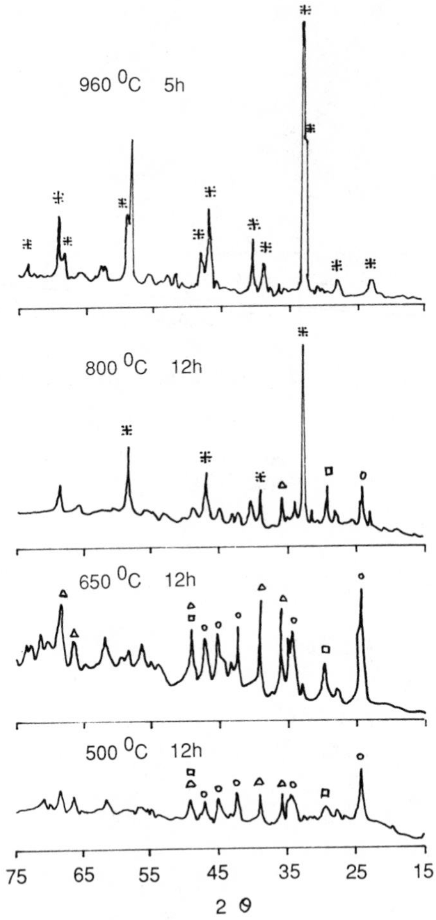

Figure 3. XRD patterns of the citric acid/ethylene glycol precursor system after heating in O_2 at different temperature. (□ Y_2O_3, ○ $BaCO_3$, △ CuO, ✳ $YBa_2Cu_3O_{7-x}$)

5h indicated a T_{onset} of 90K whereas electrical conductivity measurements of a pressed pellet of this powder after annealing in O_2 at 950 °C for 5h indicated a T_{onset} of 97K (Figure 4). These results are consistent with those of previous studies of 1-2-3 superconductor samples derived from solid-state reaction (17).

Figure 5 compares the XRD patterns of an intimate mixture of CuO, Y_2O_3 and $BaCO_3$ prepared by grinding the component compounds together in a mortar and the powder obtained from the polymeric precursor after heating both samples in oxygen at 850 °C for 15h. The superconducting phase was found to be only a minor product of the inorganic mixture whereas the polymer precursor formed pure $YBa_2Cu_3O_{7-x}$ under the same conditions. This observation is consistent with previously reported work(6) which showed that 900 °C was needed for the complete formation of the superconductor phase when this same inorganic mixture was employed. Therefore it appears that the citric acid/ethylene glycol precursor forms the 1-2-3 superconducting phase at a temperature at least 50 °C lower than that needed for the inorganic mixture precursor, presumbly due to the fine grained, intimately mixed microstructure of the latter material.

Thin films of the 1-2-3 superconductor were prepared on zirconia and alumina substrates by adding *ca.* 10% solutions of the citric acid/ethylene glycol precursor to the surface of the substrate followed by drying and pyrolysis in oxygen to 950 °C. XRD patterns indicate that the 1-2-3 orthorhombic phase is the only crystalline phase formed under these conditions. Detailed microstructural and physical property studies of these films are in progress.

Further insight into the pyrolysis chemistry involved in the conversion of this precursor to the 1-2-3 superconductor was obtained from infrared measurements. FTIR spectra were obtained on this precursor, the polyester derived from citric acid and ethylene glycol alone, and on the separate CuO and BaO precursor systems after heating in air for an extended period at various temperatures (Figure 6). The spectrum of the polyester (Fig. 6a) shows a strong, broad peak near 1200 cm^{-1} and a weaker peak at 1050 cm^{-1} which can be assigned to the asymmetric stretching vibrations of the ester C=O(OR) group (18). In the range of the C=O stretching vibration, only one peak was observed at 1740 cm^{-1}, suggesting that all of the citric acid COOH groups had reacted with the ethylene glycol to form esters.

In the spectrum of the 1-2-3 precursor (Fig. 6b), a new peak appears at 1610 cm^{-1} in addition to the ester C=O and C-O peaks at 1740 cm^{-1} and 1200 cm^{-1}. This peak is consistent with the presence of COO^- groups coordinated in a monodentate fashion to a metal ion (19) In addition, two sharp peaks are observed at 1075 and 880 cm^{-1}, which appear in the spectrum of ethylene glycol and are attributed to the presence of excess ethylene glycol in the precursor. In the spectrum of the precursor after heating to 250 °C (Fig. 6c) these peaks are lost, as is expected from the pyrolysis g.c. results. At this point the ester peaks are largely unperturbed but the peak at 1610 cm^{-1} in the original precursor has shifted to 1580 cm^{-1} and changed in appearance suggesting that a change in the coordination of some of the metal ions has occurred.

Upon heating to 350 °C (Fig. 6d), the ester C=O and C-O peaks have decreased substantially in relative intensity and a set of new

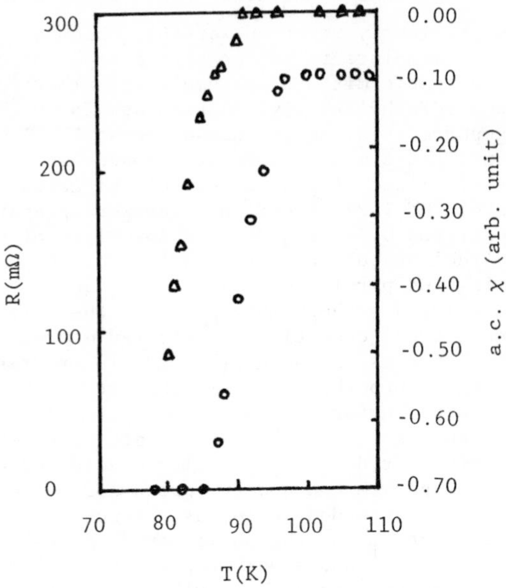

Figure 4. Results of conductivity and magnetic susceptibility measurements on the precursor-derived 1-2-3 superconductor. (o R, Δ χ)

bands at 1560 and 1405 cm^{-1} have grown in. These bands are typical of a COO^- group acting as a bidentate ligand toward a metal ion, and are assigned to the $\nu_{as}(COO^-)$ and $\nu_s(COO^-)$ stretches, respectively (18,20). By 450 ^0C (Fig. 6e) the ester peaks have completely disappeared and the large split peak at *ca.* 1460 cm^{-1} is suggestive of bidentate COO^- groups along with possibly the coordinated CO_3^{2-} ion. The latter is suggested in particular by the comparison with the spectrum of $BaCO_3$ (Fig. 6f) which has a single strong peak in this region plus a sharp peak at around 950 cm^{-1} similar to that in spectrum 6e. In contrast to this precursor spectrum, both the $BaCO_3$ and the BaO precursor (Fig. 6g) spectra after heating to 400 ^0C show little or no sign of an OH stretch, suggesting that this feature in spectrum 6e is associated with one or both of the other two metal ions in the precursor. These results, along with the XRD spectra, are consistent with the formation of $BaCO_3$ beyond about 450 ^0C.

In order to monitor the coordination of the Cu^{2+} ions during the pyrolysis of the precursor, the IR spectra of the CuO precursor was also determined as a function of temperature. Comparison of spectrum 6h with that of the original 1-2-3 precursor (Fig. 6b) reveals many common features with the exception of the strong bands at 1200 and 950 cm^{-1} which arise from the SO_4^{2-} ion used as the counterion in the CuO precursor preparation and the C=O band at 1610 cm^{-1} which is split in the CuO precursor spectrum. This may indicate more than one (COO^-) coordination environment involving the Cu^{2+} ions or H-bonding to some of the COO^- groups (21).

The splitting of the 1610 cm^{-1} band is apparently lost on heating the CuO precursor to 250 ^0C (Fig. 6i) and a shoulder has appeared on the C=O peak at 1775 cm^{-1}, as might arise from a free COOH group. After heating to 450 ^0C (Fig. 6j) the ester C=O peaks have disappeared but the band attributed to the monodentate coordinated COO^- group at 1650 cm^{-1} is still apparent, as was observed for the 1-2-3 precursor after heating to this temperature; however, there are also major differences in the C=O stretching region of these spectra which can be resolved by the addition of the CuO and BaO precursor spectra (Fig. 6j and Fig. 6g) suggesting that the mode of coordination to the COO^- groups is substantially different for the Cu^{2+} and Ba^{2+} ions in these samples. In particular these results suggest that the COO^- groups coordinate in a bidentate fashion to the Ba^{2+} ion and end up as CO_3^{-2} groups after heating in air to 450 ^0C. On the other hand the Cu^{2+} ions appear to remain associated in a monodentate fashion with the COO^- groups which, along with the presence of the OH^- groups, may contribute to the ready formation of CuO on heating to >450 ^0C.

In the case of the Y_2O_3 CVD precursor, g.c./FTIR studies carried out on the gaseous products of pyrolysis in nitrogen or vacuum indicate that isobutylene is the major hydrocarbon produced with propylene as the second highest component. Lesser amounts of various C_1, C_2, C_3, and C_4 saturated and unsaturated hydrocarbons were also observed suggesting radical cleavage of the tertiary butyl groups. The mass spectrum of this compound also shows a major peak at m/e = 57 which corresponds to M + 1 for butylene.

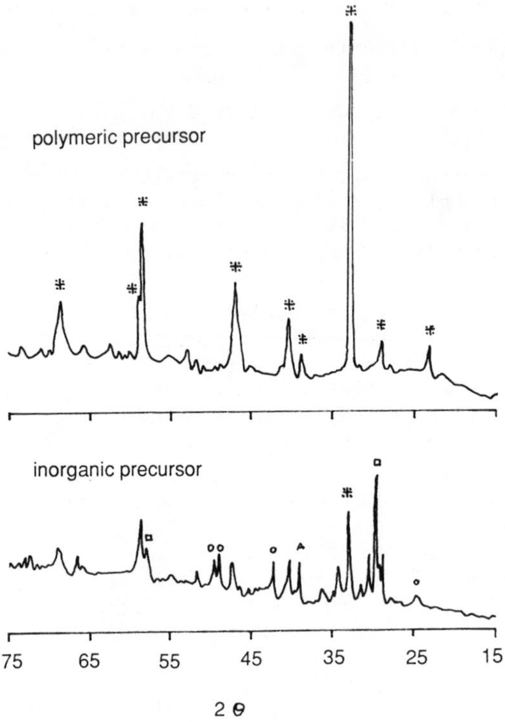

Figure 5. XRD patterns of both the inorganic mixture and polymeric precursor after heating in O_2 at 850 ^0C for 15h. (□ Y_2O_3, ○ $BaCO_3$, Δ CuO, ⁜ $YBa_2Cu_3O_{7-x}$)

Preliminary efforts to use this compound as precursor for the CVD of Y_2O_3 have centered on the production of dense, polycrystalline films on SiC fibers and Si wafers and have led to uniform, high quality films both with and without added oxygen. Auger spectra obtained for such films deposited on single crystal silicon show the presence of Y and O with varying amounts of carbon depending on the carrier gas used and the deposition rate. It was found that even in the absence of added oxygen the carbon incorporation in these films could be reduced to on the order of 1% or less by slowing down the deposition rate to *ca.* 0.1 μm/h.

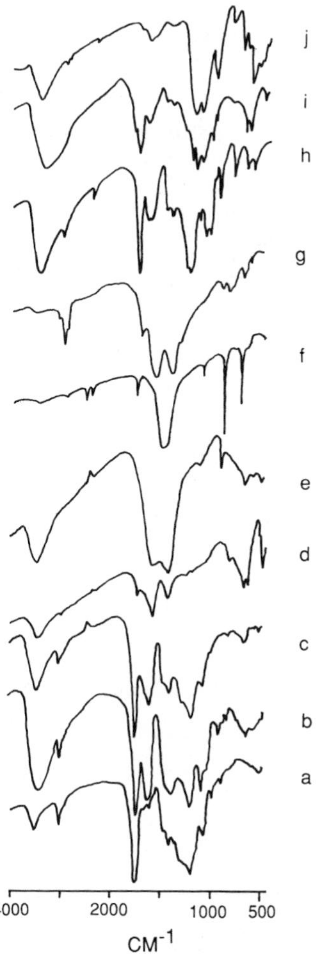

Figure 6. FTIR spectra of precursors and solid pyrolysis products after heating in air at various temperatures.
(a) polyester from citric acid and ethylene glycol.
(b) 1-2-3 precursor. (c) 1-2-3 precursor, 250 ^{0}C, 2h.
(d) 1-2-3 precursor, 350 ^{0}C, 1h. (e) 1-2-3 precursor, 450 ^{0}C, 1h.
(f) $BaCO_3$ (g) BaO precursor, 400 ^{0}C, 1h. (h) CuO precursor.
(i) precursor, 250 ^{0}C, 2h. (j) CuO precursor, 450 ^{0}C, 1h.

Acknowledgments

The authors are grateful to Prof. R. MacCrone for the measurement of the magnetic susceptibility and Lydie Valade for assistance with the conductivity measurements. We thank the Office of Naval Research for financial support of this research.

Literature Cited

1. Wu, M. K.; Ashburn, J. R.; Torng, C. J.; Hor, P. H.; Meng, R. L.; Meng, R. L.; Gao, L.; Huang, Z. J.; Wang Y. Q.; Chu, C.W. Phys. Rev. Lett. 1987, 58, 908.
2. Takagi, H.; Uchida, S.; Kishio, K.; Kitazawa, K.; Fueki, K.; Tanaka, S. Jpn. J. Appl. Phys. 1987, 26, L320.
3. Cava, R. J.; van Dover, R. B.; Batlogg, B.; Riteman, E. A. Phys. Rev. Lett. 1987, 58, 408.
4. Capone, D.W.; Hinks, D. G.; Jorgensen, J. D.; Zhang, K. Appl. Phys. Lett. 1987, 50, 543.
5 Jorgensen, J. D.; Schuttler, H. B.; Hinks, D.J.; Capone, D. W.; Zhang, K.; Brodsky, M. B.; Scalapino, D. J. Phys. Rev. Lett. 1987, 58, 1024.
6. Chu, C.T.; Dunn, B. J. Am. Ceram. Soc. 1987, 70, C357. Chaudari, P. et al., Science 1987, 238, 342.
8. Laibowitz, R. B.; Koch, R. H.; Chaudhari, P.; Gambino, R. J. Phys. Rev. B. 1987, 35, 882.
9. Hong, M.; Liou, S.H; Kwo, J.; Davidson, B. A. Appl. Phys. Lett. 1987, 51, 694.
10. Enomoto, Y.; Murawaki, T.; Suzuki, M.; Moriwaki, M. J. Appl. Phys. 1987, 26, L1248.
11. Rice, C. E.; van Dover, R.B.; Fisanick, G. J. Final program and Abstracts, The Mat. Res. Soc. 1987 Fall Mtg. p. 50
12. Kumagai, T.; Yokota, H.; Kawaguchi, K.; Kondo, W.; Mizuta, S. Chem. Lett. 1987, 1645.
13. Pechini, M. P., U.S. Patent 3 330 697, 1967.
14. Eror, N. G.; Anderson, H. U. Mat. Res. Soc. Symp. Proc. 1986, 73, 571.
15. Chiang. Y. M.; Ikeda, J. A. S.; Furcone, S.; Rudman, D. A. Final Program and Abstracts, The Mat. Res. Soc. 1987 Fall Mtg. P. 26.
16. Sievers, R. E.; Eisentraut, K. J.; Springer, Jr. C. S. Lanthanide/Actinide Chemistry; Adv. in Chem.,71, p. 141.
17. Hor, P. H.; Meng, R. L.; Wang, Y. Q.; Gao, Z.; Huang, Z. J.; Bechtold, J.; Forster, K.; Chu, C.W. Phys. Rev. Lett. 1987, 58, 1891.
18. Silverstein, R. M.; Bassler, G. C.; Morrill, T.C. Spectrometric Identification of Organic Compounds; 4th. Edition, John Wiley & Sons Inc., 1981, p. 121.
19. Kirschner, S.; Kiesling, R. J. Am. Chem. Soc. 1960, 82, 4174.
20. Livage, J. Mat. Res. Soc. Symp. Proc. 1986, 73, 717.
21. Nakamoto, K. Infrared Spectra of Inorganic and Coordination Compounds; John Wiley & Sons Inc., 1963, p. 210.

RECEIVED July 6, 1988

Chapter 15

High-Temperature Processing of Oxide Superconductors and Superconducting Oxide–Silver Oxide Composite

M. K. Wu[1,4], B. H. Loo[1], P. N. Peters[2], and C. Y. Huang[3]

[1]University of Alabama–Huntsville, Huntsville, AL 35899
[2]Marshall Space Flight Center, Space Science Laboratory, Huntsville, AL 35812
[3]Lockheed Research Center, Palo Alto, CA 94088

High temperature processing was found to partially convert the green 211 phase oxide to 123 phase. High T_c superconductivity was observed in Bi-Sr-Cu-O and Y-Sr-Cu-O systems prepared using the same heat treatment process. High temperature processing presents an alternative synthetic route in the search for new high Tc superconductors. An unusual magnetic suspension with enhancement in critical current density was observed in the 123 and AgO composite.

Immediately after the discovery (1) of the first 90 K oxide superconductor, consisting of the green insulating Y_2BaCuO_5 (211) and the black superconducting $YBa_2Cu_3O_7$ (123), we found that Bi-Sr-Cu-O (2) when prepared under extreme conditions, exhibited an anomaly indicative of superconductivity with an onset temperature at about 60 K. However, an equilibrium phase was found to have a Tc of only 15 K. While making the first 90K Y-Ba-Cu-O material we observed that samples reacted at a temperature higher than 950°C, but for a short time, exhibited sharper T_c transitions (3). We also learned that extremely careful heat treatment is required to prepare a single phase bulk polycrystalline 123 compound. These results suggested that the presence of the 211 phase may be thermodynamically favorable to the formation of the 123 phase. The results also suggested that high temperature processing may stabilize some phases that are otherise unstable at lower processing temperatures. In fact, using high temperature processing, we have successfully prepared the high temperature superconductors $BiSrCuO_{4-y}$ (4) and $YSrCuO_{4-y}$ (5). In addition, an unusual magnetic suspension and enhanced critical current density was observed in a 123 and AgO composite (6). In this paper we report the detailed results of these studies.

----

[4]After September 1988: Department of Materials Science, Henry Krumb School of Mines, Columbia University, New York, NY 10027

0097–6156/88/0377–0181$06.00/0

EXPERIMENTAL

The compounds used in high temperature processing were prepared in the following manner. Appropriate amounts of metal oxides were mixed, pressed into pellets, heated at 950°C for 12 hours, and then quenched to room temperature (RT). Annealing procedures depended on the particular study being conducted and are described in detail in appropriate sections. Electrical resistivity measurements were made with the conventional 4-probe technique. The DC magnetic moment measurements were made with a SQUID at the National Magnetic Laboratory at MIT. A standard 4-probe using pulse current was used to determine the critical current density at zero field. Structural and phase determinations were made by x-ray diffraction and Raman microprobe analysis.

RESULTS AND DISCUSSIONS

1. Study of Green 211 Phase

Sintered 211 phase was used as the starting material for this work. Firing the sintered disc at 1300°C for 15 minutes converted some of the green phase into black phase (4). Raman microprobe analysis indicated this black phase is that of $YBa_2Cu_3O_7$. The sintered sample had low resistivity at room temperature and behaved like a semiconductor, as shown in Figure 1(a). The sample became superconducting, as evidenced in Figure 1(b), after it was oxygen-annealed at 950°C for several hours and furnace-cooled. Other rare earth 123 phases have been produced from their corresponding 211 phases using the same processing methods.

The thermodynamics of the system at temperatures higher than 950°C have not been thoroughly investigated. Our results suggest that the 123 phase is more stable than 211 at higher temperature. In view of these observations, a detailed study of the equilibrium phase diagram at these temperatures would provide better control for synthesizing the superconducting compound.

2. Superconductivity in Bi-Sr-Cu-O

Early in April 1987 (2), we observed a resistance anomaly at 60K in the Bi-Sr-Cu-O system prepared at 900°C with 30 minutes sintering time. However, a longer processing time at 850°C showed a complete superconducting transition at 15 K. We used a high temperature processing technique to reexamine the system (4). A material with nominal composition of BiSrCuO4-y was prepared from Bi2O3, SrO, and CuO, pressed into a pellet, and heated to 800°-850°C for 12 hours. The sample was quenched, reground, and annealed in oxygen for two hours at 1200°C and slow-cooled in the furnace. The samples melted during heating but needle-like crystals were found inside the melt. Figure 2 shows the temperature dependence of the magnetic moment, m, for a sample (18 mg) cooled in fields of 10 and 100 G. Clearly the moment deviates around 60 K, suggesting the superconducting onset at 60 K. This onset temperature is in agreement with the onset of the resistance measurement as shown in the inset of Figure 2. Figure 3 shows the field dependence when the sample was cooled in zero

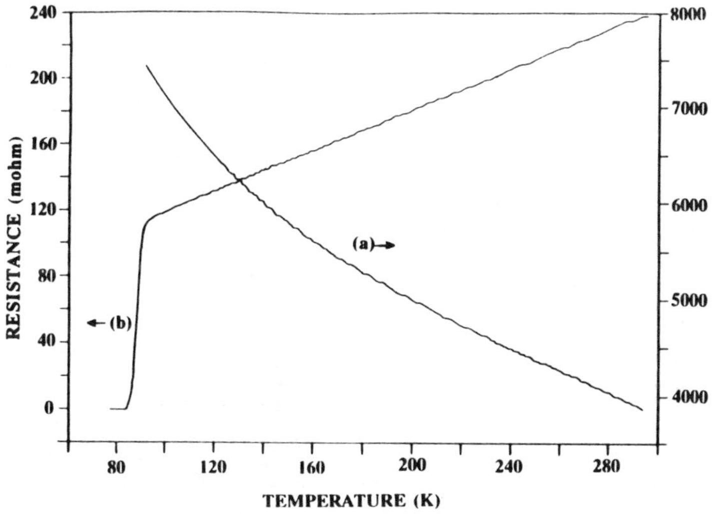

Figure 1: Resistance curve of a sintered Y_2BaCuO_5 disc after fired at 1300oC and (a) without O_2 anneal and (b) with O_2 anneal. (Reproduced with permission from Ref. 4. Copyright 1988 American Physical Society.)

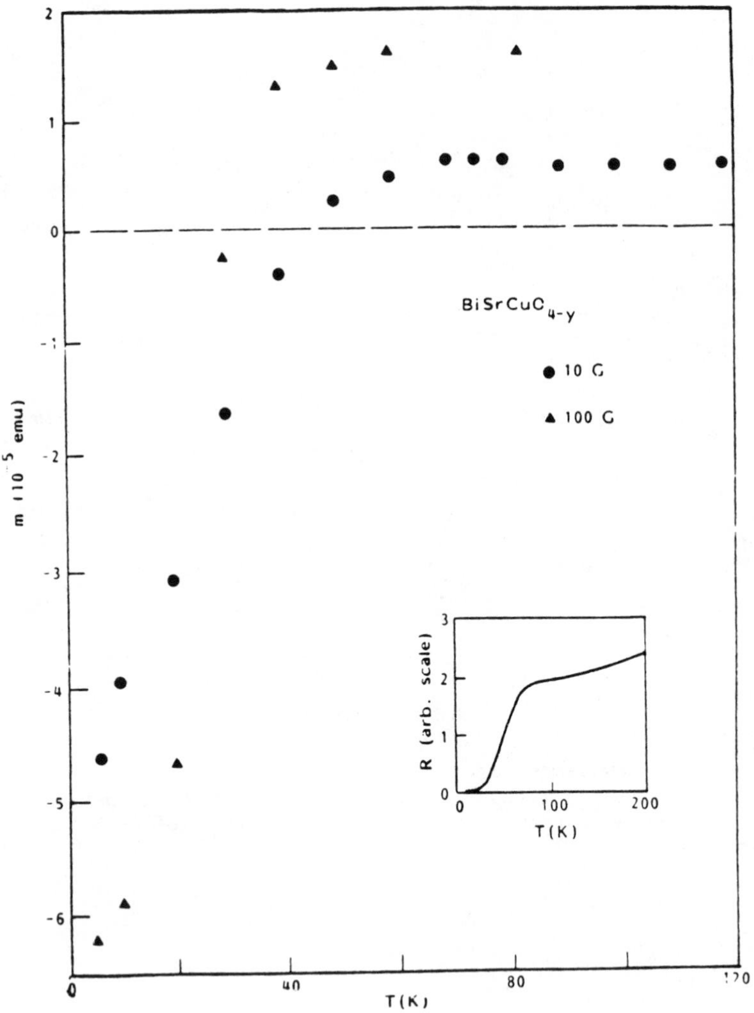

Figure 2: Magnetic moments of $BiSrCuO_{4-y}$ at 10 G and 100 G. The Inset shows the temperature dependence of resistance R.

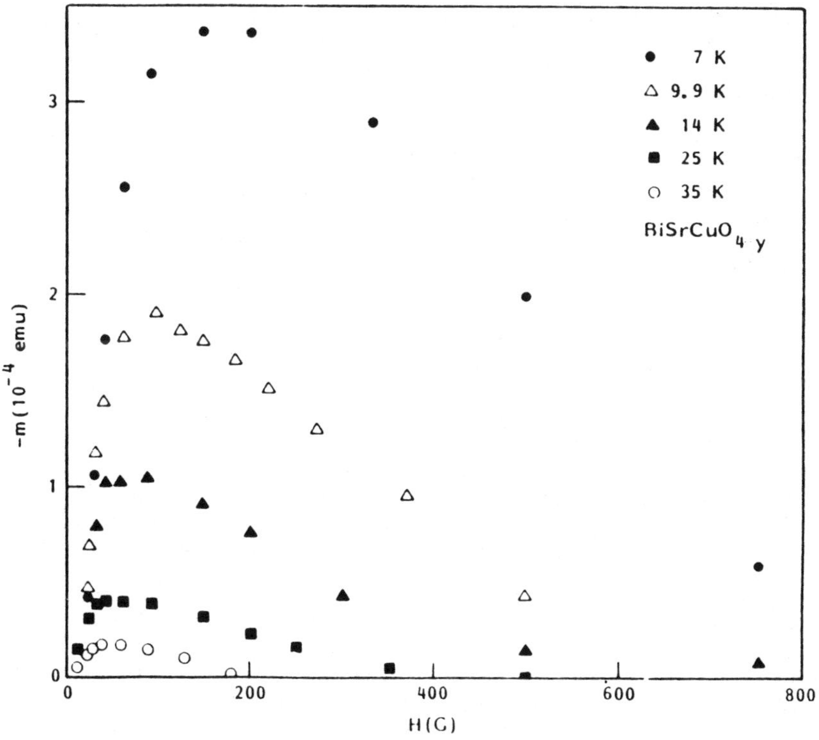

Figure 3: Field dependence of the magnetic moment for $BiSrCuO_{4-y}$ at various temperatures.

magnetic field. At low fields, m is linear in H, but starts to deviate at higher fields, thus defining H_{c1}. The H_{c1} value at 7 K is 20 G. As shown in Fig. 2, m is positive above T_c, indicating the presence of magnetic moments which are presumably caused by the presence of Cu^{2+} from unknown phases. These moments might be the reason that m crosses zero at relatively low field values, as shown in Fig. 3. X-ray and Raman microprobe analysis (Fig. 4) of this compound differed from both those of the 214 and 123 phases, indicating the possible existence of a new phase. Structure determination of this system is currently underway.

Two important conclusions can be drawn from this study. First, ions other than rare earth metals can form new copper oxide compounds that still exhibit high temperature superconductivity. Second, high temperature processing may be favorable to the formation of certain high temperature copper oxide systems.

3. Superconductivity in Y-Sr-Cu-O

Complete replacement of the Ba^{+2} ions with the smaller alkaline earth metal ions has not produced high-T_c superconductors. Thus far only $Y_{0.3}Sr_{0.7}CuO_{3-y}$ with a T_c of about 40 K has been reported (7). Based on the successful results on Bi-Sr-Cu-O we decided to reexamine the Y-Sr-Cu-O (5) system using similar high temperature processing conditions.

A sample with a nominal composition $YSrCuO_{4-y}$ was prepared by mixing appropriate amounts of Y_2O_3, SrO, and CuO. The mixture was ground and pressed into pellets, heated to 1300oC for 2h, and then quenched to RT. The material was then reground, repressed, reheated to 1200°C for 6 h in O_2, and then slowly cooled to RT. High purity alumina crucibles were used in the sample preparation.

The temperature dependence of the resistance for an Y-Sr-Cu-O sample is displayed in Figure 5. The superconducting onset is about 80 K, which is consistent with that of the magnetic moment measurement (sample mass is 25 mg) cooled in 10 G shown in the inset. A linear temperature dependence of R before the onset of superconductivity was observed. The resistance curve of our sample does not indicate the presence of a second superconducting phase. However, the I-V curve clearly reveals the existence of two superconducting phases, as shown in Figure 6. In addition, magnetic moment data also indicate one superconducting transition at 80 K and another at 40 K. Figure 7 shows the field dependence of magnetic moment at various temperatures. These curves are similar to those for $BiSrCuO_{4-y}$.

The 80 K superconducting phase did not form when our samples were prepared at a lower temperature. This indicates that the 80-K phase is thermodynamically stable only at higher temperatures. This is consistent with the conversion of the green 211 phase to the corresponding 123 phase with good superconducting characteristics by processing at 1300°C.

In addition to the two superconducting transitions, the magnetic moment curve shows an anomaly at low temperatures. Figure 8 shows details of the low temperature data for the sample cooled in 20, 50, 100, 200, 1000 and 2500 G. The magnetic moment is diamagnetic for $T < T_c$, but there is a sharp reduction in the diamagnetic moment at 14 K, indicating a sharp drop in superconductivity. It is not likely

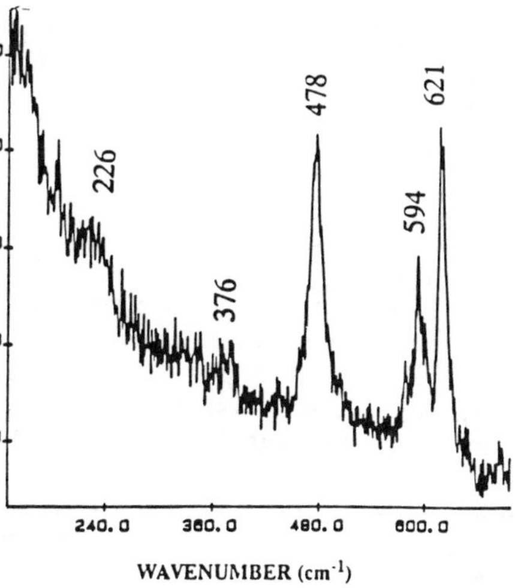

Figure 4: Raman spectra of $BiSrCuO_{4-y}$.

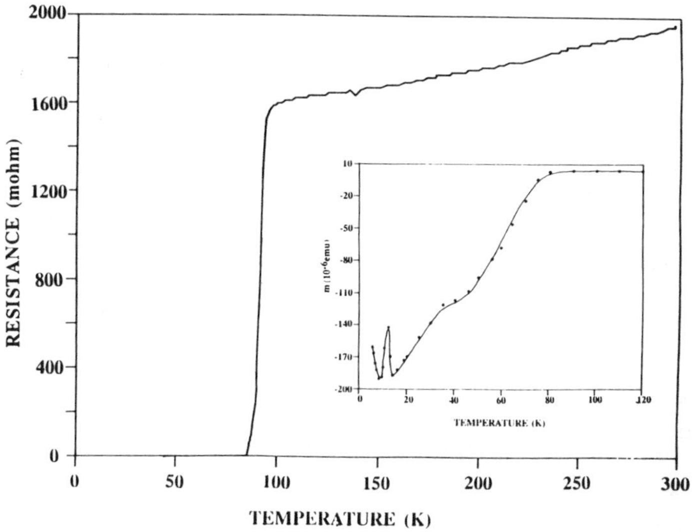

Figure 5: Resistance as a function of temperature of $YSrCuO_{4-y}$. Inset illustrates its magnetic moment at 10 G. (Reproduced with permission from Ref. 5. Copyright 1988 American Physical Society.)

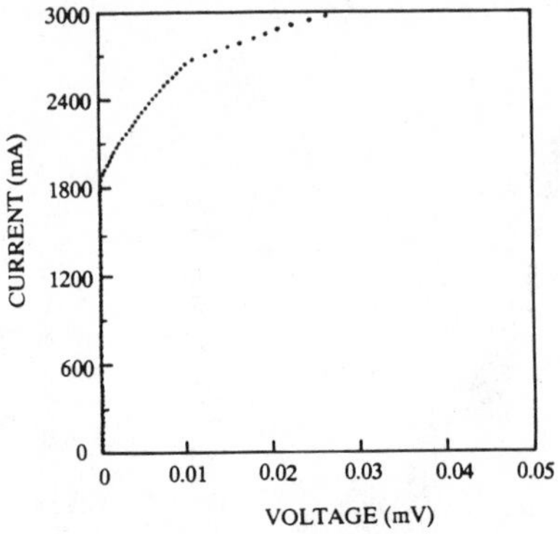

Figure 6: I-V characteristics of $YSrCuO_{4-y}$. (Reproduced with permission from Ref. 5. Copyright 1988 American Physical Society.)

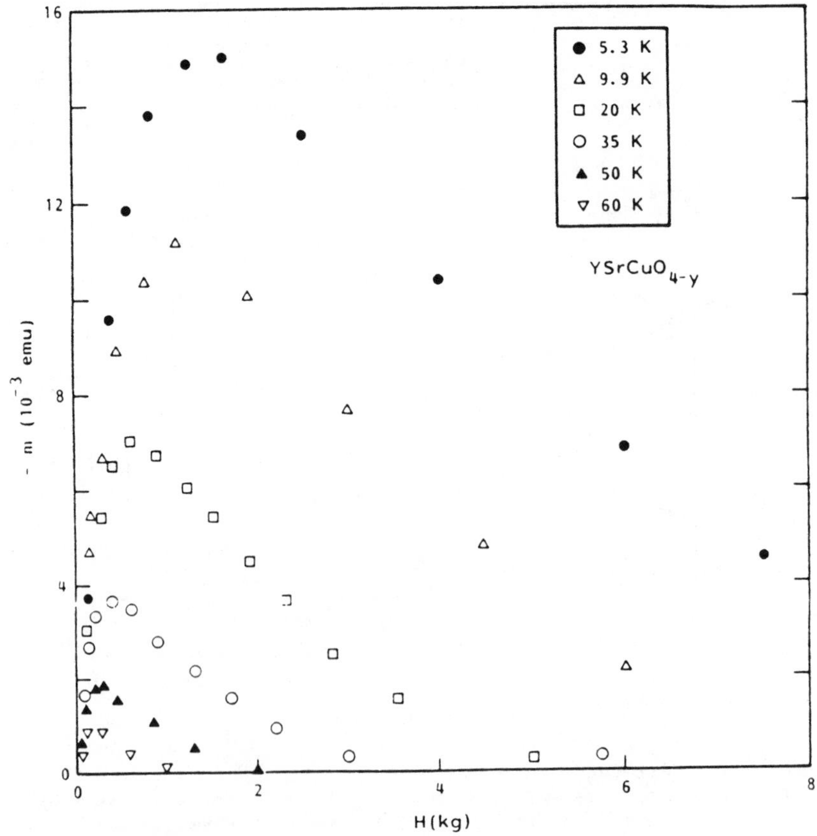

Figure 7: Field dependence of magnetic moment at various temperatures.

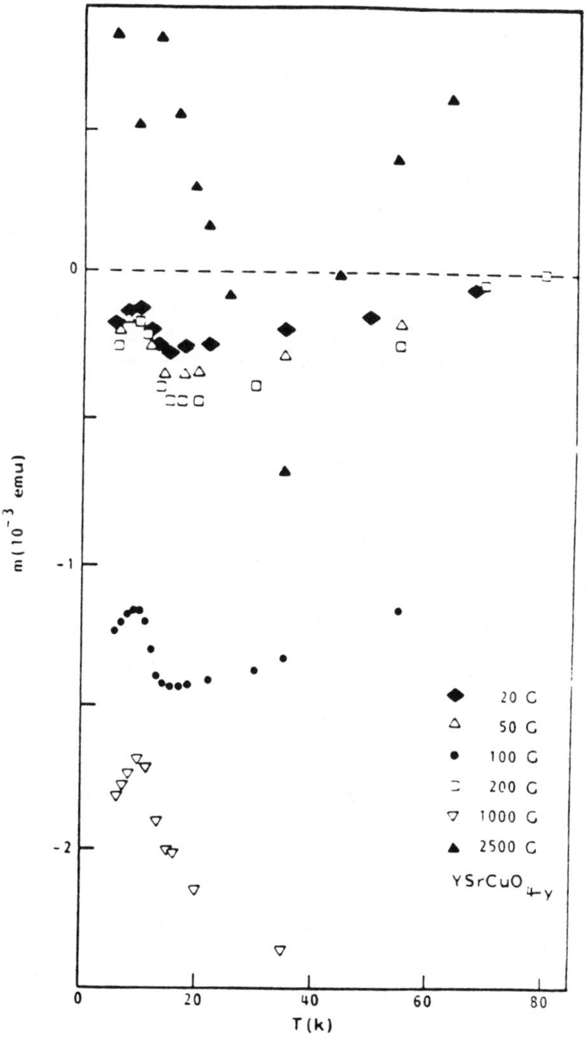

Figure 8: Magnetic moment of $YSrCuO_{4-y}$ at different fields.

that this is caused by the onset of magnetic ordering, because the moment above T_c is small. It is not presently clear whether there is a structural phase transition at this temperature or whether an unidentified phase present.

4. Y-Ba-Cu-O (123) and AgO Composite

AgO decomposes when it is heated above 200°C and does not react with high T_c superconductors. Therefore, a $Y_1Ba_2Cu_3O_7$ (123) composite can be formed by firing the mixture of 123 and AgO powders. Because of the high conductivity of the silver component the 123-AgO composite may exhibit better superconducting characteristics than the plain 123 material.

We have successfully synthesized the 123-AgO composites in several weight ratios. It should be noted that the room temperature conductivity increases as the silver content increases. Direct measurements of critical current density showed that J_c is approximately 250 to 350 A/cm2 at 77K, about two orders of magnitude higher than for our undoped samples at the same temperature. We have also observed a stable suspension (9) of the samples in the divergent field of magnets. While superconducting, they can be suspended above (the normally observed case) or below the magnet. When suspended below, attractive forces equal to the weight plus any repulsion is provided by flux pinning. A total weight of 2.07 times the sample weight was successfully lifted from a 77K surface with a magnet having a maximum field of 3.3 kG and maximum gradient of 2.6 kG/cm.

The M-H hysteresis loop for a sample (weight ratio 3:1) measured at 77K (8) is shown in Figure 9 to a maximum field H_{max} = 0.9 kOe. The measurement was taken after the sample was cooled in zero field and was independent of the sweep rate between 0.03 and 0.4 kOe/min. The critical current density estimated from the magnetization at 0.1 kOe is 3.3×10^3 A/cm^2, based on Bean's critical state model (9). This value of J_c is an order of magnitude larger than that measured from transport properties. The residual magnetization is 3 emu/g, and is not sensitve to the maximum field larger than 0.5 kOe. A slight decrease of the residual magnetization with time was observed up to 10 minutes. A crucial feature of the data as shown in Fig. 9 is the development of a positive magnetization when the sweep direction is reversed at H_{max}. In fact, M becomes positive at a high field which is only about 0.1 kOe below H_{max}. As discussed later, this explains the magnetic suspension.

A hysteresis loop was also obtained at 87 K by employing liquid Ar, as shown in Figure 10. The result was nearly independent of the sweep rate in the range between 0.22 and 1.08 kOe/min. Note that these hysteresis loops are for temperatures only a few degrees below T_c, indicating the presence of extremely strong pinning centers, presumably due to silver. The residual magnetization for the Sample at 87 K is smaller by a factor of about five than at 77 K, but is still reasonably large.

The change in sign of the magnetization when H is decreased only slightly, after reaching H_{max}, gives rise to the unusual magnetic suspension of a superconductor in the field gradient below a permanent magnet. As described in reference 6: when a strong permanent magnet (P) was moved toward one of these 123-AgO superconductors (S) at 77 K, S was first repelled by P. This is

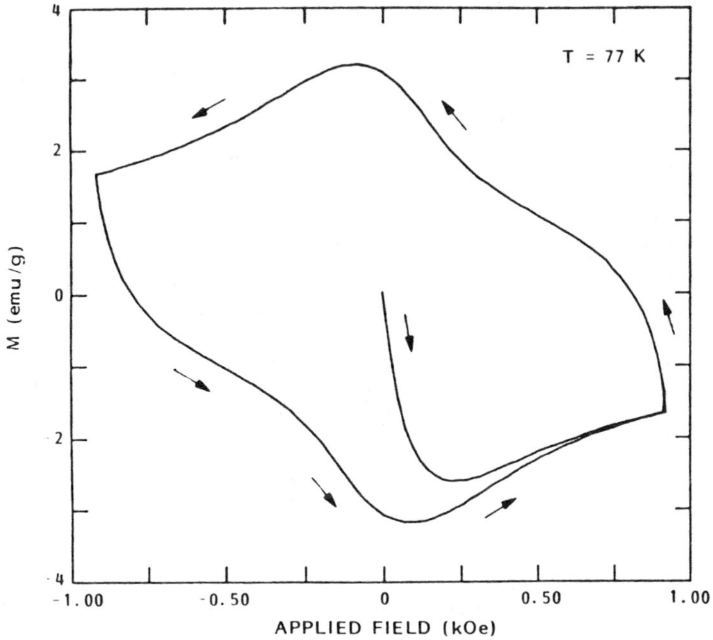

Figure 9: M-H hysteresis loop for 123 + AgO (3:1 weight ratio) at 77 K.

consistent with Fig. 10, which shows that M < 0 (diamagnetic) during the initial increase of H from 0 to H_{max}. However, if P was first moved close to S (field at S > 0.3 kOe), then pulled away from S, the field at S reduced such that M became positive, resulting in S being attracted to P. As a result of this attraction, S stayed stably suspended below P.

A simple model calculation (8) suggests that for the sample at 77 K with a magnet (H_{max} = 3.3 kG, $\overline{dH/dz}$ = 2.6 kG/cm), stable magnetic suspension is possible for heights in the range of $|z_o|$ < 9 mm, corresponding to H > 0.4 kOe. This is in rough agreement with the observation that the superconductor was suspended only when P was lowered to S, resulting in an initial applied field greater than 0.3 kOe. For $|z_o|$ > 13.5 mm (H < 0.18 kOe), suspension definitely cannot be achieved because the gradient force will require the magnetization to exceed the maximum residual magnetization, 3 emu/gram.

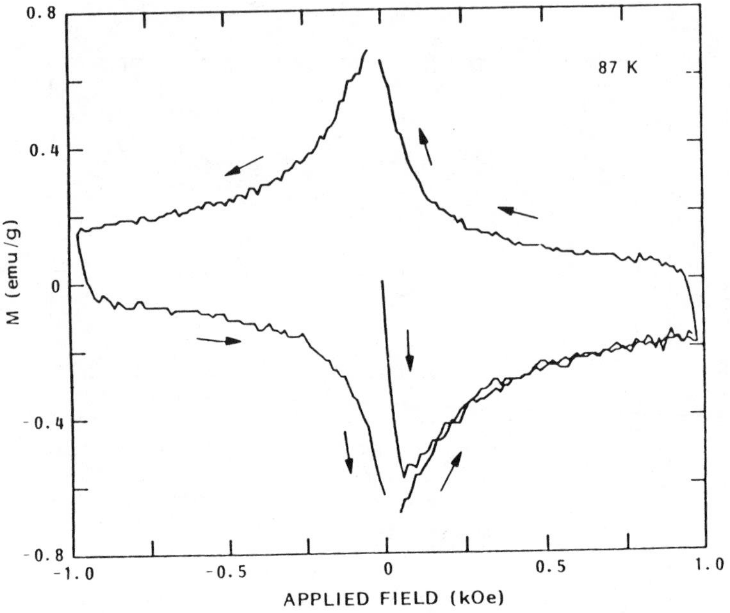

Figure 10: M-H hysteresis loops for 123 + AgO (3:1 weight ratio) at 87 K.

SUMMARY

The studies presented here indicate that proper processing procedures are critical to the formation of high temperature copper oxide superconductors. A new high Tc copper oxide compound with non-rare earth elements was prepared using high temperature processing. High temperature processing presents an alternative synthetic route in the search for new high Tc superconductors. A 123-AgO composite with improved electrical conductivity and strong flux pinning was also prepared.

ACKNOWLEDGMENTS

The work at the University of Alabama in Huntsville was supported by NASA grants NCC8-2, NAGW-812, and NAG8-089, and by NSF Alabama EPSCoR RII-8610669, and at Marshall Space Flight Center by the Center Director's Discretionary Fund. The authors would like to thank J.R. Ashburn, C.A. Higgins, I.E. Theodoulou, W.E. Carswell, D.H. Burns, A. Ibrahim, T. Rolin, and R.C. Sisk for technical assistance and Professor N.P. Ong and the Harvard Superconducting group for stimulating discussions.

REFERENCES

1. Wu, M.K.; Ashburn, J.R.; Torng, C.J.; Hor, P.H. Ashburn, C.J. Torng, P.H. Hor, R.L. Meng, L. Gao, Z.J. Huang, Y.Q. Wang, and C. W. Chu, Phys. Rev. Lett. 58, 908 (1987).
2. Reported at International Conference on Cryogenic Materials, Chicago, May 1987, and also at Japan-U.S. Symposium on High High Temperature Supercondctivity, Tokyo, October 1987.
3. Wu, M. K.; Ashburn, J.R.; and Torng, C. J., unpublished results.
4. Wu, M.K.; Ashburn, J.R.; Higgins, C.; Loo, B.H.; Burns, D.H.; Ibrahim, A. Rolin, T.; Peters, P.N.; Sisk, R.C.; and Huang, C.Y.; Appl. Phys. Lett., June 1, 1988.
5. Wu, M.K.; Ashburn, J.R.; Higgins, C.; Loo, B.H.; Burns, D.H.; Ibrahim, A.; Rolin, T.; Chien, F.Z.; and Huang, C.Y.; Phys. Rev. B, June 1, 1988.
6. Peters, P.N.; Sisk, R.C.; Urban, E.W.; Huang, C.Y.; and Wu, M.K., Appl. Phys. Lett. (June 20, 1988).
7. Mei, Y. Green; S.M. Jiang, C.; and Luo, H.L., in Novel Superconductivity, edited by Wolf, S.A. and Kresin, V.Z. (Plenum, New York, 1987), p. 1041.
8. Huang, C.Y.; McNiff, E.J.; Peters, P.N.; Schwartz, B.; Shapira, Y.; Wu, M.K.; Shull, R.D.; and Chiang, C.K.; to be published.
9. Bean, C.P. Phys. Rev. Lett., 1962, 8, 250.

RECEIVED July 7, 1988

Chapter 16

Photogeneration of Self-Localized Polarons in $YBa_2Cu_3O_{7-\delta}$ and La_2CuO_4

Y. H. Kim[1], C. M. Foster[1], A. J. Heeger[1], S. Cox[2], L. Acedo[2], and G. Stucky[2]

[1]Department of Physics and Institute for Polymers and Organic Solids, University of California, Santa Barbara, CA 93106

[2]Department of Chemistry, University of California, Santa Barbara, CA 93106

Photoinduced infrared absorption measurements of $YBa_2Cu_3O_{7-\delta}$ (δ=0.75) and La_2CuO_4 are reported. We have observed photoinduced infrared active vibrational modes and associated phonon bleachings which indicate the formation of a localized structural distortion in the Cu-O plane around the photogenerated carriers. In both materials, an associated electronic transition indicates that this structural distortion causes the formation of a self-localized electronic state in the energy gap. The photoinduced distortion and the associated self-localized gap state demonstrate that the photoexcitations are relatively long-lived polarons (or bipolarons). The dynamic mass associated with the distortion is smaller in $YBa_2Cu_3O_{7-\delta}$ than in La_2CuO_4 (smaller dynamic mass implies a longer range distortion). Since these features are not observed in the isostructural compound, La_2NiO_4, we suggest that polaron (or bipolaron) formation may play an important role in the high temperature superconductivity.

Research on the superconducting copper oxides has focused on identifying the mechanism(s) responsible for high T_c. Determining whether the mechanism involves pairing via the indirect attractive interaction mediated by phonons, pairing via electronic excitations, or whether a totally new mechanism is involved is of fundamental importance. The absence of an isotope effect[1] in $YBa_2Cu_3O_{7-\delta}$ and the positive (but relatively weak) isotope effect[2] in $La_{2-x}Sr_xCuO_4$ have been interpreted as evidence that the electron-lattice interaction is not the dominant pairing mechanism in these high T_c superconducting materials. Nevertheless, direct experimental evidence of the importance of the electron-lattice interaction has been demonstrated through photoinduced infrared absorption measurements[3].

In the limit of strong electron-lattice coupling, charge carriers (introduced by doping or by photoexcitation) will form polarons which attract one another via overlap of their structural distortions. It has been

0097–6156/88/0377–0194$06.00/0

proposed that this attractive interaction can lead to BCS-like superconductivity with an enhanced T_c (due to the relatively long range of the localized distortion)[4]; or to formation of polarons or bipolarons which Bose condense[5] or pair[6] at relatively high temperatures. Even if other pairing interactions dominate (e.g., resonating valence bond[7], electronic exciton[8], or magnetic "bag"[9] mechanisms etc.), the electron-lattice interaction could still play an important role.

Structural sensitivity to chemical doping in the cuprate superconductors is well-known. In both $La_{2-x}Sr_xCuO_4$ and $YBa_2Cu_3O_{7-\delta}$, a structural phase transition occurs as the number of carriers is increased either by doping (increasing x in $La_{2-x}Sr_xCuO_4$ causes a transition from orthorhombic to tetragonal) or by adding oxygen (decreasing δ in $YBa_2Cu_3O_{7-\delta}$ causes a transition from tetragonal to orthorhombic), respectively. In both cases, the correlation of the orthorhombic-tetragonal phase transition with T_c demonstrates the importance of structural distortions (caused by changes in x or δ) to the superconductivity.

Experimental studies of optical properties of ceramic samples of superconducting $La_{2-x}Sr_xCuO_4$ have demonstrated a dopant-induced optical transition at ~ 0.5 eV[10] which correlates with the doping concentration, x. In the case of $YBa_2Cu_3O_{7-\delta}$ a corresponding electronic feature at ~ 0.3 eV was observed[11]. In both cases, this electronic oscillator strength evolves toward lower energies and merges with the free carrier absorption for high quality (non-ceramic) samples[12]. Whether this doping-induced infrared oscillator strength is due entirely to the free carriers or whether there is an additional (overlapping) electronic transition at a finite frequency is an important issue which remains to be resolved. In the case of an additional electronic transition at finite frequency, one must inquire into the origin of the transition. Does the oscillator strength originate solely from the electronic interactions (excitons), or is it associated with a self-localized gap state formed by a structural distortion (due to electron-lattice interaction)? Our results[3] demonstrate that even when charge carriers are injected by photoexcitation, a localized distortion forms around the carrier, and that associated with this localized distortion there is a localized electronic state in the energy gap. These spectral features demonstrate the formation of self-localized polarons (or bipolarons), and they imply that the semiconducting phase is sufficiently close to the structural instability that even a single carrier locally changes the structure to that characteristic of the metallic (and superconducting) phase.

In the previous study[3a] of the photoinduced infrared absorption in La_2CuO_4, evidence was presented of polaron formation resulting from the electron-lattice interaction. Photoinduced infrared active vibrational (IRAV) modes and phonon bleachings were observed, together with a broad photoinduced electronic absorption which peaked at ~0.5 eV. The former indicated the formation of a localized structural distortion, while the latter indicated the formation of a localized electronic state in the energy gap. The association of this gap state with the localized structural distortion formed around the photoinjected charge carriers was implied by the common temperature and intensity dependences of all photoinduced spectral features. The electronic transition is the photoinduced analogue of the doping-induced absorption in the superconducting materials; a demonstration of the essential equivalence of charge injection by chemical doping and by photoexcitation (as is the case in conducting polymers) in two-dimensional Cu-O plane.

In an effort to probe more deeply into these phenomena, we have recently expanded this study to the $YBa_2Cu_3O_{7-\delta}$ system[3b]; the presence of photoinduced IR absorption in $YBa_2Cu_3O_{7-\delta}$ (δ=0.75) implies the existence of a tetragonal to orthorhombic structural distortion around the photogenerated charge carriers and an associated photoinduced electronic absorption peaked at ~0.13 eV. These data provide direct evidence of the importance of the electron-lattice interaction in the $YBa_2Cu_3O_{7-\delta}$ system. Moreover, the smaller dynamic mass (~11.4m_e) and the more weakly bound gap state (~0.13 eV) found in $YBa_2Cu_3O_{7-\delta}$ as compared with La_2CuO_4 (~23.8m_e and ~0.5eV, respectively) suggest a longer range distortion around the carriers in the higher T_c system. These relatively large photoinduced absorption signatures appear to be unique to the superconducting oxide systems; any photoinduced absorption signals in the isostructural compound, $LaNi_2O_4$, are essentially at the noise level and thus more than an order of magnitude weaker than in the copper-oxide systems[13].

Sample Preparation

The single phase La_2CuO_4 samples were prepared by the solid state reaction of dried La_2O_3 and CuO by following the method of Longo and Raccah[14]. La_2O_3 and CuO were ground together in an aluminum mortar with a pestle for 30 min. The mixture was slowly heated in an alumina boat to 1100 °C over a period of 4 hours, held for 15 hours at 1100 °C, and then slowly cooled down to 600 °C during 4 hours in an oxygen atmosphere. The sintered mixture was reground and compressed into a 13 mm dia. pellet. The pellet was heat treated in an oxygen environment in the same way done for the initial mixture. The sintered pellet was reground, pressed, and again heat treated under the same conditions. The regrinding-pressing-sintering cycle was repeated 4 times.

The tetragonal semiconducting $YBa_2Cu_3O_{7-\delta}$ (δ=0.75) samples were prepared by heating orthorhombic superconducting material (T_c= 92 K: a=3.8170 Å, b=3.8888 Å, c=11.6721 Å, V=173.26 Å^3) in air at 850 °C in an alumina boat for 3 hours followed by rapid quenching to room temperature. The resulting sample was not a superconductor; resistivity measurements gave the high dc value and characteristic temperature dependence of $\delta \approx 0.75$. X-ray powder patterns indicated a tetragonal unit cell: a=3.8590 Å, b=3.8590 Å, c=11.7949 Å, V=175.65 Å^3. The oxygen content was determined by comparing these values with the lattice constants given by Tarascon et al.[15].

The La_2NiO_4samples were obtained from P. Odier. Both semiconducting and metallic samples were prepared and fully characterized by techniques previously described in the literature[16].

For photoinduced absorption measurements, sintered pellets of the different materials were reground and mixed at a concentration of 2 wt. % with KBr powder (for the spectral range from 400-4000 cm^{-1}) or 1 wt. % with CsI powder (120 - 500 cm^{-1}). The mixture was pressed into thin dark grey semi-transparent pellets, which were then reground and repressed to achieve greater homogeneity. This process was repeated until satisfactory transparency and homogeneity was achieved.

Measurements

An IBM/98 (Bruker) Fourier-transform infrared spectrometer, modified to allow access onto the sample of the external beam from a Ar^+ laser, was used to cover the spectral range from 120 cm^{-1} (0.015 eV) to 8000 cm^{-1} (1 eV); the CsI cutoff set the low-frequency limit. For temperature control, the sample was placed on the cold finger of an Air Products Heli-tran system. The sample was optically pumped by the Ar^+ laser, and the fractional changes in infrared transmission were measured[17] (with resolution of 2 cm^{-1} in the far-infrared, 4 cm^{-1} in the mid-infrared, and 8 cm^{-1} in near-infrared). The net change in absorption coefficient ($\delta\alpha$) was determined from the photoinduced change in transmission (ΔT) without chemically changing the system; for small ΔT, $(-\Delta T/T) = \delta(\alpha d)$ where d is the sample thickness. To obtain an acceptable signal-to-noise ratio, long-time signal averaging was required (typically two hours).

Results and Discussion

Figure 1 shows the photoinduced absorption spectra for semiconducting $YBa_2Cu_3O_{7-\delta}$ (δ=0.75) obtained in the spectral range from 120 cm^{-1} to 4000 cm^{-1} at 15K (pumped at 2.7 eV with 30 mW/cm^2), with the inset showing the photoinduced absorption spectrum of La_2CuO_4 obtained at 15 K (pumped at 2.56 eV with 50 mW/cm^2). For $YBa_2Cu_3O_{7-\delta}$ (δ=0.75), we found a number of photoinduced IRAV modes, photoinduced phonon bleachings, and an associated broad electronic absorption peaked at 1050 cm^{-1} (~0.13 eV). A similar characteristic photoinduced absorption spectrum was obtained for La_2CuO_4, with the electronic absorption located at ~ 0.5 eV .

In Figure 2a, we show the photoinduced IRAV modes and phonon bleachings of $YBa_2Cu_3O_{7-\delta}$ (δ=0.75) in greater detail. We found three new major IRAV modes at 396, 436, and 520 cm^{-1} which have one-to-one correspondence with Raman active (symmetric) modes in the tetragonal structure and four major photoinduced bleachings at 192, 219, 362, and 598 cm^{-1} which are directly related to the characteristic infrared modes of the orthorhombic structure. Such specific changes in the infrared spectra imply a strucural rearrangement in the Cu-O plane around photogenerated charge carriers. We show the corresponding photoinduced phonon spectrum of La_2CuO_4 in Figure 2b; photoinduced absorption bands were found at 285, 398, 486, 560, and 640 cm^{-1} with associated bleaching at 214 and 706 cm^{-1}. These imply a localized structural change from the orthorhombic to the tetragonal phase around the photogenerated charge carriers[3a]. Therefore, the data lead to the following conclusion: charge injection through photoexcitation locally changes the structure just as charge injection through chemical doping changes the structure. Thus, the photogeneration of charge carriers is qualitatively equivalent to increasing the number of carriers by reducing δ or increasing x.

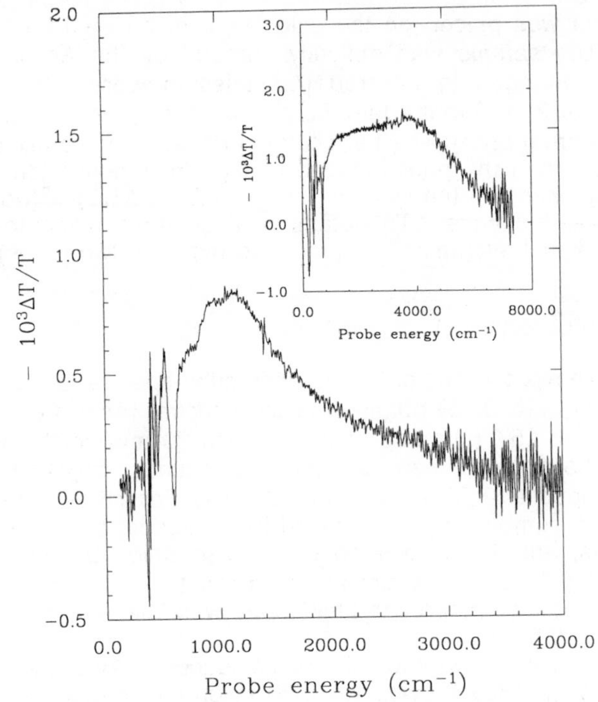

Figure 1. Photoinduced absorption spectrum of $Y_1Ba_2Cu_3O_{7-\delta}$ (δ=0.75) at 15K (2.7 eV pump at 30 mW/cm^2); inset shows photoinduced spectrum of La_2CuO_4 obtained at 15 K (2.56 eV at 50 mW/cm^2).

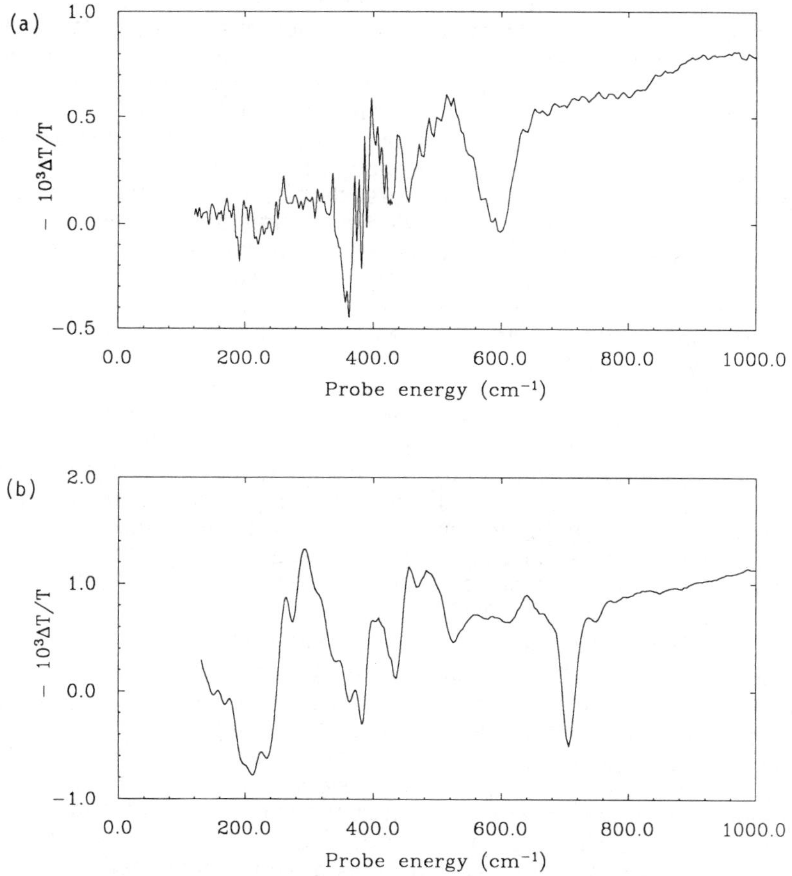

Figure 2. (a) Photoinduced IRAV features of $YBa_2Cu_3O_{7-\delta}$ (δ= 0.75); (b) Photoinduced IRAV modes and corresponding phonon bleachings of La_2CuO_4.

Experimental artifacts associated with the photoinduced absorption technique can be generated by sample heating caused by the incident laser pump beam. In systems such as oxide superconductors, small changes in the "dark" IR absorption caused by a strong temperature dependence can be incorrectly interpreted as photoinduced absorptions. Small shifts in frequency with changes in temperature would appear as derivative shaped signals in the photoinduced spectra; changes in linewidth would appear as second-derivative shaped signals. Similarly, a decrease in intensity accompanying an increase in temperature will appear as photoinduced bleaching. That the photoinduced features observed in Fig. 1 result from the photogenerated charge carriers (and do not arise from sample heating) can, therefore, be inferred from the lack of a derivative shape to the photoinduced bleachings and from the absence of second derivative features even though the tetragonal phase infrared modes exhibit a strong temperature dependence. Nevertheless, as confirmation, we carried out detailed measurements of the "dark" IR spectrum as a function of temperature for both materials and generated from these data numerical ($-\Delta T/T$) spectra that would arise from heating. The heating-induced spectra are different from that shown in Fig. 1, and they do show all the spectral features expected from the sample heating artifacts. We conclude that the spectrum of Fig. 1 is a genuine indication of the change in IR absorption due to photogenerated carriers.

It is known[18] that small metallic particle effects arising from random depolarization factors in highly anisotropic metallic grains in ceramic samples can lead to artifacts in the infrared reflection (and absorption) spectrum. Since the samples used in the photoinduced absorption experiments are insulating micron sized crystallites embedded in an insulating medium, such small particle effects will not affect either the infrared spectrum or the photoinduced IR absorption spectrum where the measured $\Delta T/T$ is of order 10^{-3}.

In Figure 3, we compare the temperature dependences of the IRAV mode at 520 cm^{-1}, the phonon bleaching at 598 cm^{-1}, and the broad electronic absorption peaked at 1050 cm^{-1} (~0.13 eV) in the photoinduced absorption spectrum of $YBa_2Cu_3O_{7-\delta}$. The common temperature dependence indicates that all originate from the same charged species (the decrease in signal strength with increasing temperature shown in Fig. 3 probably results from the temperature dependence of the carrier lifetime). This was confirmed through measurements of the intensity dependence; an $I^{0.5}$ dependence (possibly indicative of a bimolecular recombination mechanism) was observed for all of the spectral features. In the case of La_2CuO_4 (see Figure 4), the IRAV modes and the electronic absorption at ~ 0.5 eV again have the same $I^{0.5}$ dependence. We conclude that the IRAV modes, the mode bleachings, and the broad electronic transition arise from the same photogenerated charged excitation. This implies that the gap state is self-localized and originates from the carrier-induced structural distortion in the Cu-O plane as a result of electron-lattice coupling.

The origin of the photoinduced IRAV modes in conducting polymers and their relation to resonant Raman active modes has been clearly established[19]. A one-to-one correspondence between photoinduced or doping induced IRAV modes and Raman-active modes arises because

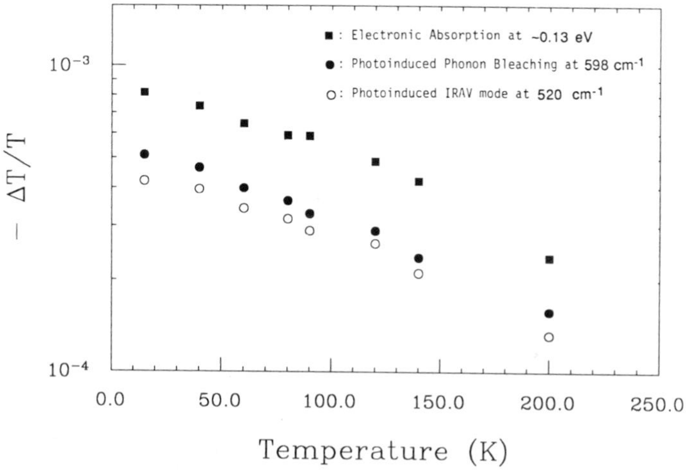

Figure 3. Temperature dependence of the photoinduced electronic absorption at ~ 0.13 eV, the photoinduced bleaching at 598 cm^{-1}, and the photoinduced IRAV mode at 520 cm^{-1}.

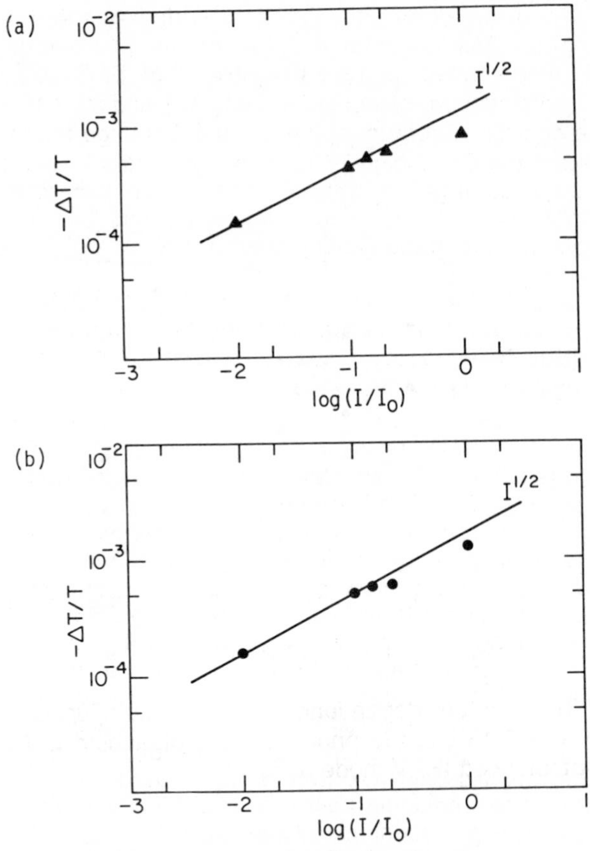

Figure 4. Pump intensity dependence of the photoinduced absorption: (a) the photoinduced electronic absorption centered at ~ 0.5 eV (the measurement of the intensity dependence was done at 3000 cm^{-1}); (b) the photoinduced IRAV mode at 486 cm^{-1}.

photogenerated charges change the local symmetry and make those Raman-active modes which are strongly coupled to the π-electrons infrared active. Thus, by analogy, we expect to observe new photoinduced infrared active localized phonon modes associated with those symmetric Raman modes which are coupled to the injected carriers. This point has been confirmed in the photoexcitation studies of La_2CuO_4 and $YBa_2Cu_3O_{7-\delta}$ (δ=0.75). We note that since only those Raman modes which strongly couple to the injected electrons will be affected, these data provide important information on the relative electron-phonon coupling strengths. Assignments of some of the photoinduced IRAV features can be made based upon the IR studies[20] of $SmBa_2Cu_3O_{7-\delta}$. For example, the photoinduced bleaching at 598 cm^{-1} and the IRAV mode at 520 cm^{-1} are a specific fingerprint of the structural change from tetragonal to orthorhombic of the Cu-O plane. The bleachings at 190 and 219 cm^{-1} can be assigned to deformation of the Cu-O plane coupled to the Y-sites in the tetragonal phase. Thus, when charge carriers are added, they couple to specific modes of the Cu-O planar structure, which then locally distorts.

An estimate of the dynamic mass of the photoinduced structural distortions was obtained from the ratio of the integrated oscillator strength of all the IRAV modes (I_{IRAV}) to that of the subgap electronic transition (I_{el}), $m_d = (I_{el}/I_{IRAV})\, m_e$, where m_e is the electronic band mass. For $YBa_2Cu_3O_{7-\delta}$ (δ=0.75), we find $m_d \sim 11.4 m_e$, which is considerably smaller than the corresponding dynamic mass obtained for La_2CuO_4 of ~$23.8 m_e$. This result and the smaller energy of the electronic transition (0.13 eV as compared with 0.5 eV for La_2CuO_4) suggest that the distortion around a charge carrier is more widely spread in $YBa_2Cu_3O_{7-\delta}$. Based on our experience obtained from analogous studies of conductive polymers[21], in order to transfer so much oscillator strength into the IRAV modes, the distortion must extend over several lattice units.

The large photoinduced infrared absorption signatures observed for $YBa_2Cu_3O_{7-\delta}$ and La_2CuO_4 appear to be unique to the superconducting oxide systems. Our initial infrared photoinduced absorption studies of non-superconducting nickel oxide based materials showed quite different photoinduced activity in contrast to the case of La_2CuO_4. The metallic La_2NiO_4 (tetragonal phase) sample gave null results while the semiconducting La_2NiO_4 (orthorhombic phase) yielded extremely weak signals (essentially at the noise level) which are much smaller in magnitude than in La_2CuO_4. This implies that the dynamic mass of the self-localized defect (if any) in La_2NiO_4 is an order of magnitude larger than that of copper oxide systems[13].

Summary and Conclusion

In summary, we have demonstrated that the charge carriers in La_2CuO_4 and $YBa_2Cu_3O_{7-\delta}$ are polarons (or bipolarons) which have a relatively small dynamic mass, $m_d \sim 23.8\, m_e$ and $11.4 m_e$ respectively. Photoinduced IRAV modes and associated phonon bleachings were observed, indicative of a local distortion from tetragonal to orthorhombic in the vicinity of the charge carrier. The subgap electronic transition in $YBa_2Cu_3O_{7-\delta}$ (δ=0.75) implies a self-localized electronic state and occurs at a significantly lower energy (~0.13 eV) than that observed for La_2CuO_4 (~0.5 eV). The reduced dynamic mass, the smaller gap state binding energy, and the implied spatially extended distortion are important components to an understanding of the high transition temperature.

Based on the results obtained from the dilute concentration of photogenerated carriers, we speculate on what happens at a higher concentration of holes corresponding to the superconducting samples ($x>0.05$ or $\delta<0.2$). At these high carrier concentrations, there is no well-defined peak in the IR absorption (nor in the frequency dependent conductivity) for high quality samples[12]; the oscillator strength associated with the electronic transition shifts toward zero frequency and merges with the free carrier absorption (see the discussion in the Introduction regarding the possibility of a separate finite frequency contribution to the electronic oscillator strength). This implies that the hole polaron (or bipolaron) level broadens and shifts, and probably merges into the valence band. On the other hand, the IRAV modes remain, with enhanced intensity[22]; a situation remarkably similar to the metallic state in conducting polymers[21]. The enhanced IRAV modes suggest that in the superconducting phase, the holes carry overlapping distortions (like polarons or bipolarons) leading to an average distortion plus large local fluctuations about that average. Whether the implied attractive interaction is sufficient to overcome the direct Coulomb repulsion and thereby to yield high T_c superconductivity remains to be seen. However, since these photoinduced infrared absorption signatures are not observed in the isostructural compound, $LaNi_2O_4$, we conclude that the polaron (or bipolaron) formation may play an important role in the high temperature superconductivity.

Acknowledgments

We thank N. Basescu and Z.-X. Liu for the transport measurements on the samples and we would like to acknowledge Prof. D.B. Tanner and Dr. S. Etemad for helpful discussions regarding the optical data on $YBa_2Cu_3O_{7-\delta}$. We are grateful to Professor P. Odier for providing the La_2NiO_4 samples. This research was supported by ONR through N00014-83-K-0450 and N00014-87-K-0457.

References

1. a) B. Batlogg, R.J. Cava, A. Jayaraman, R.B. van Dover, G.A. Kourouklis, S. Shine, D.W. Murphy, L.W. Rupp, H.S. Chen, A. White, K.T. Short, A.M. Mujsce, and E.A. Rietman, Phys. Rev. Lett. **58**, 2333 (1987).
b) L. Bourne, M.F. Crommie, A. Zettle, Hans-Conrad zur Loye, S.W. Keller, K.L. Leary, A.M. Stacy, K.J. Chang, M.L. Cohen, and D.E. Morris, Phys. Rev. Lett. **58**, 2337 (1987).

2. a) B. Batlogg, G. Kourouklis, W. Weber, R.J. Cava, A. Jayaraman, A.E. White, K.T. Short, L.W. Rupp, and E.A. Rietman, Phys. Rev. Lett. **59**, 912 (1987).
b) T.A. Faltens, W.K. Ham, S.W. Keller, K.J. Leary, J.N. Michaels, A.M. Stacy, Hans-Conrad zur Loye, D.M. Morris, T.W. Barbee, L.C. Bourne, M.L. Cohen, S. Hoen, and A. Zettle, Phys. Rev. Lett. **59**, 915 (1987).
3. a) Y. H. Kim, A.J. Heeger, L. Acedo, G. Stucky, and F. Wudl, Phys. Rev. B**36**, 7252 (1987).
b) Y.H. Kim, C. Foster, A.J. Heeger, S. Cox, and G. Stucky, submitted to Phys. Rev. Lett.
4. E. Sigmund, R. Ruckh, and K.W.H. Stevens, Physica B (in press).
5. P. Prelovsek, T. M. Rice, and F. C. Zhang., J. Phys. C: Solid State Phys. **20**, L229 (1987).
6. M. J. Rice and Y. R. Wang, Phys. Rev. B**36**, 8794 (1987).
7. a) P. W. Anderson, Science **235**, 1196 (1987).
b) S. A. Kivelson, Phys. Rev. B **36**, Nov. (1987).
8. C. M. Varma, S. Schmitt-Rink, and E. Abrahams, Solid Stat. Comm. **62**, 681 (1987).
9. J.R. Schrieffer, X.-G. Wen, and S.-C. Zhang, Phys. Rev. Lett. **60**, 944 (1988).
10. a) S. Etemad, D.E. Aspnes, M.K. Kelly, R. Thompson, and G.W. Hull, (preprint).
b) S. Etemad, D.E. Aspnes, P. Barboux, G.W. Hull, S.L. Herr, K. Kamaras, C.D. Porter, and D.B. Tanner, to be published in Mat. Res. Soc. Proc. (1987).
11. K. Kamaras, C.D. Porter, M.G. Doss, S.L. Herr, D.B. Tanner, D.A. Bonn, J.E. Greedan, A.H. O'Reilly, C.V. Stager, and T. Timusk, Phys. Rev. Lett. **59**, 919 (1987).
12. a) I. Bozovic, D. Kirilov, A. Kapitulnik, K. Char, M.R. Hahn, M.R. Beasley, T.H. Geballe, Y.H. Kim, and A.J. Heeger, Phys. Rev. Lett. **59**, 2219 (1987).
b) Z. Schlesinger, R.T. Collins, D.L. Kaiser, and F. Holtzberg, Phys. Rev. Lett. **59**, 1958 (1987).
13. Y.H. Kim, C.M. Foster, A.J. Heeger, S. Cox, L. Acedo, G. Stucky, and P. Odier, (to be published).
14. J.M. Longo and P.M. Raccah, J. Solid State Chem. **6**, 526 (1973).
15. J. M. Tarascon, P. Barboux, B.G. Bagley, L.H. Greene, W.R. McKinnon, and G.W. Hull, in "Chemistry of High-Temperature Superconductors" ed. by D. L. Nelson, M. S. Whittingham, and T. F. George, Chapter 20, (American Chemical Society, Washington, DC, 1987).
16. a) P. Odier, M. Leblanc, and J. Choisnet, Mat. Res. Bull. **21**, 787 (1986).
b) M. Sayer and P. Odier, J. Solid State Chem. **67**, 26 (1987).
17. G. B. Blanchet, C.R. Fincher, T.C. Chung, and A.J. Heeger, Phys. Rev. Lett. **50**, 1938 (1983).

18. a. J. Orenstein and D.H. Rapkine, Phys. Rev. Lett. 60, 968 (1988).
b. K. Kamaras, C.D. Porter, M.G. Doss, S.L. Herr, D.B. Tanner, D.A. Bonn, J. E. Greedan, A. H. O'Reilly, C.V. Stager and T. Timusk, Phys. Rev. Lett. 60, 969 (1988).
c. G.A. Thomas, J. Orenstein, D.H. Rapkine, M. Capizzi, A.J. Millis, R.N. Bhatt, L.F. Schneemeyer and J.V. Waszczak, Phys. Rev. B (in press).
19. B. Horovitz, Solid Stat. Comm. **41**, 729 (1982).
20. a) G. Burns, F.H. Dacol, P. Freitas, Sol. State Comm. **64**, 71 (1987).
b) G. Burns, F.H. Dacol, P.P. Freitas, W. König, and T.S. Plaskett, Phys. Rev.B **37** (Apr. 1, 1988).
21. For a review see: A. J. Heeger *et al.* in "Solitons is Conducting Polymers", Rev. Mod. Phys. (in press).
22. D. B. Tanner and S. Etemad, private communication.

RECEIVED July 12, 1988

Chapter 17

Electron-Transfer Studies at $YBa_2Cu_3O_7$, $Bi_2Sr_{2.2}Ca_{0.8}Cu_2O_8$, and $Tl_2Ba_2Ca_2Cu_3O_{10}$

John T. McDevitt[1], Robin L. McCarley[1], E. F. Dalton[1], R. Gollmar[1], Royce W. Murray[1], James Collman[2], Gordon T. Yee[2], and William A. Little[3]

[1]Department of Chemistry, University of North Carolina, Chapel Hill, NC 27599–3290
[2]Department of Chemistry, Stanford University, Stanford, CA 94305
[3]Department of Physics, Stanford University, Stanford, CA 94305

We have developed the methodology for fabricating well behaved electrodes utilizing high temperature superconductors $YBa_2Cu_3O_7$ (T_c = 94K), $Bi_2Sr_{2.2}Ca_{0.8}Cu_2O_8$ (T_c = 90K) and $Tl_2Ba_2Ca_2Cu_3O_{10}$ (T_c = 112K). Electrochemical corrosion studies have been completed with these electrodes and relative reactivity with water of each phase has been assessed. Freshly surfaced electrodes display facile room temperature electron transfer between superconductor phases and both solution redox species and surface confined electroactive polymers. This work should lay the grounds for construction of novel molecular film/superconductor junctions which can be studied at a variety of temperatures.

The recent discoveries of superconductivity in materials lacking rare earth elements (1-5) have added new excitement to the already active field of high temperature superconductivity. Studies of heterogeneous electron transfers at interfaces between these superconductors and other conductors both above and below T_c are essential towards understanding these novel materials. The ability of a superconductor to pass charge via bound electron pairs, the proximity effect, anomalous magnetic effects (i.e., perfect diamagnetism) and the ability to tunnel electron pairs over large distances are properties of superconductors that make these materials of interest both from theoretical and practical points of view. Previous studies have found that the properties of high temperature superconductor junctions are highly variable (6-12). Anisotropic gaps, high reactivity of these phases to atmospheric components, lack of chemical purity, variable oxygen content and a remarkably short coherence length are complicating factors in the analysis of such junctions.

0097–6156/88/0377–0207$06.00/0

Electrochemical methods provide additional means by which the transfer of charge across an interface between two conductive chemical systems can be studied (13). The fundamental events which support electron transfers from one phase to the other typically occur in close proximity (<10Å) to the interface between the two phases. Electrochemical studies are similar to tunneling experiments in that both are highly sensitive to the nature of the interface. For this reason, electrochemical experiments should be useful in probing the chemical purity and energetics of the surfaces of these superconductors, as well as the likely anomalous electrochemical behavior of a superconducting electrode operating at or below its critical temperature. In spite of such possibilities, the cuprate ceramics have received little electrochemical study. Obstacles are the tendency of high T_c phases towards corrosion in the presence of water and acids (14), and their high porosity, which causes large capacitive background currents that can mask the faradaic electron transfer currents of interest.

In a recent report, we described (15) how encapsulating the $YBa_2Cu_3O_7$ pellets in an epoxy matrix effectively fills the outer pores and diminishes the problem with large background currents. These electrodes undergo reversible electron transfers with electroactive solutes in a variety of nonaqueous solvents, as observed by cyclic voltammetry. This reversible voltammetry was used as a phenomenological method for estimating the lifetime of such electrodes, in a variety of corrosive media, based on monitoring the voltammetric response as a function of exposure time to corrosive media.

In this paper, we elaborate on methods for further improving the electrochemical performance of $YBa_2Cu_3O_7$ electrodes, and extend our efforts to obtain the first electrochemical information on the two new high T_c phases $Bi_2Sr_{2.2}Ca_{0.8}Cu_2O_8$ and $Tl_2Ba_2Ca_2Cu_3O_{10}$. Details of the corrosion of these two phases, as well as for $YBa_2Cu_3O_7$, are described. Finally, descriptions of electrochemical polymerization and deposition of conductive molecular films onto the high T_c phases, and fabrication of conductive polymer/high T_c superconductor microstructures, are provided. All of these experiments are conducted at room temperature.

EXPERIMENTAL

Chemicals and Equipment. Acetonitrile (Burdick and Jackson) was dried over 4A molecular sieves and used without further purification. Pyrrole (Aldrich) was vacuum distilled, stored under N_2 and refrigerated prior to use. Preparations of LiTCNQ, $[Os(bpy)_2(vpy)_2](PF_6)_2$, Et_4NClO_4 and Bu_4NClO_4 were performed according to standard literature procedures (16-18). Ferrocene (Aldrich); bis(pentamethylcyclopentadienyl)iron(II), Cp^*_2Fe (Strem); LiTCNQ (Aldrich); Y_2O_3 (Aldrich, 99.99); $BaCO_3$ (Aldrich, 99.999); CuO (Aldrich, 99.999); Bi_2O_3 (Aldrich, 99.999); $CaCO_3$ (Aldrich, 99.995); $SrCO_3$ (Aldrich, 99.995) and Tl_2O_3 (Aldrich, 99.99) were used as received. Electrochemical experiments were performed in normal three electrode mode. Aside from the corrosion studies, the majority of the voltammetric experiments were conducted in an N_2 filled glove box (Vacuum Atmospheres) employing as reference electrodes, either $Ag/AgNO_3$ (.01 M), Ag wire (pseudo-reference) or SSCE (NaCl saturated calomel electrode). All results are reported vs. SSCE based on ferrocene as an internal standard.

Electrode Fabrication. Sintered pellets 13 mm in diameter and 1 mm thick of $YBa_2Cu_3O_7$ (19), $Bi_2Sr_{2.2}Ca_{0.8}Cu_2O_8$ (3) and $Tl_2Ca_2Ba_2Cu_3O_{10}$ (4,5) were prepared according to previously described methods. All samples were characterized by four point probe measurements (see Table I), X-ray powder diffraction, scanning electron microscopy (SEM), energy dispersive X-ray spectroscopy (EDX) and X-ray photoelectron spectroscopy (XPS). Whereas samples of $YBa_2Cu_3O_7$ were found to be high purity single phase materials, the Bi and Tl-based ceramics possessed multiple components. Epoxy encapsulated electrodes were prepared as described previously (15). Electrodes were resurfaced using either Buehler 300 or 600 grit sand paper or by cleaving the electrode surface with a sharp blade immediately prior to use.

Corrosion Studies. The lifetimes of electrodes soaked in water were estimated as detailed in an earlier publication (15), using electrodes each having ca. 0.5 cm^2 exposed surface area. To monitor the formation of the layer that passifies the electrode and decreases the voltammetric currents for oxidation of Cp^*_2Fe solutions, pellets of the three high T_c phases were bathed in distilled deionized water at room temperature for two days. The pellets were then removed from the bath and washed with additional water to remove any loosely adhering precipitate. EDX and XPS measurements before and after the water exposure were employed to monitor the formation of corrosion products.

Electropolymerization of $[Os(bpy)_2(vpy)_2](PF_6)_2$ and Pyrrole. In an inert atmosphere box, poly-$[Os(bpy)_2(vpy)_2](ClO_4)_2$ films were electropolymerized from a 1.3 mM solution of the monomer in 0.1 M Et_4NClO_4/acetonitrile, by scanning the electrode potential ten times between limits of -0.75 and +1.8V vs SSCE (20). In a similar fashion, films of poly(pyrrole) were grown from neat pyrrole with 0.25 M Bu_4NClO_4 electrolyte (21), by scanning the electrode potential at 100 mV/sec four times between 1.05 and -0.85 V vs. SSCE. Identical conditions were used for the electropolymerization onto Pt and the three ceramic electrode materials. Electrodes were rinsed with copious amounts of acetonitrile before studying the voltammetry of the films in monomer-free 0.1M Et_4NClO_4/acetonitrile.

For microscopy, a film of poly(pyrrole) was coated onto a sample of $YBa_2Cu_3O_7$ that was partially masked with a strip of insulating Apeizon wax which was then removed with methylene chloride. In another experiment, a miniature whisker of poly(pyrrole) was grown onto a ca. 40μm diameter $YBa_2Cu_3O_7$ electrode prepared as described previously (15). A 300 Å overlayer of gold was deposited onto the whisker structure before the SEM work.

RESULTS AND DISCUSSION

Voltammetry of Redox Solutes. We have studied the voltammetric response of the cuprate ceramic electrodes in electrolyte solutions

Table I

Corrosion, Electrochemical and Conductive Properties of Electrode Materials

item	Pt	$YBa_2Cu_3O_7$	$Bi_2Sr_{2.2}Ca_{0.8}Cu_2O_8$	$Tl_2Ba_2Ca_2Cu_3O_{10}$
2mM Cp^*_2Fe		"cleaved"		
E'_o vs. SSCE	-0.10	-0.09	-0.10	-0.10
ΔEp(volts)	0.095	0.130	0.105	0.130
		"sanded"		
E'_o vs. SSCE		-0.11	-0.09	-0.05
ΔEp(volts)		0.250	0.165	0.248
2mM LiTCNQ				
E'_o vs. SSCE	0.21	0.22	0.22	0.22
ΔEp(volts)	0.076	0.115	0.090	0.120

poly-[Os(bpy)$_2$(vpy)$_2$](ClO$_4$)$_2$				
E_o' vs. SSCE	0.79	0.84	0.81	0.84
ΔEp(volts)	0.030	0.090	0.070	0.090
Electrode lifetime in water (days)	--	4	7	14
Porosity	--	2%	4%	18%
σ-(ohm-cm)$^{-1}$	$1X10^5$	1400	625	10-100
T_c(onset)(^{o}K)	--	94	90	112
T_c(zero)(^{o}K)	--	91	68	88

containing no redox couple, seeking insights into the chemical purity and intrinsic reactivity of these materials. Aside from somewhat larger capacitive currents, epoxy encapsulated $YBa_2Cu_3O_{7}$ and $Bi_2Sr_{2.2}Ca_{0.8}Cu_2O_8$ electrodes provide background voltammetric responses in 0.1M Et_4NClO_4/acetonitrile as featureless as that obtained with a highly polished Pt electrode in the window of +1.4 to -1.2V vs. SSCE. Exceeding these potentials leads to a degradation of the electrode surface evidenced by an increase in the background current in the absence of electroactive species and larger peak splitting in the presence of redox couples during subsequent voltammetry. Resurfacing the electrode restores the original response. A highly irreversible feature with an anodic wave centered at 0.9V and a cathodic one at -0.6V vs. SSCE superimposed on a large capacitive current is observed for the $Tl_2Ba_2Ca_2Cu_3O_{10}$ electrode. An overnight soaking in acetonitrile dramatically improves this electrode's response to the point where it is comparable to that of the Y and Bi-based electrodes. This behavior is possibly due to the presence of thallic and/or thallous oxide surface films which are leached out of the electrode by acetonitrile. The X-ray powder pattern for $Tl_2Ba_2Ca_2Cu_3O_{10}$ is consistent with the presence of these contaminants. The bismuth phase also contains multiple impurities, but none appear to be electroactive in acetonitrile.

In order to study the intrinsic charge transfer properties of these electrodes, the electrochemical responses in acetonitrile solutions of well behaved redox couples were studied. Figure 1 displays cyclic voltammetry recorded at the three high T_c phases for 0.2mM solutions of [bis(pentamethylcyclopentadienyl)iron(II)], Cp^*_2Fe, and the lithium salt of the radical anion of tetracyanoquinodimethane, LiTCNQ. These voltammetric responses closely parallel those obtained at Pt: the voltammograms are stable and reproducible, depend on potential scan rate as expected for a diffusion controlled process ($i_p \propto v^{1/2}$), and exhibit cathodic and anodic peak currents of equal magnitude. The peak potential separations, ΔE_p, are modestly larger than the theoretical value, 59mV (see Table I), and are definitely larger than for Pt. This effect will be discussed subsequently in further detail. For now it is important to emphasize that chemically reversible charge transfers between the surfaces of all three high T_c phases and dissolved redox species do indeed occur.

Electrochemically Polymerized Films. Electrochemically polymerized molecular films have been used extensively to modify the properties of electrode surfaces (22,23). We have employed this methodology to deposit films of poly-$[Os(bpy)_2(vpy)_2](ClO_4)_2$ and poly(pyrrole) onto high T_c phases. Reductive polymerization of the vinyl containing complex $[Os(bpy)_2(vpy)_2](PF_6)_2$ onto $Bi_2Sr_{2.2}Ca_{0.8}Cu_2O_8$ and Pt is displayed in Fig. 2. The two bpy$^{0/-}$ reduction waves gradually and continually grow in size as greater amounts of electroactive film are deposited. The formal potentials for both bpy waves at $Bi_2Sr_{2.2}Ca_{0.8}Cu_2O_8$ are the same as at Pt, but there is a markedly larger ΔE_p for the former electrode. The behavior of the other two ceramic electrodes is very similar.

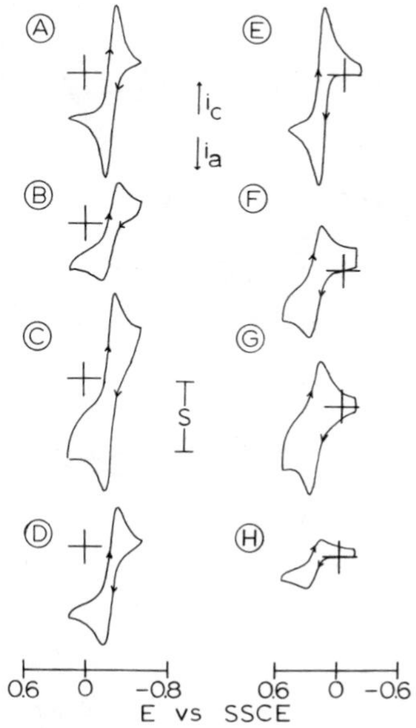

Figure 1. Cyclic voltammetry of (A-D) 2mM Cp^*_2Fe and (E-H) 2mM LiTCNQ in 0.1M Et_4NClO_4/acetonitrile recorded in a glove box at 50mV/sec scan rate. A,E at Pt; B,F at $YBa_2Cu_3O_7$; C,G at $Bi_2Sr_{2.2}Ca_{0.8}Cu_2O_8$; D,H at $T_2Ba_2Ca_2Cu_3O_{10}$; (A,H) S = 20μA, (B,C) S = 10μA, (D) S = 4μA, (E) S = 0.8μA, (F,G) S = 40μA.

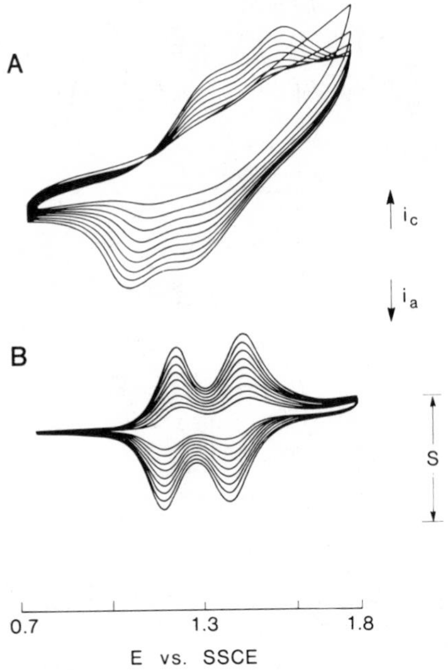

Figure 2. Reductive polymerization of $[Os(bpy)_2(vpy)_2]^{2+}$ at (A) $Bi_2Sr_{2.2}Ca_{0.8}Cu_2O_8$ and (B) Pt electrodes; (A) S = 200µA; (B) S = 20µA.

That deposition of the polymer onto the electrodes had indeed occurred was confirmed by washing the films with acetonitrile and transferring them to monomer free solutions of 0.1M Et_4NClO_4/acetonitrile. The voltammetric response for the Os(II/III) couple shown in Figure 3B-D confirms, by comparison to this well-known behavior (22) on Pt (Fig. 3A), the presence of surface localized films. Films exhibit the expected scan rate dependence for a surface localized species ($i_p \propto V$), and exhibit, as do most films of this type including that on polished Pt, full-width-half maximum peak widths larger than that for simple Nernstian response (i.e., 90.6 mV).

Poly(pyrrole) films were grown electrochemically from neat pyrrole containing 0.25M Bu_4NClO_4. Using the pure monomer as a source of the reactant enables growth of relatively thick poly(pyrrole) films in a matter of minutes (21). Voltammetry of such films in the absence of monomer in acetonitrile is shown in Figure 3F-H and that on Pt by Fig. 3E. Poly(pyrrole) is a neutral non-conductive polymer at potentials more negative than 0.1 V vs. SSCE and a highly conductive polymer at potential positive of this value. This, combined with a high surface area to volume ratio, results in film voltammetry which is complicated by a combination of faradaic and capacitive effects (24). The kinetics of ion motion into and out of the films also play an important role in determining the appearance of the voltammetry (25). In spite of these factors, the responses of the films on the ceramic electrodes are quite comparable to that obtained at Pt. As was the case with solution electrochemistry, facile charge transport between the high T_c ceramics and the electroactive films seems to occur.

Electron conduction in redox polymers such as poly-$[Os(bpy)_2(vpy)_2](ClO_4)_2$ occurs via site-to-site electron self exchanges and requires concurrent ionic motion for charge compensation (23). Conductivities of such polymers are relatively low and, therefore, the thickness of electrochemically grown films is limited. In contrast, poly(pyrrole) has high electronic conductivity and can function as its own electrode. Relatively thick films can be obtained with polymers of this type. This was confirmed by examination of scanning electron micrographs of films grown on pellets of the ceramics. Shown in Fig. 4A is a poly(pyrrole) coated $YBa_2Cu_3O_7$ electrode that was partially masked with an insulator film prior to the polymerization; the film was removed before the microscopy to reveal the original electrode surface. A ca. 40 micron thick, textured film of poly(pyrrole) (the dark region in Fig. 4A) evenly coats the surface of $YBa_2Cu_3O_7$. EDX (energy dispersive X-ray spectroscopy) measurements of the poly(pyrrole) film detect only the Cl signal from the incorporated anion (EDX is not sensitive to C, N or O), and only Y, Ba and Cu in the region which was initially masked. Examination of the latter region reveals that the ceramic pellet is highly dense and has an approximately 10μm diameter average grain size. The topology of the film does not follow that of the underlying electrode, but the film is composed of interconnected 1μm spherical globules of polypyrrole. We have obtained similar structures on polished Pt electrodes. Previous studies (25) have shown that the morphology of poly(pyrrole) films is more dependent on nucleophilicity of the

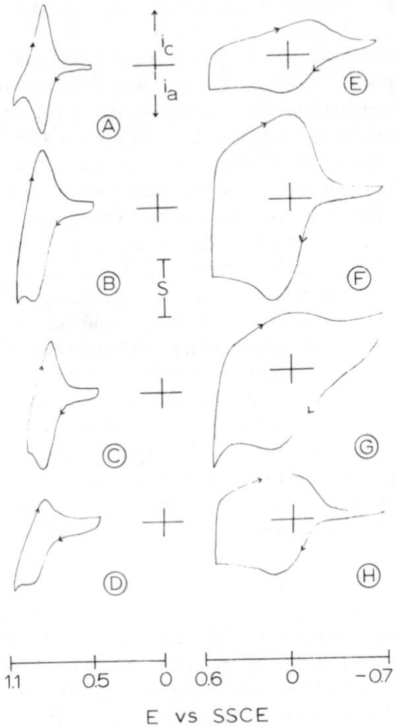

Figure 3. Cyclic voltammetry of surface confined films of (A-D) poly-$[Os(bpy)_2(vpy)_2](ClO_4)_2$ and (E-H) poly(pyrrole) recorded in a glove box in the absence of monomer for 0.1M Et_4NClO_4/acetonitrile at scan rate of 25mV/sec; A (S = 2.0μA), E (S = 100μA) at Pt; B (S = 10μA), F (S = 40μA) at $YBa_2Cu_3O_7$; C (S = 20μA), G (S = 40μA) at $Bi_2Sr_{2.2}Ca_{0.8}Cu_2O_8$; D (S = 20μA), H (S = 40μA) at $Tl_2Ba_2Ca_2Cu_3O_{10}$.

solvent, the electrolyte and solvent viscosity than on the texture of the underlying electrode.

Poly(pyrrole) films have been used recently in conjunction with lithographic techniques to generate chemical microstructures (26) such as molecular diodes and transistors. We have recently developed the methodology to fabricate microelectrodes of micron dimensions employing the high T_c phases and as part of our plans to devise chemical microstructures on superconductors have grown poly(pyrrole) structures on top of the microelectrodes. Figure 4B shows for example a whisker of poly(pyrrole) grown on an inverted 40μm diameter $YBa_2Cu_3O_7$ microelectrode. The whisker's radial dimension has expanded about 6 fold over that of the underlying electrode and protrudes 300μm downward from the surface of the electrode. Similar results are achieved using Pt microelectrodes (21). We have taken this notion one step further to produce $YBa_2Cu_3O_7$/poly(pyrrole)/$YBa_2Cu_3O_7$ molecular junctions having about 200μm separation between the slabs of superconductor. The fabrication and responses of such junctions will be discussed in future publications.

Corrosion Studies. Reaction of $La_{1.8}Sr_{0.2}CuO_4$ and $YBa_2Cu_3O_7$ phases with water is now well documented (14,27-29). Reactions of this type have been shown to result in resistance anomalies some of which have been mistakenly attributed to the occurrence of extraordinarily high temperature superconductivity (30). Surface reactivity with atmospheric moisture is also possibly responsible for the variable results obtained in point contact tunnelling experiments. A better understanding of the surface corrosion processes of the high T_c phases is obviously important. Previous studies (29) of cuprate ceramics have implied that low density (high effective surface area), high level of impurities, and high electrical conductivity may enhance the rate of corrosion. These factors must be considered in any corrosion study and in comparisons of corrosion rates.

We have recently developed a phenomenological test for corrosion of electrodes composed of high T_c phases (15). The method involves monitoring the voltammetric response of a redox couple in a non-corrosive solvent at a superconductor electrode, as a function of preceding time of exposure of the electrode to a corrosive medium. The corrosion occurs with the formation of both soluble and insoluble products at the electrode surface. The insoluble products passivate the electrode altering its charge transfer characteristics, as illustrated in Fig. 5 for the corrosion of the Tl based electrode in water. Initially the electrode responds well to the presence of Cp^*_2Fe, but over a period of time a capacitive and resistive appearance begins to dominate the voltammetry, and after 46 days the faradaic Cp^*_2Fe response is virtually undetectable. Similar studies have been completed for $YBa_2Cu_3O_7$ and $Bi_2Sr_{2.2}Ca_{0.8}Cu_2O_8$, and effective electrode lifetimes are summarized in Table I along with fractional porosity (roughly estimated by SEM) and conductivity of the bulk phase. Powder X-ray diffraction showed the level of purity to be $YBa_2Cu_3O_7 > Bi_2Sr_{2.2}Ca_{0.8}Cu_2O_8 > Tl_2Ba_2Ca_2Cu_3O_{10}$. With all of these factors in mind, we speculate that the stability towards reaction with water is $Tl_2Ba_2Ca_2Cu_3O_{10} > Bi_2Sr_{2.2}Ca_{0.8}Cu_2O_8 > YBa_2Cu_3O_7$.

Figure 4. (A) Scanning electron micrograph of a poly(pyrrole)-coated $YBa_2Cu_3O_7$ electrode. The top half shows the polymer film and the lower half shows the underlying electrode. S = 70 μm.

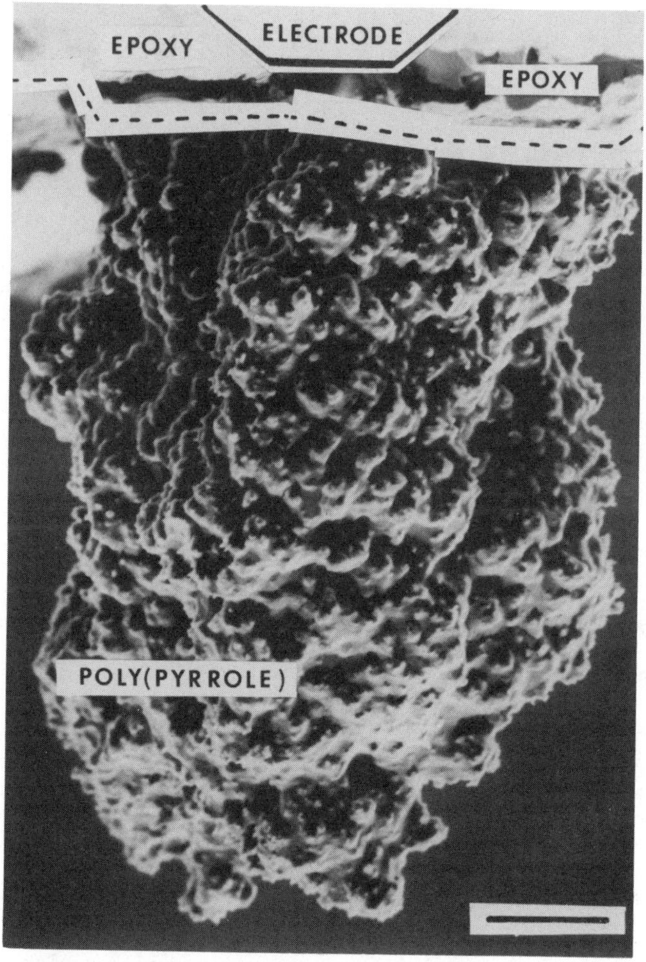

Figure 4. (B) Scanning electron micrograph of a poly(pyrrole) whisker grown underneath a $YBa_2Cu_3O_7$ microelectrode. Approximate dimensions and location of the underlying electrode are indicated in the upper central portion of the figure. The interface between the epoxy and the poly(pyrrole) is delineated by a broken line. S = 40 μm.

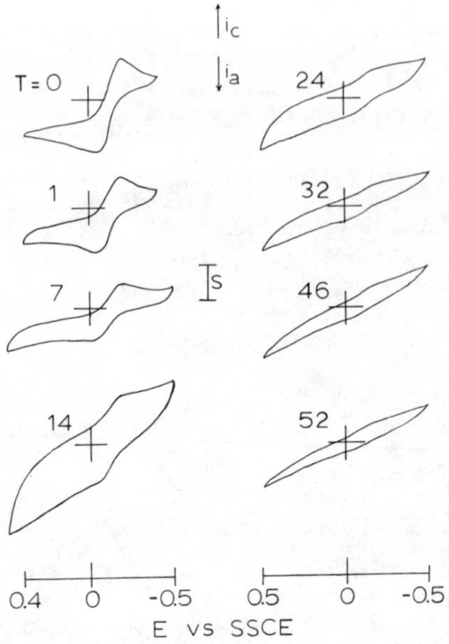

Figure 5. Voltammetry of a solution of 5mM Cp^*_2Fe in 0.1M Et_4NClO_4/acetonitrile recorded at $Tl_2Ba_2Ca_2Cu_3O_{10}$ as a function number of days exposure to water. V = 50mV/sec, S = 40μA for the left column and S = 100μA for the right column.

Close inspection of the Fig. 5 voltammetry reveals that there is no significant increase in peak splitting, ΔE_p, as surface passivation proceeds; instead the Cp^*_2Fe wave simply decreases in size. If corrosion in water were to proceed by formation of an insulating film that steadily increases in thickness, a correspondingly steady increase in ΔEp would certainly be expected. For this reason, we believe the corrosion process involves a surface passivation that initially encompasses only a small portion of the surface (probably at the grain boundaries), leaving behind regions that retain their capacity for facile electron transfers. Over a period of time the insulating layer spreads and eventually engulfs the entire electrode. SEM photomicrographs and EDX microscopy support this notion.

Unlike corrosion with water, the method of surface preparation dramatically alters ΔE_p. Table I provides representative data for the voltammetry of Cp^*_2Fe. More reversible electrochemistry (i.e., smaller ΔE_p values) is obtained with electrodes with freshly cleaved surfaces relative to those resurfaced by sanding. Further differences in ΔE_p were noted for electrodes sanded inside and outside a glove box (not included in Table I). Cleaving electrodes in an inert atmosphere provide the smallest ΔE_p. ΔE_p is a measure of the heterogeneous electron transfer rate constant; the most reversible electrode/redox couple systems produce values close to 59/n mV (13) and slow electron transfers increase ΔE_p.

Important questions remain regarding the chemical nature and morphological changes which occur during the resurfacing of these electrodes. It is tempting to speculate that corrosion effects similar to those observed in the aqueous corrosion studies, also occur during electrode resurfacing and are responsible for the variability in ΔE_p. However, the pattern of changes in ΔE_p observed during solution corrosion vs. the various re-surfacing experiments is different. The interfacial barriers to electron transfers that are introduced during resurfacing produce large variations in ΔE_p whereas solution corrosion affects surface access more than change in barrier (vide supra). The re-surfacing phenomena are not yet understood, and further work is needed to understand their details. This is a very important topic in terms of producing chemically well-defined junctions.

In summary, we have developed the methodology for fabricating well behaved electrodes using the high temperature superconductors. We have used such electrodes in the context of electrochemical corrosion studies and have found the following trend in terms of stability in water: $Tl_2Ba_2Ca_2Cu_2O_{10}$ > $Bi_2Sr_{2.2}Ca_{0.8}Cu_2O_8$ > $YBa_2Cu_3O_7$. Freshly surfaced electrodes exhibit chemically reversible electron transfers between the high Tc phases and solution redox species as well as surface confined electroactive films. This work should lay the grounds for fabrication of molecular film/high T_c ceramic junctions that further probe electron transfers over a range of temperatures.

Acknowledgments. This research was supported in part by grants from the National Science Foundation to R.W.M. and to J.P.C., from the Office of Naval Research to R.W.M., and by grant DE-FG03-H6ER45245 from the Department of Energy (J.P.C.).

Literature Cited

(1) Michel, C.; Hervieu, M.; Boiel, M. M.; Gradin, A.; Deslandes, F.; Provost, J.; Raveau, B. Z. Phys. 1987, B68, 421.
(2) Maeda, H.; Tanaka, Y.; Fukutomi, M.; Asano, T. Jpn. J. Appl. Phys. 1988, 27, L209.
(3) Chu, C. W.; Bechtold, J.; Gao, L.; Hor, P. H.; Haung, Z. J.; Meng, R. L.; Sun, Y. Y.; Wang, Y. Q.; Xue, Y. Y. Phys. Rev. Lett. 1988, 60, 941.
(4) Sheng, Z. Z.; Herman, A. M. Nature 1988, in press.
(5) Sheng, Z. Z.; Herman, A. M. Phys. Rev. Lett. 1988, in press.
(6) Wu, M. K.; Ashburn, J. R.; Torng, C. J., Hor, P. H.; Meng, R. L.; Gao, L.; Huang, Z. J., Wang, Y. O., Chu, C. W. Phys. Rev. Lett. 1987, 58, 908.
(7) Yamashita, T.; Kawakami, A.; Nishihara, T.; Takata, M.; Kishio, K. Jpn. J. Appl. Phys. 1987, 26, L671.
(8) Nishino, T.; Haseqawa, H.; Nakane, H.; Ito, Y.; Takagi, K.; Kawabe, U. Jpn. J. Appl. Phys. 1987, 26, L674.

(9) Tai, J. S.; Kubo, Y.; Tabuchi, T. Jpn. J. Appl. Phys. 1987, 26, L701.
(10) Koch, R. H.; Umback, C. P.; Clark, G. J.; Chaudhari, P.; Laibowitz, R. B. Appl. Phys. Lett. 1987, 51, 200.
(11) Iquchi, I.; Watanabe, H.; Kasai, Y.; Mochiku, T.; Sugishita, A.; Yamaka, E. Jpn. J. Appl. Phys. 1987, 26, L645.
(12) Koch, H.; Cantor, R.; March J. F.; Eickenbusch, H.; Schollhorn, R. Phys. Rev. B, 1987, 36, 722.
(13) Bard, A. J.; Faulkner, L. R. Electrochemical Methods; John Wiley and Sons: New York, 1980, p. 1.
(14) Rosamilia, J. M.; Miller, B.; Schneemeyer, L. F.; Waszczak, J. V.; O'Bryan, H. M. J. Electrochem. Soc. 1987, 134, 1863.
(15) McDevitt, J. T.; Longmire, M.; Gollmar, R.; Jernigan, J. C.; Dalton, E. F., McCarley, R.; Murray, R. W. J. Electroanal. Chem. 1988, 243, 465.
(16) Melby, L. R.; Harder, R. J.; Hertler, W. R.; Mahler, J., Benson, R.E.; Mochel, W. E. J. Am. Chem. Soc., 1962, 84, 3374.
(17) Calvert, J. M.; Schmehl, R. H.; Sullivan, B. P.; Facci, J. S.; Meyer, T. J.; Murray, R. W. Inorg. Chem. 1983, 22, 2151.
(18) Sawyer, D. T.; Roberts, J. L. Experimental Electrochemistry for Chemists, Wiley, New York, 1974.
(19) Chen, X. D.; Lee, S. Y.; Golben, J. P.; Lee, S. I.; McMichael, R. D.; Song, Y.; Noh, T. W.; Gaines, J. R. Rev. Sci. Instrum. 1987, 58, 1565.
(20) Leidner, C. R.; Murray, R. W. J. Am. Chem. Soc. 1984, 106, 1606.
(21) McCarley, R. L.; Morita, M.; Wilbourn, K. O.; Murray, R. W. J. Electroanal. Chem. 1988, 245, 321.
(22) Murray, R. W. Ann. Rev. Mater. Sci. 1984, 14, 145.
(23) Chidsey, C. E.; Murray, R. W. Science, 1986, 231, 25.
(24) Burgmayer, P.; Murray, R. W. J. Am. Chem. Soc. 1982, 104, 6139.
(25) Diaz, A. F.; Bargon, J. in Handbook of Conducting Polymers; Skotheim, T. A., Ed.; Marcel Dekker, New York, 1986; Vol. 1, p. 81.
(26) Kittlesen, G. P.; White, H. S.; Wrighton, M. S. J. Am. Chem. Soc. 1984, 106, 7389.
(27) Frase, K. G.; Liniger, E. G.; Clarke, D. R. Advanced Ceramic Materials 1987, 2, 698.
(28) Hyde, B. G.; Thompson, J. G.; Withers, R. L.; Fitzgerald, J. G.; Steward, A. M.; Bevan, D. J. M.; Anderson, T. S.; Bitmead, J.; Paterson, M. S. Nature, 1987, 327, 402.
(29) Dou, S. X.; Liu, H. K.; Bourdillon, A. J.; Tan, N. X.; Zhou, J. P.; Sorrell, C. C.; Easterling, K. E. Mod. Phys. Lett. B 1988, 1, 363.
(30) Kitazawa, K.; Kishio, K.; Haseqawa, T.; Nakamura, O.; Shimoyara, J.; Sugii, N.; Ohtomo, A.; Yaegashi, S.; Fueki, K. Jpn. J. Appl. Phys. 1987, 26, L1979.

RECEIVED July 6, 1988

Chapter 18

Microwave Absorption at Various Preparation Stages of $YBa_2Cu_3O_{7-\delta}$ Superconductor

Larry Kevan, John Bear, Micky Puri, Z. Pan, and C. L. Yao

Department of Chemistry and Texas Center for Superconductivity, University of Houston, Houston, TX 77204–5641

The microwave absorption of the $YBa_2Cu_3O_{7-\delta}$ high temperature superconductor has been studied by electron spin resonance (ESR). The responses are related to the development of the superconducting phase by examination of samples after each of three heating stages in the sample preparation process. An apparently axially symmetric g = 2 signal ($g_{\parallel}$ = 2.24, $g_{\perp}$ = 2.05) is seen which is assigned to dipole broadened Cu^{2+} in impurity phases Y_2BaCuO_5, $BaCuO_5$ and possibly others. In early stages of the sample preparation process the g = 2 signal is observable to 6 K while at later stages it disappears below the superconducting transition temperature, T_c. A low field non-resonant absorption is also seen for samples cooled in a kilogauss field which has a derivative maximum, $(d\chi''/dH)_{max}$. The low field absorption is characteristic of superconducting $YBa_2Cu_3O_{7-\delta}$ and it disappears at T_c. The field at which $(d\chi''/dH)_{max}$ occurs decreases with successive stages in the sample preparation process and suggests that the area of superconducting clusters increases during the preparation process. From the low field absorption it is estimated that nanogram quantities of a superconducting phase can be detected.

The microwave absorption of the $YBa_2Cu_3O_{7-\delta}$ high temperature superconductor as measured by electron spin resonance (ESR) has been recorded by several groups with rather diverse results (1-32). Two different types of magnetic responses have been reported. One is a g = 2 signal which is apparently due to dipolarly broadened cupric ion species. The second is a very low field absorption which is only partially understood. Our goal in this work is to investigate these two types of signals with respect to the development of the superconducting phase of the $YBa_2Cu_3O_{7-\delta}$ compound by examination of

0097–6156/88/0377–0223$06.00/0

samples after each stage in the sample preparation process. The development of the superconducting phase during the sample preparation procedure can shed light on the solid state compound formation mechanism.

Experimental

Materials used for sample preparation included yttrium oxide and barium carbonate from Alfa Chemicals and cupric oxide from Fluka Chemicals. These compounds were received as powders and the various stages of sample preparation are denoted as follows. Sample type 0, so designated, involves simple mixing of the three powders in the stoichmetric ratio needed for the title compound. The second step in the sample preparation process involves grinding the mixed powder with an agate mortar and pestle for typically 40 minutes until it becomes a uniform gray color to obtain sample type I. The next step involves the initial heating of the ground powder at 900 °C in air in a muffle furnace for 6 hours. This produced a compacted material of large particles which was ground to a powder and designated as sample type II. The next step was to take the reground material and subject it to a second heating at 900 °C in air for 12 hours in a muffle furnace. This product was ground to a powder and designated as sample type III. The final step of sample preparation involves pressing a pellet from the reground material in a press at a pressure of 5000 to 10,000 psi for 20 minutes and then reheating the pellet at 850 °C in flowing oxygen at atmospheric pressure for 6 hours in a tube furnace followed by cooling slowly at a rate of 60 °C per hour in flowing oxygen at atmospheric pressure. This product was ground to a powder and designated as sample type IV. Thus including the first sample mixing step we have five stages in the sample preparation with samples designated as types 0, I, II, III, and IV.

The samples were characterized by x-ray diffraction on a Philips Electronics Model 2500 powder diffractometer. Sample types 0 and I were about the same and characteristic of the starting materials. Sample type II shows diffraction lines characteristic of $YBa_2Cu_3O_{7-\delta}$ (123 compound) (30) and $YBaCuO_5$ (211 compound) (31) together with some unreacted CuO. Sample type III shows mainly lines of the 123 compound with a trace of 211 compound while in sample type IV the lines of the 211 compound are usually absent.

Pelletized samples of about 200 mg were also tested for bulk superconductivity at 77 K by examining whether a small pellet would leviate a small magnet in a dewar of liquid nitrogen. Sample types III and IV always strongly leviated the magnet and some preparations of type II leviated it slightly.

Microwave absorption measurements were made with a Bruker ER300 ESR spectrometer operated with 100 kHz magnetic field modulation where the derivative of the absorption part of the magnetic susceptibility with respect to magnetic field was detected ($d\chi''/dH$). Temperature was varied with an Oxford Instruments ESR 900 helium flow system from room temperature to about 6 K. The temperature at the sample position in the flow dewar was calibrated with a separate thermocouple and found to be 2-3 K higher than the instrumental readout from nominally 4 K to 100 K. Approximately 5 mg samples were evacuated and sealed into 2 mm i.d. by 3 mm o.d. Suprasil quartz tubes for the ESR measurements.

g = 2 Signal

Normal ESR conditions were used to observe the g = 2 signal in 5 mg samples with a typical microwave power of 2 mW and magnetic field modulation of 10 G. No signals in the g = 2 region were seen for sample 0 or sample I preparations. Figure 1 shows some results in sample types II, III and IV at high and low temperatures. In sample II preparations after the first 900 °C heating a signal is seen which appears to have an axially symmetric g tensor characterized by $g_{\parallel} = 2.28$ and $g_{\perp} = 2.02$. This signal is observed from room temperature to 6 K.

For type III samples after the second heating stage a similar apparently axially symmetric g = 2 signal is seen at room temperature which is somewhat stronger than the signals seen in sample type II. The signal in sample III is characterized by slightly different values of $g_{\parallel} = 2.24$ and $g_{\perp} = 2.05$. However the g = 2 signal for sample III disappeared at the superconducting transition temperature T_c of about 90 K.

Sample type IV prepared after the third heating in oxygen flow showed the same signal as in type III samples at $g_{\parallel} = 2.24$ and $g_{\perp} = 2.05$. The signal intensity in type IV samples is comparable to that in type III samples. The type IV sample signals were also observable at room temperature down to T_c and disappear below T_c.

The x-ray diffraction data shows that the phase purity of the 123 compound increases steadily as one progresses through the sample preparation steps from type II to type III to type IV samples. This is not consistent with the intensities of the g = 2 signal which indicates that this signal is not associated with the superconducting phase. It is interesting that when there is a relatively small amount of superconducting phase present in type II samples the g = 2 signal persists down to 6 K and shows no attenuation on decreasing the temperature through T_c. Thus it seems clear that the g = 2 signal is not due to pure phase 123 compound and presumably is not associated with a pure phase high temperature suerconductor. This conclusion conflicts with some earlier work but is consistent with more recent work (1,2).

It is interesting however that in both type III and IV samples that this g = 2 signal does serve as a diagnostic for T_c in that it disappears at the superconducting transition temperature. This is presumably due to the Meissner effect exclusion of magnetic flux in most of the bulk sample in a largely superconducting preparation such that the normal ESR signal is no longer observed. If only a small amount of bulk superconducting phase is present, as in type II samples, then the bulk exclusion of magnetic flux is too weak to prevent the observation of a g = 2 signal even below the superconducting transition temperature.

In order to possibly identify one or more of the nonsuperconducting phases which may give rise to this characteristic apparently axially symmetric g = 2 signal we prepared several other compounds involving cupric oxides which gave results as follows. These other compounds were prepared by a one stage heating of the ground, stoichiometric powder at 900 °C for 8 h in air in a muffle furnace followed by slow cooling and regrinding. The ESR results are shown in Figure 2.

Preparation of Y_2BaCuO_5 (211 compound) gave rise to a g = 2 signal with apparent axial symmetry analogous to that observed in the type III and IV samples described above. This signal is observable from room temperature to 6 K although it broadens somewhat below about 30 K as shown in Figure 2.

The $BaCuO_2$ compound also showed the apparent axially symmetric g = 2 signal described above at room temperature which persisted down to 80 K and then at lower temperature broadened into a symmetric line which was quite broad (about 570 G peak-to-peak linewidth) at 6 K with $g_{apparent} \sim 2.15$ (see Figure 2). In contrast, the $Y_2Cu_2O_5$ compound showed a broad symmetric line over the entire temperature range from room temperature where the peak-to-peak linewidth is about 554 G to 6 K where the linewidth is about 830 G. The signal in $Y_2Cu_2O_5$ differed from that in $BaCuO_2$ in that it was broad and symmetric over the entire temperature range and did not show the apparent axially symmetric signal at room temperature.

We conclude that the signal observed in all these samples is dipolarly broadened cupric ion. We also suggest that the 211 compound is the most likely nonsuperconducting phase that gives rise to this g = 2 signal in preparations nominally containing only the 123 superconducting compound. This contrasts with the conclusion of Vier et al. (1) who suggested $BaCuO_2$ as the principal impurity phase but they did not show the ESR spectral shapes at different temperatures. In previous studies (6) a broad relatively symmetric ESR signal has been reported in nominally superconducting materials at temperatures below the 90 K superconducting transition temperature. We suggest that this symmetric signal is most likely due to impurity phases of $BaCuO_2$ and $Y_2Cu_2O_5$ which show such a broad apparently symmetric signal at low temperature. However we caution that if such an impurity phase is only a very minor part of a relatively good preparation of the 123 compound the ESR signal should still disappear at T_c due to the Meissner effect exclusion of the magnetic field.

Low Field Absorption

A low field absorption in the range of 100 - 200 gauss was observed in our laboratory in some of the original samples prepared in C.W. Chu's laboratory in February, 1987. However the origin of such absorptions was not understood at that time. Undoubtedly, similar observations were made in other laboratories and the first published account appears to be that of Durney et al. (9). However, the first paper that offered some explanation of this phenomenon is that by Blazey et al. (11). This low field signal can be described as an ac diamagnetic susceptibility response induced by a microwave magnetic field. It has been suggested to be associated with magnetic flux penetration through weakly coupled superconducting regions. More recently it has been described as a loss in microwave conductivity due to interaction between superconducting carriers and

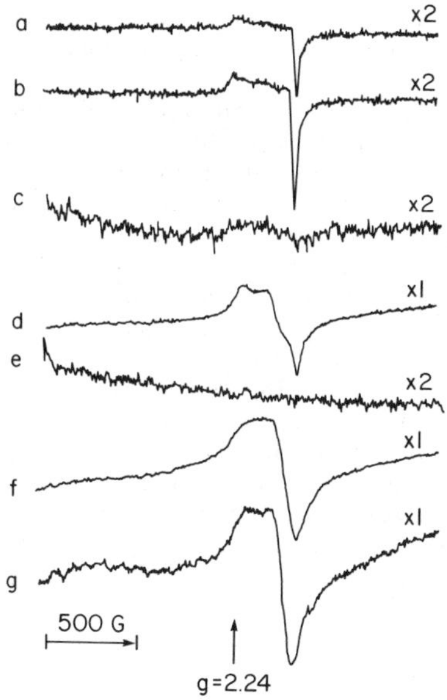

Figure 1. ESR spectra in the g = 2 region at 9.1 GHz for $YBa_2Cu_3O_{7-\delta}$ sample type II at (a) 6 K and (b) 93 K, III at (c) 98 K and (d) 293 K, and IV at (e) 88 K, (f) 98 K and (g) 293 K.

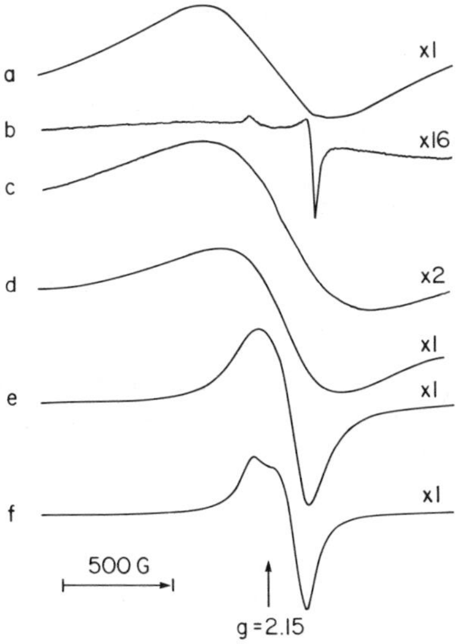

Figure 2. ESR spectra in the g = 2 region at 9.1 GHz for $BaCuO_5$ at (a) 6 K and (b) 293 K, $Y_2Cu_2O_5$ at (c) 6 K and (d) 133 K and Y_2BaCuO_5 at (e) 18 K and (f) 293 K.

damped magnetic fluxoids (21). The important result is that this low field absorption seems to clearly be characteristic of the superconducting phase.

Here our focus is to use this signal diagnostically to detect small amounts of superconducting phase in various stages of sample preparation. Figure 3 shows a dramatic example of this in a type I sample in which the component compounds are simply ground together in a mortar and pestle with no heating and show a definite low field absorption. Based on a comparison with the signal intensity in a type IV sample and optimum microwave power and modulation amplitude we estimate that a few nanograms of superconducting material can be detected showing the extreme sensitivity of this measurement for small amounts of superconducting material.

In our experiments all the samples ($\sim$ 5 mg) were cooled in a 3 kG field and the observation conditions were typically 2 μW microwave power with 1 to 10 G magnetic field modulation. This low field absorption becomes relatively strong for type II samples as is seen in Figure 4. As one progresses from type II to IV samples distinct changes are seen in the shape of the low field absorption with respect to temperature as also shown in Figure 4. Thus there is diagnostic information here on the superconducting phase formation which is currently being studied. It is also clear that the intensity of the maxima of $(d\chi''/dH)$ at a given temperature is not directly related to the volume amount of superconducting phase present.

Blazey et al. (11) suggested that the field at which $(d\chi''/dH)_{max}$ occurs is related to the area of superconducting clusters or grains which form a set of coupled Josephson junctions which maintain superconducting current loops below T_c. These superconducting current loops can trap magnetic flux which interacts with the microwave magnetic field to generate the observed signal.

Our data suggests that the field at which $(d\chi''/dH)_{max}$ occurs decreases as one progresses from type II to type III to type IV preparations. Given the above interpretation, this indicates that the area of the clusters which form pathways for superconducting current loops increases as one progresses through these preparation stages. Further work is required to unravel what this implies with respect to generation of the microscopic structure of the superconducting phase.

The intensities of the low field absorption and the g = 2 signal both decrease to zero at T_c as shown in Figure 5a. The field at which $(d\chi''/dH)_{max}$ occurs also decreases monotonically with increasing temperature and extrapolates to zero field at T_c as shown in Figure 5b. Thus T_c can be determined by monitoring the low field intensity, the g = 2 intensity or the field at which $(d\chi''/dH)_{max}$ occurs without requiring electrical contact.

Conclusions

The g = 2 signal seems characteristic of nonsuperconducting phases. It is probably associated with cupric ion spins that are dipolarly broadened. In fairly pure superconducting samples it is interesting that the g = 2 signal can be used to determine the superconducting transition temperature in that it disappears below T_c.

The low field ESR absorption seems diagnositic of the supercon-

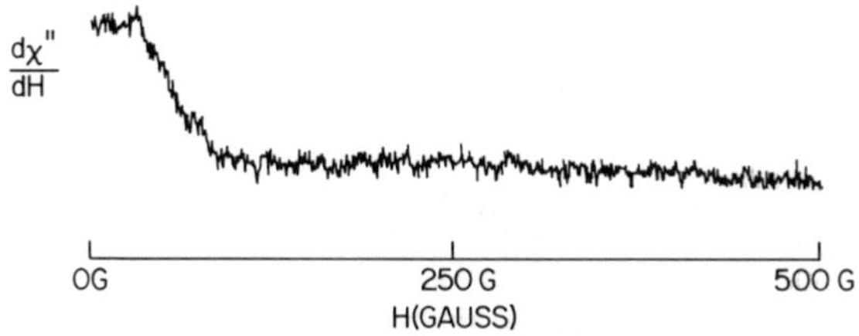

Figure 3. Low field absorption for a type $YBa_2Cu_3O_{7-\delta}$ sample cooled in a 3-kG field at 2 μW microwave power and 10-G modulation amplitude.

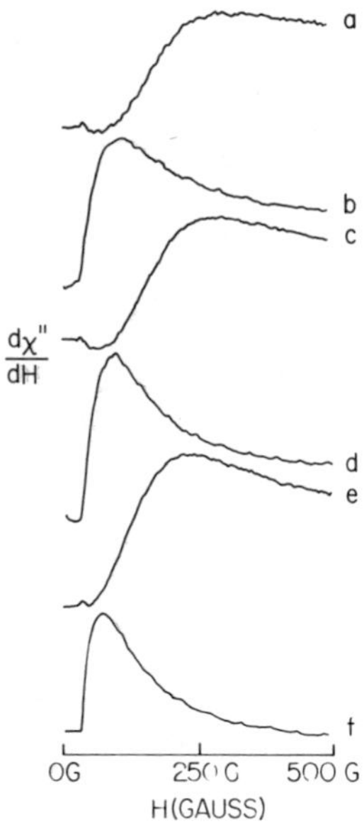

Figure 4. Low field absorption for → $YBa_2Cu_3O_{7-\delta}$ samples cooled in a 3-kG field of sample types II at (a) 18 K and (b) 73 K, III at (c) 18 K and (d) 73 K, and IV at (e) 18 K and (f) 73 K.

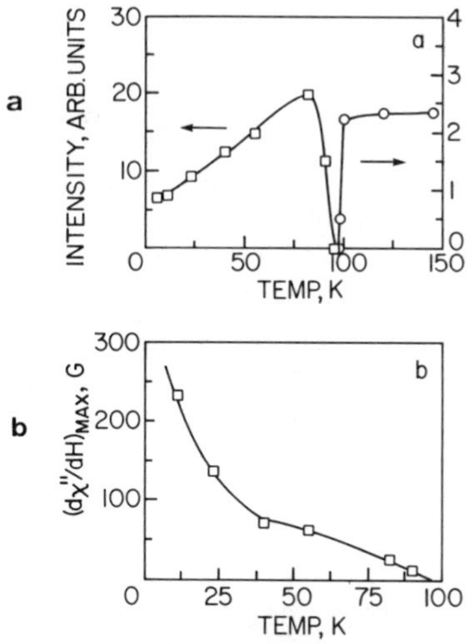

Figure 5. $YBa_2Cu_3O_{7-\delta}$ type IV sample cooled in a 3-kG field: (a) intensity of low field absorption (□) and of g = 2 signal (○) versus temperature, and (b) magnetic field at $(d\chi''/dH)_{max}$ for low field absorption versus temperature.

ducting phase. It is quite sensitive to extremely small amounts of superconducting phase materials with an extrapolated sensitivity of a few nanograms. This method is easier and faster than resistivity measurements since no contacts are needed to measure T_c. It can measure T_c quite accurately as shown in Figure 5.

The sensitivity of the shape of the low field absorption under various experimental conditions is thought to be an extremely sensitive diagnostic tool for the local structure of a superconducting phase present as a disordered powder or glassy material on a microscopic scale. Exact reproducibility of the shape and intensity of this signal under nominally analogous experimental conditions for different preparations is hard to achieve. However semiquantitative reproducibility is good. This may be an experimental problem at present, but we feel that it is more likely due to subtle differences in the solid state compound formation process in different sample preparations. Additional work and other locally microscopic techniques are needed to clarify these aspects.

Acknowledgment

This work was supported by the Texas Center for Superconductivity. Helpful discussions with R.N. Schwartz are gratefully acknowledged.

Literature Cited

1. Vier, D.C.; Oseroff, S.B.; Salling, C.T.; Smyth, J.F.; Schultz, S.; Dalichaouch, Y.; Lee, B.W.; Maple, M.B. Phys. Rev. B. 1987, 36, 8888-91.
2. Bowden, G.J.; Elliston, P.R.; Wan, K.T.; Dou, S.X.; Easterling, K.E.; Bourdillon, A.; Sorrell, C.C.; Cornell, B.A.; Separovic,F. J. Phys. C: Solid State Phys. 1987, 20, L545-52.
3. Kohara, T.; Yamagata, H.; Matsumura, M.; Yamada, Y.; Nakada, I.; Sakagumi, E.; Oda, Y.; Asayama, K. Physica 1987, 148B, 459-61.
4. Castilho, J.H.; Venegas, P.A.; Barberis, G.E.; Rettori, C.; Jardin, R.F.; Gama, S.; Davidov, D.; Felner, I. Solid State Commun. 1987, 64, 1043-45.
5. Kojima, K.; Ohbayashi, K.; Udagawa, M.; Hihara, T. Japn. J. Appl. Phys. 1987, 26, L766-67.
6. Shattiel, D.; Genossar, J.; Grayersky, A.; Kalman, Z.H.; Fisher, B.; Kaplan, N. Solid State Commun. 1987, 63, 987-90.
7. Blank, D.H.A.; Flokstra, J.; Gerritsma, G.J.; Van de Klundert, L.J.M.; Velders, G.J.M. Physica 1987, 145B, 222-26.
8. Mehran, F.; Barnes, S.E.; McGuire, T.R.; Gallagher, W.J.; Sandstrom, R.L.; Dinger, T.R.; Chance, D.A. Phys. Rev. B. 1987, 36, 740-42.
9. Durny, R.; Hautala, J.; Ducharme, S.; Lee, B.; Symke, O.G.; Taylor, P.C.; Zheng, D.J.; Xu, J.A. Phys. Rev. B. 1987, 36, 2361-63.
10. Khachaturyan, K.; Weber, E.R.; Tejedor, P.; Stacy, A.M.; Portis, A.M. Phys. Rev. B. 1987, 36, 8309-14.
11. Blazey, K.W.; Müller, K.A.; Bednorz, J.G.; Berlinger, W.; Amoretti, G.; Buluggiu, E.; Vera, A.; Mattacotta, F.C. Phys. Rev. B. 1987, 36, 7241-43.

12. Stankowski, J.; Kahol, P.K.; Dalal, N.S.; Moodera, J.S. Phys. Rev. B. 1987, 36, 7126-28.
13. Bhat, S.V.; Ganguly, P.; Ramakrishnan, T.V.; Rao, C.N.R. J. Phys. C: Solid State Phys. 1987, 20, 6559-63.
14. Rettori, C.; Davidov, D.; Belaish, I.; Felner, I. Phys. Rev. B. 1987, 36, 4028-30.
15. Müller, K.A.; Blazey, K.W.; Bednorz, J.G. Takashige, M. Physica 1987, 149-54.
16. Sastry, M.D.; Dalvi, A.G.I.; Babu, Y.; Kadam, R.M.; Yakhi, J. V.; Iyer, R.M. Nature 1987, 350, 49-51.
17. Shrivastava, K.N. J. Phys. C: Solid State Phys. 1987, 20, L789-96.
18. Dulcic, A.; Leontic, B.; Peric, M.; Rakvin, B. Erophys. Lett. 1987, 4, 1403-7.
19. Schwartz, R.N.; Pastor, A.C.; Pastor, R.C.; Kirby, K.W.; Rytz, D. Phys. Rev. B. 1987, 36, 8858-59.
20. Portis, A.M.; Blazey, K.W.; Müller, K.A.; Bednorz, J.G. Europhys. Lett. 1988, 5, 467.
21. Blazey, K.W.; Portis, A.M.; Bednorz, J.G. Solid State Commun. 1988, 65, 1153-56.
22. Bartucci, R.; Colavita, E.; Sportelli, L.; Balestrino, G.; Barbanera, S. Phys. Rev. B. 1988, 37, 2313-16.
23. Kim, B.F.; Bohandy, J.; Moorjani, K.; Adrian, F.J. J. Appl. Phys. 1988, 63, 2024-32.
24. deAguiar, J.A.O.; Menovsky, A.A.; Van den Berg, J.; Brom, H.B. J. Phys. C: Solid State Phys. 1988, 21, L237-41.
25. Tyagi, S.; Barsoum, M.; Rao, K.V. Phys. Lett. A 1988, 128, 225-27.
26. McKinnon, W.R.; Morton, J.R.; Preston, K.F.; Selwyn, L.S. Solid State Commun. 1988, 65, 855-58.
27. Bist, H.D.; Khulbe, P.K.; Shahabuddin, Md.; Chand, P.; Narlikar, A.V.; Jayaraman, B.; Agrawal, S.K. Solid State Commun. 1988, 65, 899-902.
28. Yu, J.T.; Lii, K.H. Solid State Commun. 1988, 65, 1379-83.
29. Pakulis, E.J.; Osada, T. Phys. Rev. B. 1988, 37, 5940-42.
30. Cava, R.J.; Batlogg, B.; van Dover, R.B.; Murphy, P.W.; Sunshine, S.; Siegrist, T.; Remeika, J.P.; Rietmann, E.A.; Zahurak, S.; Espinosa, G.P. Phys. Rev. Lett. 1987, 58, 1676.
31. Michel, C.; Raveau, B. J. Solid State Chem. 1982, 43, 73.

RECEIVED July 15, 1988

SURFACES AND INTERFACES

Chapter 19

Advances in Processing High-Temperature Superconducting Thin Films with Lasers

T. Venkatesan[1], X. D. Wu[2], A. Inam[2], M. S. Hegde[2], E. W. Chase[1], C. C. Chang[1], P. England[1], D. M. Hwang[1], R. Krchnavek[3], J. B. Wachtman[2], W. L. McLean[2], R. Levi-Setti[4], J. Chabala[4], and Y. L. Wang[4]

[1]Bell Communication Research, Inc., 331 Newman Springs Road, Red Bank, NJ 07701–7020
[2]Rutgers University, Piscataway, NJ 08854
[3]Bell Communication Research, Inc., 435 South Street, Morristown, NJ 07906
[4]University of Chicago, Chicago, IL 60637

The stringent requirements for the preparation of high T_c superconducting thin films, based on the demands for device fabrication, pose a major challenge to the number of thin film deposition and processing techniques. In this paper we examine the generic problems of the various techniques and expand on the capabilities of a pulsed laser deposition process. We show that using suitable processing steps the laser deposition technique for preparation of high-T_c thin films is emerging as a strong contender among the various thin film deposition techniques.

High temperature superconducting (HTSC) materials are metal oxides and the metal oxide system has been very important for micro- and opto- electronics for properties other than superconductivity [1]. Properties such as ferroelectricity, optical nonlinearities, high optical transparency, relatively large controllable refractive indices, etc., have made metal oxides very useful for technological applications (table 1). The metal oxides could be doped with transition metal ions which significantly affect their optical properties; eg., Ti doping of $LiNbO_3$ to form waveguides and Cr doping of Al_2O_3 to form light emitters. Technologies such as ion implantation could be effectively utilized to modify the surface properties of metal oxides [2]. With the discovery of HTSC in metal oxide systems, the importance of these materials has significantly escalated. This is probably the only system where, by modifying the oxygen composition, the film could be tailored from a perfect dielectric to a superconductor. As a result, films of these materials have potential novel applications in advanced technologies such micro- and opto- electronics.

0097–6156/88/0377–0234$08.75/0

Table 1. A brief list of the properties and examples of metal oxides that have applications in micro- and opto-electronics

Property	Example of metal oxide system	Application
High optical transparency	MgO, ZrO	Mirror Coatings
Low loss diectric with electro-optic effect	$LiNbO_3$, $LiTaO_3$	Optical wave guides and integrated optics
Piezoelectricity	$BaTiO_3$, $PbTi_{.46}Zr_{.54}O_3$	Transducers
Ferromagnetism	γ-Fe_2O_3	Magnetic tape memories
Optical norlinearity	Nb_2O_5-SiO_2-Na_2O-Ba_2O_3-TiO_2	All optical switching devices
High optical gain	Nd^{3+} doped $Y_3Al_5O_{12}$	Lasers
Tansparent conductors	$InSnO_x$	Novel device coatings
Superconductivity	Y-Ba-Cu-O Bi-Sr-Ca-Cu-O, Tl-Ba-Ca-Cu-O	SQUID Superconducting electronics

Since one of the potentially important applications is microelectronics it may be a worthwhile digression to speculate on how electronics, photonics and HTSC devices could coexist in a single device or system. The electron-electron interaction is strong, and this qualifies electronics for efficient switching devices, whereas the photon-photon interaction is weak, which qualifies photonics for information transmission with minimum cross-talk (further low absorption and dispersion media for photons exist as well). However, superconductors exhibit properties of both electrons and photons; there is a strong interaction between the basic quanta, and superconducting materials exhibit low absorption losses and dispersion in propagating signals. Hence, while superconductors will not replace photons in terms of their attractiveness for data transmission with low cross-talk, a hybrid evolution of superconductors coexisting with electronics and photonics seems a high likelyhood for eventual high bandwidth applications. An example of a futuristic high bandwidth hybrid system is shown in fig. 1, where high bit rate information arrives at a compound semiconductor interface where photons get converted to electrons; the electronic signal is processed by VLSI Si based chips as well as specialized HTSC chips before being retransmitted as photons via a compound semiconductor interface.

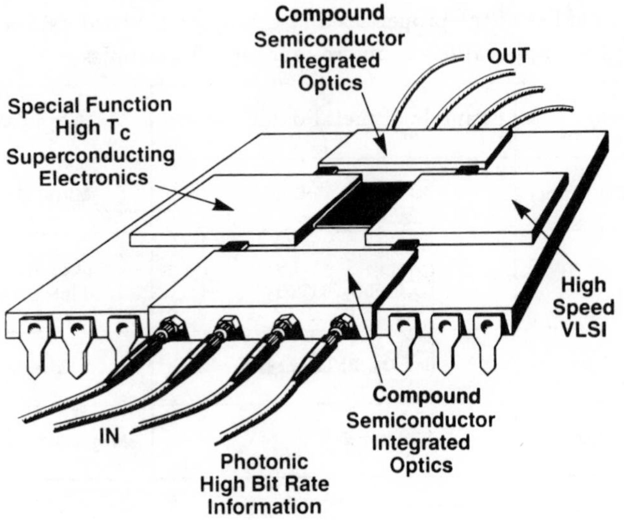

Figure 1. A futuristic chip combining opto-, micro-and super-conductor electronic technologies.

CHALLENGES

Table 2 illustrates the requirements for thin film HTSC materials in order to integrate the superconductors with micro- and opto-electronics. In fig. 2 is shown the molecular and complex crystal structure for the various HTSC materials and the increasing complexity of the materials with increasing T_c illustrates the difficulties inherent in the process for synthesizing these films in order to accomplish the requirements listed in table 2. With a viable film fabrication technique:

(a) one must be able to produce smooth films with the appropriate composition,
(b) get the right crystal phase, and
(c) produce the right oxygen stoichiometry in the film.

Table 2. Needs for device fabrication

1. High T_c(R=0)
2. Small ΔT (transition width)
3. High J_c (critical current density)
4. Surface smoothness
5. Stable film on substrates such as Si
6. Sharp interfaces

The number of thin film deposition techniques demonstrated to date vary in terms of their ease in meeting the above criteria in producing good quality films. The various deposition techniques could be generically divided into two classes: multiple sources or a

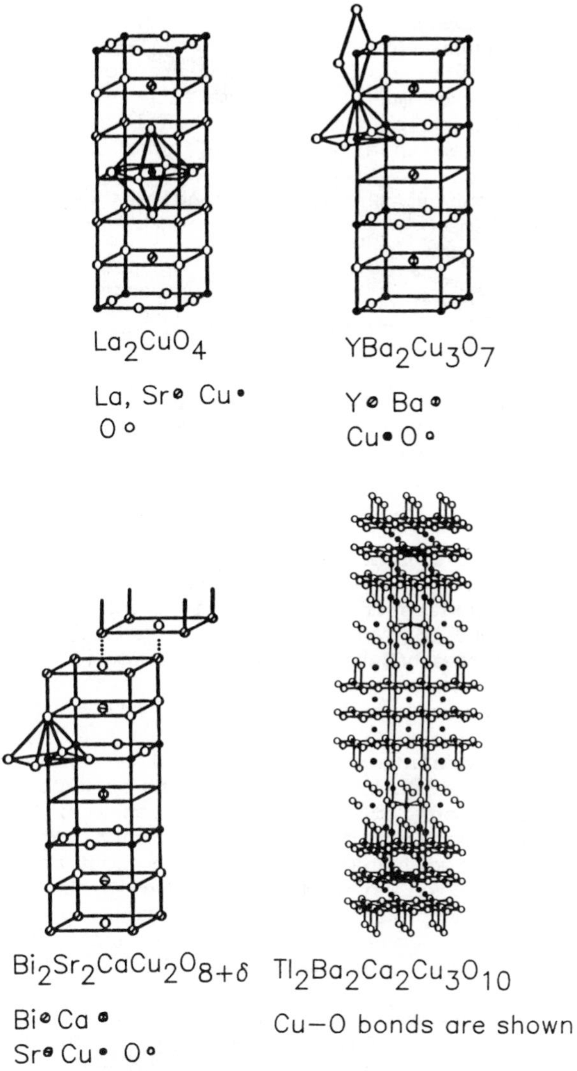

Figure 2. Approximate crystal structures of the various high T_c superconductors (courtesy of J. M. Tarascon).

single source for depositing the different elements. In fig. 3a is shown a typical geometry for deposition from different sources. The elements are ejected from these sources using heat, electrons or ions. The generic problems with these systems are:

1. the relative ejection rate of the elements must be monitored (in mTorr oxygen ambients) and kept a constant,
2. the non-overlap of the atomic trajectories from the three sources must be overcome either with planetary manipulation of the substrate holder or mounting the sources in a radial configuration.

On the other hand, the use of a single target (fig. 3b) has its associated problems:

1. the ejection rate of the three elements is not the same (eg. ion sputtering yields are different for the three species),
2. the sticking coefficients of these elements on the substrate are also different.

As a result, the target will not have the same composition as that of the deposited film, and further the optimum target composition will also depend upon the deposition parameters. In ion sputtering, effects such as negative ion bombardment of the films must be overcome.

While the above problems have been overcome and good films have been demonstrated by e-beam and thermal evaporation from multiple sources [3-5], sputtering from multiple and single targets [6-8], MBE deposition [9,10], sol-gel techniques [11], one of the most versatile techniques has been laser deposition [12].

LASER DEPOSITION

The technique consists of firing a pulsed excimer laser at a stoichiometric pellet of the material to be deposited and under suitable conditions of laser energy density, oxygen partial pressure, substrate temperature and deposition angle, high quality films are deposited. What is remarkable about the process is the stoichiometric deposition of films achieved by this technique. The composition of the pellet is closely reproduced in the films. A schematic of the deposition system is shown in fig. 4. The deposition and annealing parameters are shown in table 3.

In fig. 5a is shown a Rutherford backscattering spectrum of a film deposited from a pellet of $Y_1Ba_2Cu_3O_{7-x}$. The data agrees with a simulation of $Y_1Ba_2Cu_3O_6$! In fig. 5b and 5c are shown the spectra for films deposited from pellets where Y was substituted with Eu and Gd, and the composition is still preserved. The message is very clear: the laser deposition process is a straightforward technique to produce a complex film starting from a pellet of the desired film composition. Since the fabrication of pellets of complex materials is relatively easy, the laser deposition technique becomes increasingly attractive for the deposition of complex films.

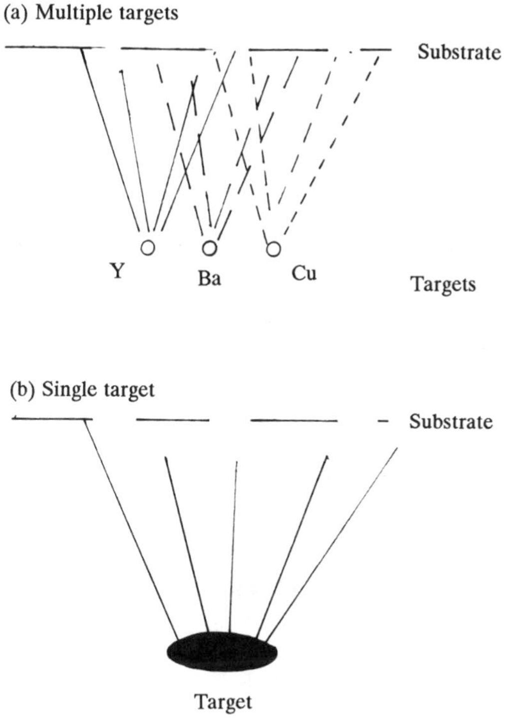

Figure 3. Generic deposition systems: (a) multiple sources and (b) single source.

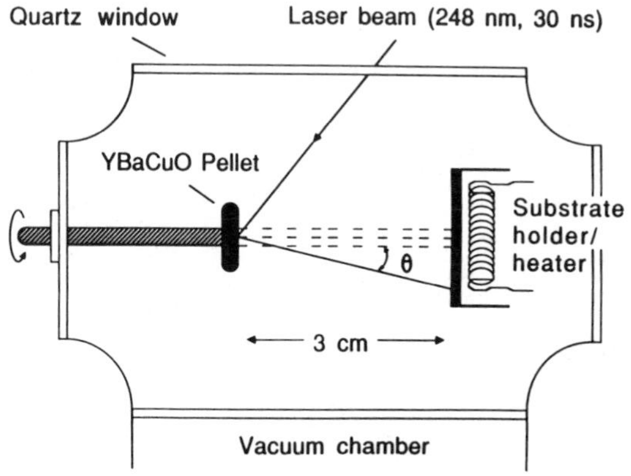

Figure 4. Schematic of the laser deposition system.

Table 3. Parameters for depositing and processing the films

Laser wavelength	248 nm
Pulse length	30 ns
Energy density	1 - 2 J/cm^2
Target	$RE_1Ba_2Cu_3O_x$
Target-substrate distance	3 cm

High Temperature Processing (HTP)	Low Temperature Processing (LTP)
$T_{substrate}$ ~ 400 C Background pressure ~ 10^{-6} Torr	~ 600 C ~ 5 mTorr O_2
Post Anneal in Oxygen 800-900 C ~ 1 h and slow cool	Post Anneal in Oxygen 450 C ~3 h and slow cool

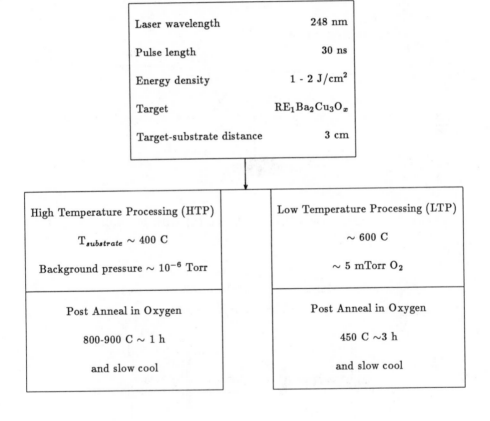

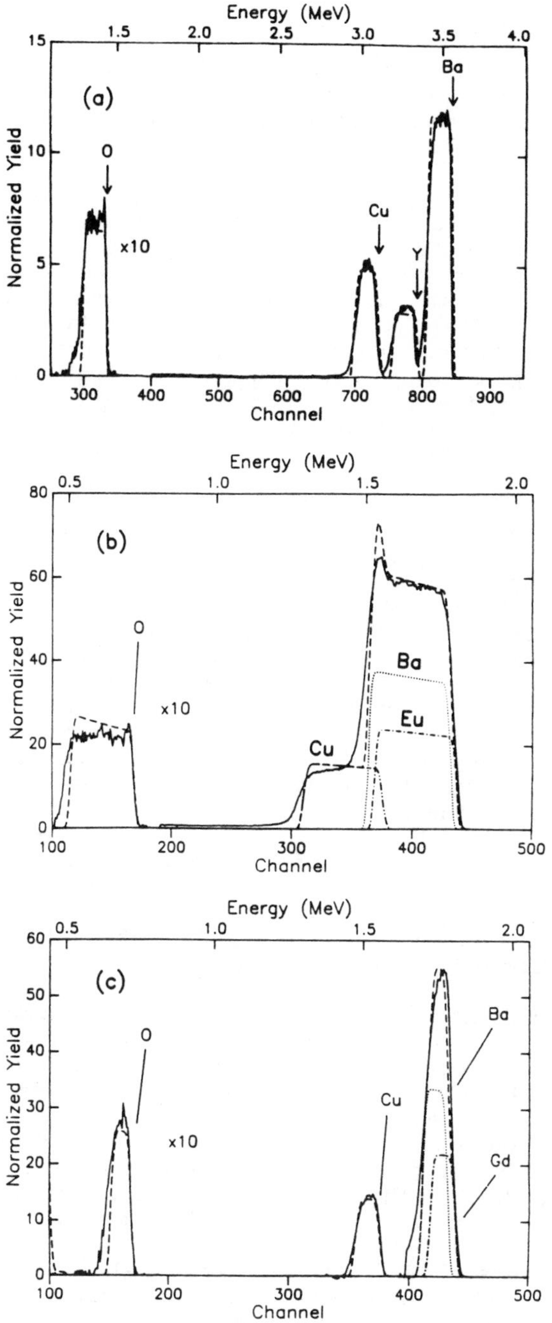

Figure 5. Rutherford backscattering spectra of films deposited from various pellets. The solid line is the data and the dashed line is a simulation for $RE_1Ba_2Cu_3O_{7-x}$ where RE is Y in (a), Eu in (b) and Gd in (c).

FILM CRYSTALLIZATION

There are two more obstacles to be overcome before one can have good superconducting films. The atoms in the film must be in the right crystallographic phase and the right amount of oxygen needs to be incorporated into the structure.

There are two approaches to accomplish this as shown in table 3: one is to deposit the film in some random phase at a relatively low temperature and follow this up with a high temperature (800-900 C) anneal in oxygen [12]. However, a more elegant way to accomplish this is to deposit the films directly in the orthorhombic phase at a temperature of 600-700 C in an oxygen ambient (a few mTorrs) [13]. From a technological point of view, in order to produce sharp interfaces, and minimize film-substrate interaction and stresses in the film, a low temperature process is absolutely essential. To this extent, the latter process is the preferred one. We will illustrate the advantages of the latter process with experimental data.

In fig. 6 are shown the SEM pictures of two HTSC surfaces prepared by the two processes [14]. Figs. 6a and 6b illustrate the deleterious effects of higher temperature processing (HTP). The surface is extremely grainy and one can even see cracks on the film surface. The substrate is sapphire and this is a typical consequence of the high temperature anneal on such a substrate. The thermal expansion mismatch between the film and substrate is minimum for $SrTiO_3$ and the least amount of surface cracks are observed on this surface. On the other hand, in figs. 6c and 6d we show the surface morphology of the low temperature processed (LTP) film on sapphire which seems to show no surface roughness what so ever. The small defect at the center was used to focus the electron beam. Using cross sectional specimens the surface roughness was estimated to be 100 A over a dimension of 1 um length of the specimen.

This result is further illustrated in the TEM cross sections in fig. 7 where in 7a one sees the surface of the high temperature processed film [15] to be rough, on the order of > 1000 Å. Further, the orientation of the crystallites seems to be quite random, though the crystalites in close proximity to the substrate do show a c axis orientation normal to the surface (fig. 7b). However, the low temperature processed films, shown in fig. 7c, exhibit a crystallographic structure with a smooth surface, where close to 90% of the crystallites have a c axis orientation normal to the surface, though there is some mosaicity with a few degree distribution of the c axis orientation with respect to the surface normal (fig. 7d). In this sense the crystal layers are analogous to those found in highly oriented pyrolytic graphite.

The results of TEM cross sectional analysis clearly show that for the HTP films the grain boundaries develop with the excess elements segregated at the grain boundaries. Even though the films do start with the right stoichiometry the high temperature annealing causes some interdiffusion at the interface resulting in the formation of off-stoichiometric phase boundaries. In fig. 8a one

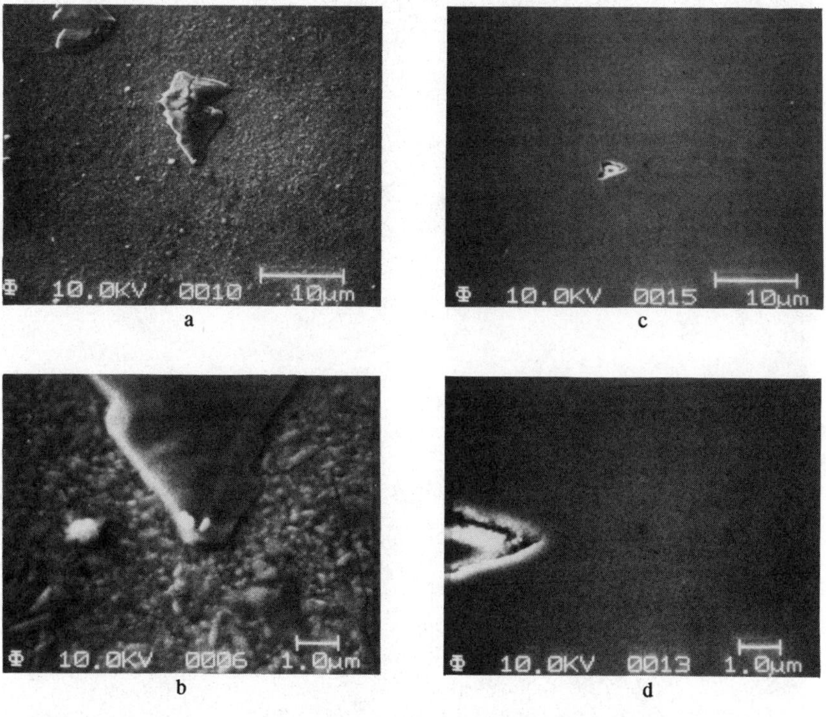

Figure 6. SEM micrographs of films deposited by (a) and (c) high temperature processing (HTP), (b) and (d) low temperature processing (LTP) (ref. 14).

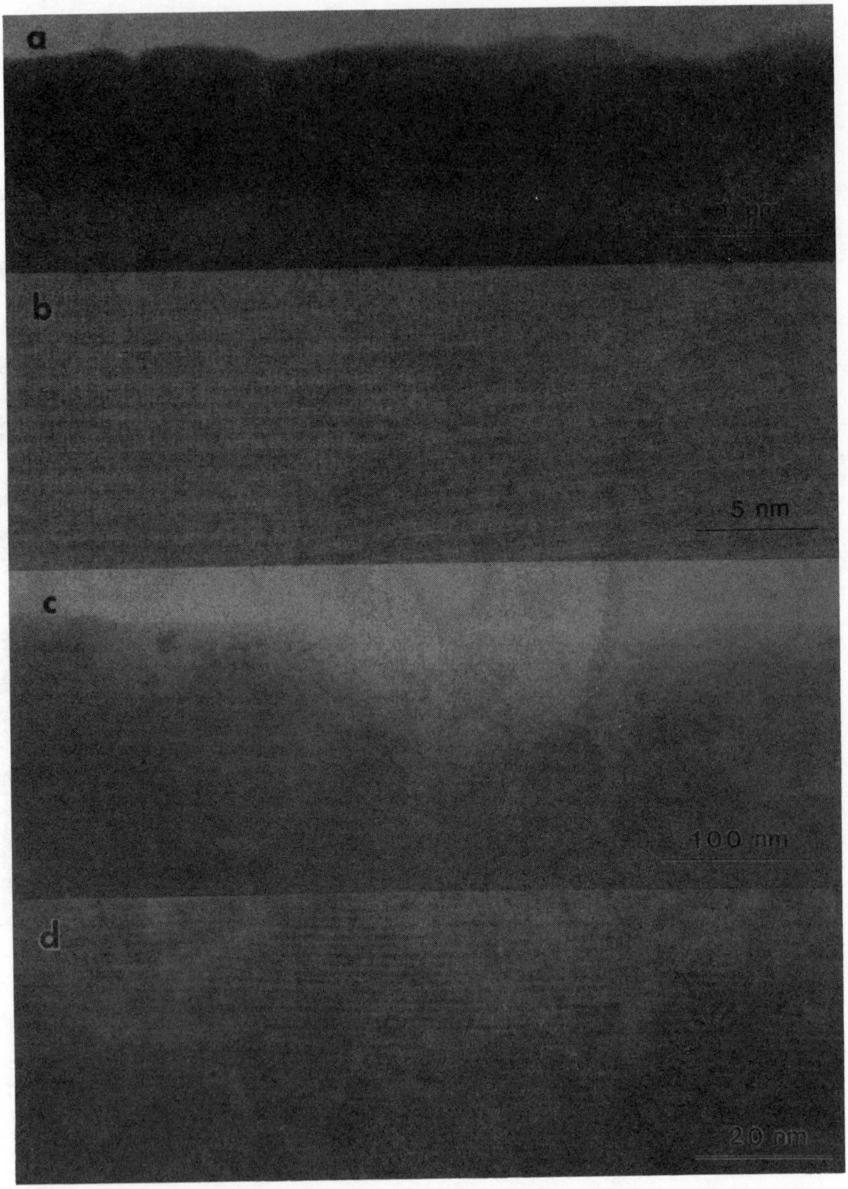

Figure 7. TEM cross sectional view of specimens prepared on $SrTiO_3$ by (a) and (b) HTP, (c) and (d) LTP, at two different magnifications.

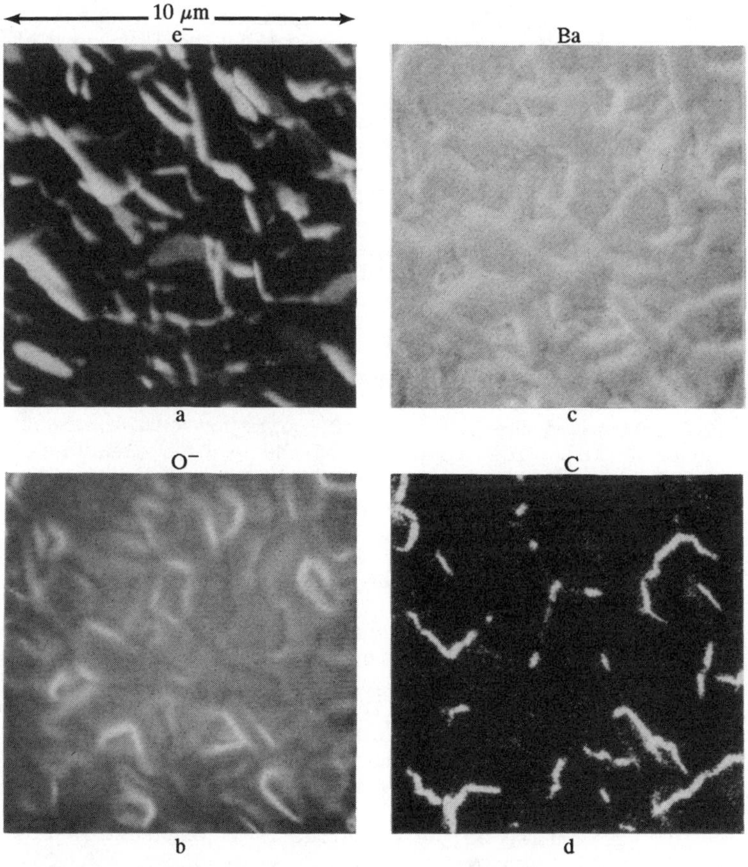

Figure 8. SIMS images generated on a HTP film using a focussed ion beam of gallium; the beam size is about 30 nm and the scales are shown in the figure. The images are for (a) secondary electrons, (b) oxygen, (c) barium and (d) carbon ions.

sees a focussed ion beam induced secondary electron image and the surface roughness is quite evident. The secondary ion images in figs. 8b and 8c illustrate the O and Ba segregation at grain boundaries which show a good correlation. The C image (fig. 8d) on the other hand shows limited correlation with the O and Ba signals. This rules out $BaCO_3$ as the possible segregant in these grain boundaries. The most likely form in which Ba is segregated is likely to be Barium hydroxide. On the other hand the LTP films showed no phase boundaries at all and the elemental composition was very homogeneous. The grain boundaries in the LTP films consisted of crystal defects such as dislocations or stacking faults.

The reduction of the process temperature results in very little reaction between the film and the substrate as shown by the Auger electron spectroscopy (AES) results of figs. 9a and 9b. In fig. 9a the AES results indicate a reacted layer on the order of 100 nm in a HTP film of the order of 350 nm. However, in the LTP film shown in fig. 9b the interface reaction extends over a region of less than 12 nm, close to the instrument resolution of the technique. This result is very important for the fabrication of structures containing abrupt junction.

The resistance versus temperature curve for the LTP film on $SrTiO_3$ shown in fig. 10 indicates a T of < 1 K and the critical current densities in the LTP films are quite reasonable, about $1x10^5$ A/cm^2 at 82 K [16]. The advantages of the process clearly point to the need for further lowering of the process temperature. As a result of using low temperature processing, good films have been deposited directly (T_{co} (R=0) of 67 K) as well as with a very thin ZrO_2 buffer layer on silicon [17]. In figs. 11a and 11b are shown the transport data on a $Y_1Ba_2Cu_3O_{7-x}$ film deposited on Si with only a 500 Å ZrO_2 buffer layer. The zero resistance transition temperature of 80 K is quite respectable though the voltage versus current characteristics shown in fig. 11b indicate a temperature dependence characteristic of granular superconductivity [18-19]. The critical current in this film was about $5x10^4$ A/cm^2 at 10 K, which may be adequate for some novel devices such as detectors, while for high current applications the critical current density needs to be a couple of orders of magnitude larger.

The versatility of the technique is illustrated in figs. 12a and 12b showing stoichiometric deposition of $Bi_4Sr_3Ca_3Cu_6O_x$ films, and the resistance versus temperature curve of an annealed HTP film showing a slight drop in resistivity at 110 K.

RECENT PROGRESS

Low temperature deposition. Recently in addition to those in refs. 13, 20 and 21 other groups have developed low temperature processes using deposition techniques such as activated and microwave-assisted reactive evaporation [22-23], electron-enchanced laser evaporation [24] to fabricate superconducting thin films at temperatures below 600 C. It is important at this stage to point out an inconsistency in the literature with regards to the definition of deposition temperature. Most groups including us, now report the substrate

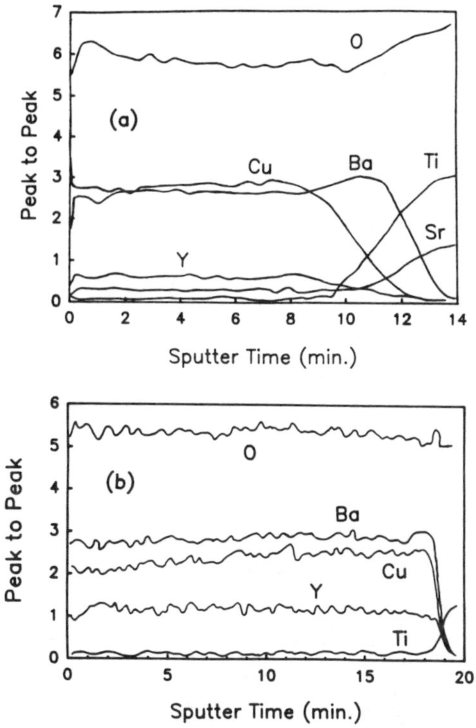

Figure 9. Auger depth profiling of (a) HTP and (b) LTP film (ref. 13).

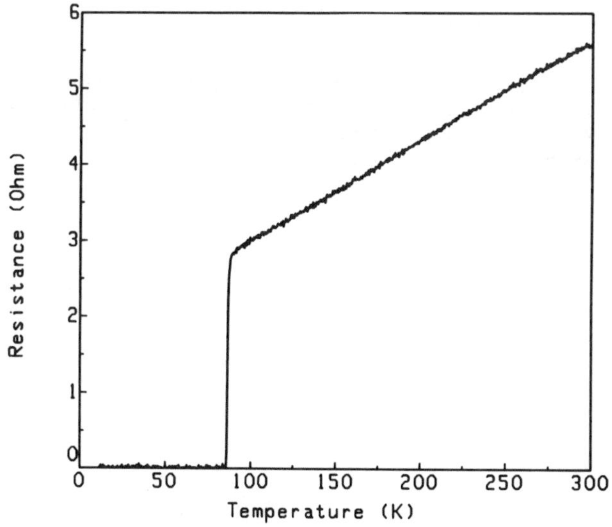

Figure 10. Resistance versus temperature characteristic of a LTP film on $SrTiO_3$ (ref. 16).

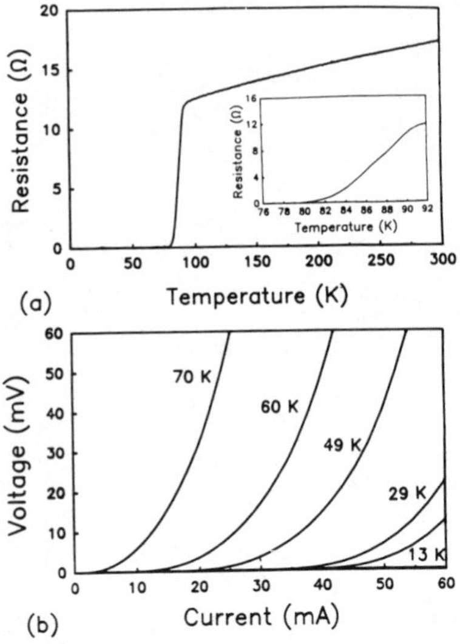

Figure 11. (a) Resistance versus temperature characteristic of an LTP film prepared on Si with a 50 nm ZrO_2 buffer layer; (b) Voltage versus current characteristic of the same film as a function of different temperatures (ref. 17).

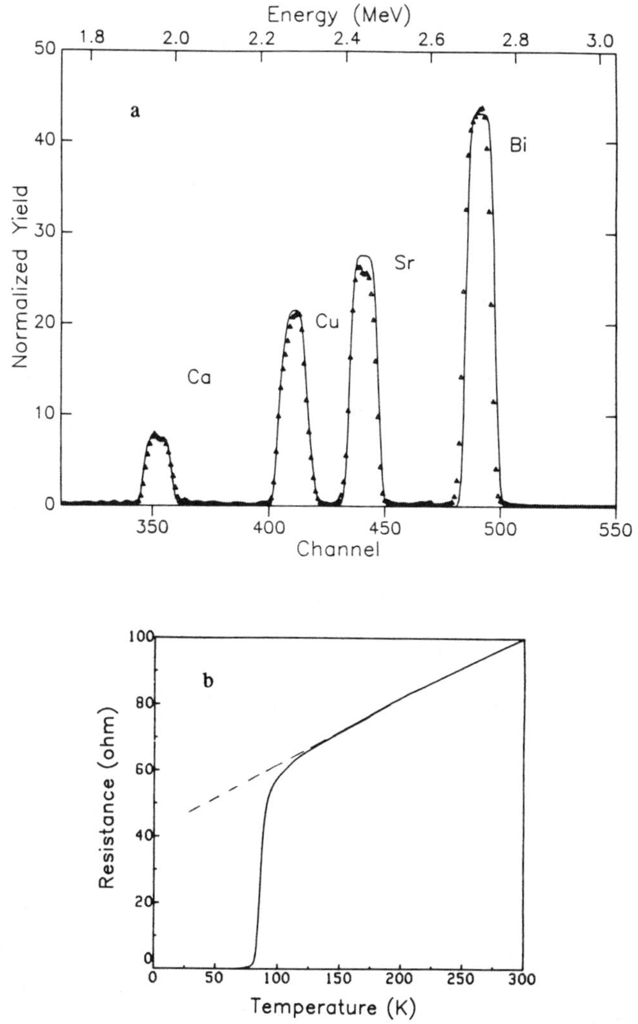

Figure 12. (a) RBS spectrum of a film deposited at room temperature with the composition $Bi_4Sr_3Ca_3Cu_6O_x$ from a pellet of similar composition; (b) Resistance versus temperature characteristic of the film subsequent to annealing in oxygen at 830 C for 5 hours (ref. 16).

holder temperature because of the difficulty of measuring the actual substrate surface temperature during each deposition run. The holder temperature is higher than the surface temperature by 50 to 150 C in our system but we prefer to report the holder temperature rather than an estimated surface temperature that is less accurate. Using the pulsed laser deposition technique (PLD), we have already demonstrated that stable thin films with surface smoothness of less than 10 nm and interfaces sharper than 15 nm can be fabricated at substrate holder temperatures below 650 C [13,16]. In this section, we report on an improved low temperature process that yields high J_c (~10^6 A/cm^2 at 77 K) and high T_c (R=0 at ~89 K) superconducting thin films without any post-deposition anneal.

Details of the PLD process have been published elsewhere [12,13,16]; briefly, pulses from a 248 nm KrF excimer laser are fired at a rate of one hertz onto a rotating $Y_1Ba_2Cu_3O_{7-x}$ target inside the deposition chamber. The ensuing plume of ejected material from the target is collected onto a substrate mounted on a resistively heated holder (the substrate holder temperature, as measured by a thermocouple, may be varied from 25 C to 650 C), which is held at a distance of ~6 cm from the target. Equilibrium oxygen pressure of a few millitorrs is maintained in the chamber during deposition. It was shown earlier that as-deposited superconducting films, with the correct "123" stoichiometry and a zero resistance transition temperature T_{co}~30 K are produced using this process. Subsequent to annealing at 450 C in flowing oxygen for three hours, films on $SrTiO_3$ had improved T_{co}'s of up to 86 K and J_c's at 82 K [16] of greater than 1.5 x 10^5 A/cm^2. X-ray diffraction analysis of as-deposited films on (100) $SrTiO_3$ showed a highly c-axis oriented orthorhombic phase, but the films probably contained oxygen deficient material possibly explaining the initially low T_c's. Post deposition annealing at 450 C in flowing oxygen improved the T_c's significantly, but the maximum T_{co} was 86 K and often lower. Since bulk $Y_1Ba_2Cu_3O_{7-x}$ material can have T_{co}'s over 90 K, these films appear to contain defects that cannot be eliminated by low temperature (450 C) post deposition anneal in oxygen.

It has been clear from our understanding of the high T_c thin film formation process, that incorporation of sufficient oxygen was important for formation of the correct orthorhombic phase. Therefore, we introduced oxygen directly into the plume emanating from the target by surrounding the plume with an open ended box-like copper enclosure measuring about 2 cm^3, and injecting a jet of oxygen gas directly into the center of this box. The oxygen was fed in at a rate of 5 sccm while the background pressure in the chamber, as measured by a pressure guage mounted about 25 cm from the nozzle, was maintained at 8 mTorr. We found that the plume passing through the oxygen column changed its color from bluish-white to red, indicating a direct reaction between the ions in the plume and molecular oxygen. We speculate that metal ions react with oxygen forming metal-oxide ions thus compensating for the oxygen deficiency. We are presently pursuing a spectroscopic study to understand the reactions between the laser induced plume and the oxygen. Apart from these modifications, the deposition procedure was the same as outlined in the preceding paragraph. Substrates of single crystal (100) $SrTiO_3$ and Al_2O_3 were used. Substrate holder temperatures, as measured via a thermocouple, were varied from 500 C to 650 C. Typically, films with a thickness of 2000 to 5000 A were deposited in about an hour. Immediately following the deposition, the substrate holder was allowed to cool down to room temperature in 250 Torr of oxygen.

Transport measurements. The as-deposited films were black and had a smooth mirror-like surface. Four probe DC resistivity measurements were made using wires attached to the films with silver ink. Figure 13 shows the resistivity versus temperature curves for as-deposited 2000 Å films on (a) Al_2O_3 and on (b) (100) $SrTiO_3$. The film on sapphire was deposited at a substrate holder temperature of 580 C and has a zero resistance temperature of 78 K. The film grown on $SrTiO_3$ was deposited at a substrate holder temperature of 650 C and has zero resistance below 88.6 K. The resistivity of the as-deposited film on $SrTiO_3$ just before the onset of transition is below 30 uohms-cm, which we believe to be among the lowest resistivities reported in the literature for a Y-Ba-Cu oxide superconducting thin film. The normal to superconducting transition for this sample is very sharp, with a width ΔT_c(10-90 %), less than 1 K.

A portion of the film on $SrTiO_3$ shown in fig. 13(b) was patterned into a 16 um wide line by reactive ion etching to measure the critical current density and for other magnetotransport measurements. Figure 14 shows the temperature dependence of the critical current density, J_c. It can been seen that at 77 K and in zero field, a current density of 0.69 x 10^6 A/cm^2 is measured. The film critical current density is greater than 4 x 10^5 A/cm^2 at 50 K and in a field of 14 Tesla. More detailed magnetotransport results will be mentioned later.

Film crystallinity. A Scintag four circle diffractometer using Cu-K_a radiation was used to perform x-ray diffraction studies on the films. Figure 15 shows the diffraction data from an as-deposited film on (100) $SrTiO_3$ grown at a substrate holder temperature of 650 C. The very strong (00L) reflections (where L=1,2,4,5,7,10; L=3,6,9 are obscured by the (100), (200), and (300) substrate peaks) indicate the film to be oriented with its c-axis normal to the substrate surface, and the absence of any impurity peaks shows that the film is predominantly single phase. The high degree of orientation of the film with respect to the substrate is confirmed by a sharp peak in a scan transverse to the (100) substrate direction, across the $Y_1Ba_2Cu_33O_{7-x}$ (005) reflection (figure 15 inset), where the peak width (FWHM) is only 0.22 degrees. Axial ion channeling measurements, to be reported below, indicate a very similar value and reveal that these films are highly crystalline. Raman measurements [25] further confirm the single crystalline nature of the film by the conspicuous absence of the 500 cm^{-1} vibration mode in the spectra. This would be expected if the c-axis of the crystallites were oriented in the direction of the laser beam with the Raman signal detected in a backscattering mode.

Surface perfection. For electronic applications of the new high

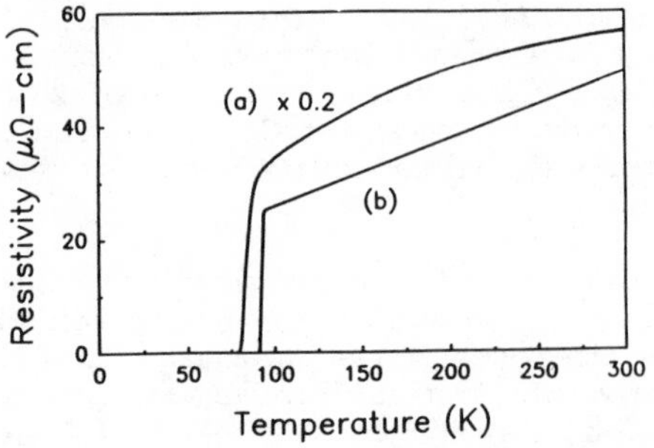

Figure 13. Resistivity vs. T curve for an as-deposited 0.2 um $Y_1Ba_2Cu_3O_{7-x}$ film on (a) Al_2O_3 made at 580 C, and (b) on (100) $SrTiO_3$ deposited at 650 C. R=0 is achieved at 78 K and 88.6 K respectively.

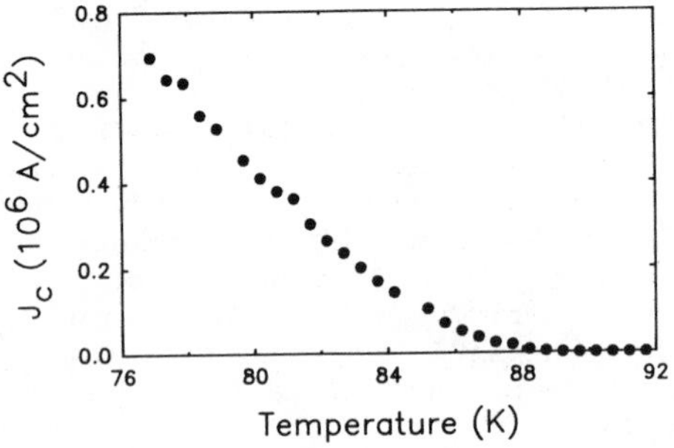

Figure 14. Critical current density as a function of temperature for an as-deposited film on (100) $SrTiO_3$. J_c at 77 K is 0.69×10^6 A/cm^2 in zero field.

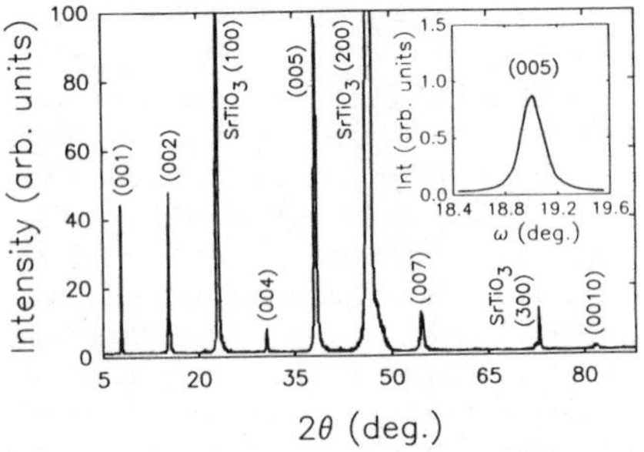

Figure 15. Radial x-ray diffraction pattern of as-deposited $Y_1Ba_2Cu_3O_{7-x}$ film on (100) $SrTiO_3$. Transverse scan across (005) reflection has a width (FWHM) of 0.22 degrees.

T_c superconductors it is crucial to have films with smooth superconducting surfaces, in order to delineate micron and submicron features and to fabricate junction devices consisting of superconducting layers sandwiching an insulator or a normal metal. In these junctions, the superconducting material must have a high T_c layer right up to the interface to within the superconducting coherence length of the material, which is ~ 0.43 nm along the c axis and ~ 3.1 nm in the a-b plane [26]. By utilizing surface sensitive techniques such as X-ray Photoelectron Spectroscopy (XPS) and Rutherford backscattering spectrometry (RBS) in the channeling mode, it is possible to obtain information about the film surfaces and interfaces. The channeling technique has been widely used to characterize various crystalline materials, single crystals with some disordered regions, polycrystalline films on single crystal substrates and so on [27]. Recently, Stoffel et. al.[28] showed that single crystals of $YBa_2Cu_3O_x$ have excellent crystallinity and stoichiometry to within about 1 nm of the surfaces by using RBS in the channeling mode. XPS has also been utilized to obtain information on the chemical state of the surface layer of the superconductors [29]. Here, we report the results of axial ion channeling and XPS studies on as-deposited high T_c Y-Ba-Cu oxide superconducting thin films on (100) $SrTiO_3$, and show that the crystallinity and composition of the material is good up to the surface to within 1 nm, which is comparable to the superconducting coherence length.

RBS and channeling measurements were made on a dual-axis goniometer using 3 MeV He^{++} ions with a 1 mm beam size. XPS spectra were taken on as-deposited films without any cleaning steps. The spectra were recorded in a KRATOS XSAM800 instrument equipped with a multichannel detector. The resolution of the XPS system was set to yield a peak width of 0.85 eV for the $Ag(3d_{5/2})$ line.

In Fig 16. we show a random RBS spectrum, and an aligned channeling spectrum for a ~ 4100 Å as-deposited Y-Ba-Cu oxide superconducting thin film on (100) $SrTiO_3$. The solid line in the figure is a simulation of $Y_1Ba_2Cu_3O_x/SrTiO_3$ using the RUMP program [30]. The result shows that the film has a composition close to ideal stoichiometry through the entire thickness. The aligned channeling spectrum shows a large reduction in the backscattering yield in the film and about 50% reduction (which depends on the film thickness) in the yield from the substrate. The minimum yield for Ba, measured near the surface, is ~7% compared to 3.5% for channeling of 1.66 Mev He^+ on single crystals [28]. Although the ion energy for channeling is different from that of ref. 28, the comparison shows that the as-deposited films have good long-range crystalline order. The minimum yield of 7% is the smallest value reported for the high T_c superconducting thin films on any substrate. As mentioned earlier the X-ray diffraction study shows that the as-deposited films on (100)$SrTiO_3$ are oriented with the c-axis perpendicular to the substrate surface. The 7% minimum yield suggests that at least 95% of the c-axis in the film is oriented. The rapid increase in the channeling yield from the film indicates that in the film there are a large number of defects, which cause dechanneling of the He ions. Preliminary transmission electron

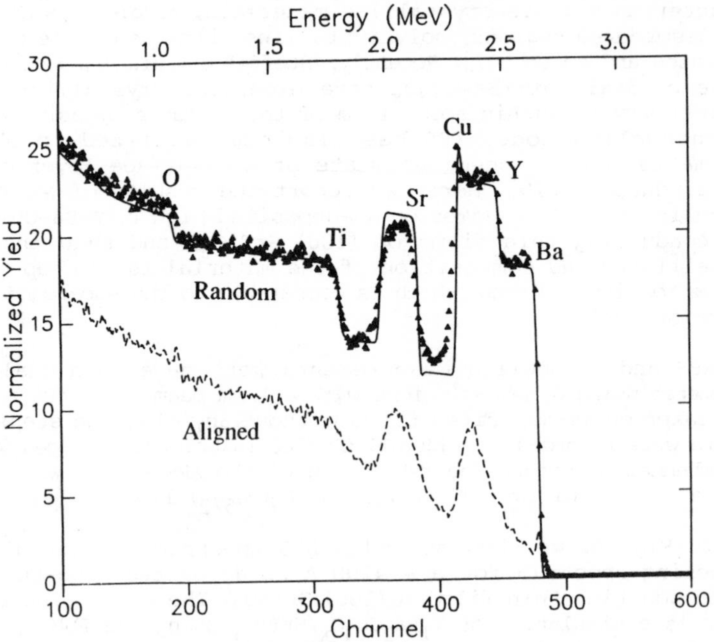

Figure 16. Random and aligned RBS (3Mev He^{++}) spectra for an as-deposited Y-Ba-Cu oxide superconducting thin film on $SrTiO_3$. The solid line is a simulation of 4100 Å $Y_1Ba_2Cu_3O_{7-x}/SrTiO_3$ (ref. 31).

microscopy (TEM) studies show existance of stacking faults in the films.

From the channeling measurements [31], the mosaicity is estimated to be less than 0.24 degrees (consistent with the X-ray data) showing that the film has grown epitaxially on the $SrTiO_3$ substrate during deposition. The most interesting feature for us in the channeling spectrum is the Ba surface peak. The area under the surface peak is proportional to the density of atoms on the surface and is a measure of the crystalline order at the surface[5]. A disordered surface layer of ~ 1 nm was estimated at the surface based on the surface peak assuming the "123" composition.

The surface layer composition and thickness can also be determined using XPS. We found that the surface layer of as-deposited films is barium enriched. In figure 17a, MgKa XPS of $Ba(3d_{5/2})$ region for "123" film at two different take-off angles, 15° (curve A) and 85° (curve B) are given. For camparison, $Ba(3d_{5/2})$ region from an in-situ, freshly scraped high T_c superconducting $Y_1Ba_2Cu_3O_{7-x}$ pellet is also shown (curve C). The Ba(3d) region shows two peaks, one at 778 eV and the other at 780 eV. These two peaks are resolved into two Gaussian distributions of equal width (1.7 eV) shown in fig. 17b. The peak at 778 eV is due to Ba in the "123" phase [32]. At a low angle of collection when only the surface region is examined, the peak at 780 eV is the largest (curve A) and, therefore, the 780 eV peak is due to Ba at the surface. Corresponding O(1s) region is shown in fig. 18. Clearly, we see three O(1s) peaks at 528.5, 531, and 532.7 eV in the "123" pellet as well as "123" film at a collection angle of 85°. These three peaks are assigned to O^{2-}, O^{1-}, and O^{2-} types of oxygen. However, in the film, the intensity of the 531 eV peak is higher which correlates with the surface Ba. At 15°, the 531 eV peak intensity relatively increases (curve A) and this further confirms its association with surface Ba ions. Even at the low collection angle, significant intensities at 528.5 and 532.7 eV are seen because the mean escape depth for the O(1s) photoelectron is larger than that for the Ba(3d) one. The XPS spectrum for C(1s) shows a small peak at 288.5 eV attributable to carbonate ions. The intensity of oxygen due to CO_3^{--} calculated from the C(1s) signal does not account for more than 5% of the total intensity in the 531 eV region. Further, $Ba(3d_{5/2})$ peaks in the case of $BaCO_3$ and $Ba(OH)_2$ appear at 782 eV and we do not see a peak in this region. Therefore, the 780 eV $Ba(3d_{5/2})$ peak is associated with the 531 eV O(1s) peak and represents some form of Ba oxide. It is known that Cu (which is also present on the surface layer) with Ba gives O(1s) and Ba(3d) at 531 and 780 eV respectively [33]. It should be noted that in films with Y:Ba:Cu in the 1:2:3 ratio but in which the "123" superconducting phase is not formed, mainly the 531 eV and 780 eV peaks in O(1s) and $Ba(3d_{5/2})$ are seen and the 528.5 and 778 eV peaks develop only after an oxygen anneal [34]. We therefore conclude that these two peaks are associated with the superconducting "123" phase. The surface composition of the films was estimated form the intensities of Ba(3d), Cu(2p), and Y(3d) peaks. At the 85° angle the composition was $Y_{0.88}Ba_{2.05}Cu_3O_x$ while at the low collection angle, 15°, the composition was $Y_{0.48}Ba_{2.06}Cu_3O_x$.

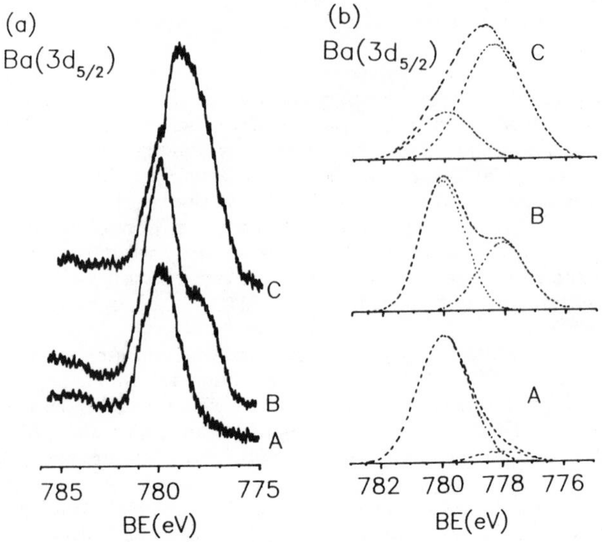

Figure 17. XPS spectra for Ba($3d_{5/2}$) region: (a) for an as-deposited superconducting film at two different photoelectron take-off angles: 85° (curve A) and 15° (curve B), and for a bulk superconductor (curve C); (b) two Gaussian distribution fitting for (a) with peak energies of 778 and 780 eV, and equal width of 1.7 eV (ref. 31).

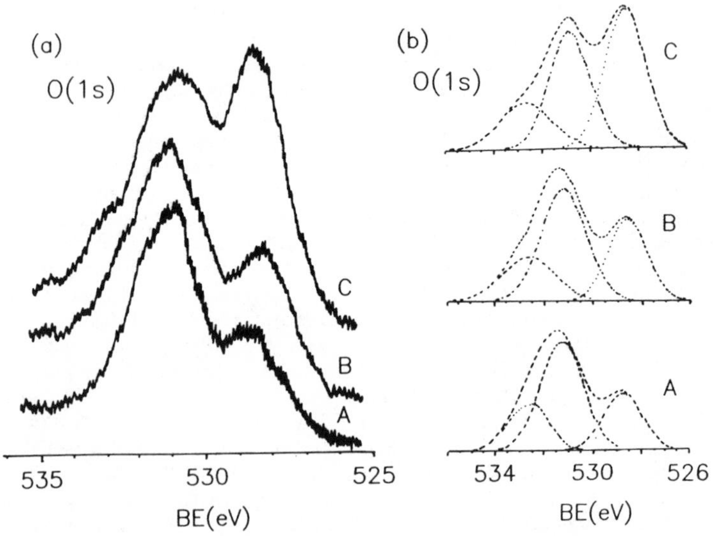

Figure 18. O(1s) XPS spectra: (a) for an as-deposited superconducting thin film at two different photoelectron take-off angles: 85° (curve A) and 15° (curve B), and for a bulk superconductor (curve C); (b) deconvolution of the XPS spectra in (a) using three Gaussian distributions at peak energies of 528.5, 531, and 532.7 eV repectively (ref. 31).

The mean escape depth for the Ba(3d) photoelectron is of the order of 29 A [35] and at this escape depth significant superconducting phase can be detected. Only at a take-off angle of 15° which corresponds to a sampling depth of 7 A, do we lose the bulk contribution. Hence we estimate the thickness of the surface barium enriched layer to be about 8 A, which is consistent with the result from the ion channeling study. This experiment therefore clearly demonstrates that the as-deposited superconducting films (grown by the laser deposition technique) have a nonstoichiometric surface layer of a thickness ~ 10 Å.

PATTERNING

The films have been successfully patterned by a number of techniques (table 4) and there is no real problem in making sub-micron structures in the smooth LTP films. Pulsed laser etching [36,37] of these films was shown to be an efficient way to make patterns and the results are shown in figs. 19a and 19b. The results indicate the potential of the technique and its eventual use in in-situ processing systems. A very reproducible and controlled dry etching is possible using pulsed laser etching.

Table 4. Patterning Techniques

1. Coventional photolithography + wet etching (10 % HNO_3 solvent)
    - Resolution ≈ 2-4 μm; grain boundary limit resolution
2. Electron beam lithography + ion milling
    - Resolution ≤ 1 μm
3. Pulsed & CW laser etching
    - Resolution ≤ 1 μm
4. Focussed ion beam etching
    - Resolution ≈ 0.1 μm (?)

One could also delineate patterns in these films using a cw focused Argon ion laser beam. In an unannealed film, direct writing with the Argon laser in an oxygen atmosphere produces a metallic film, and with higher laser intensities patterns could be directly written with micron resolution. Such a pattern is shown in fig. 20, where trenches as narrow as 1 um were made by a scanning cw argon laser. However, the etched trenches show a significantant deposit of debris on the sides, which must be eliminated to make the process useful.

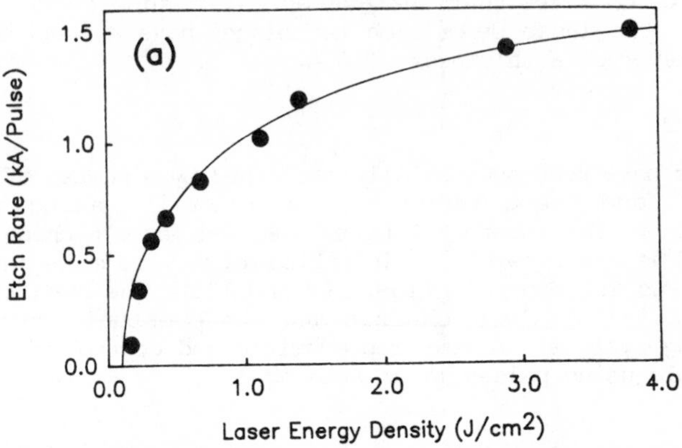

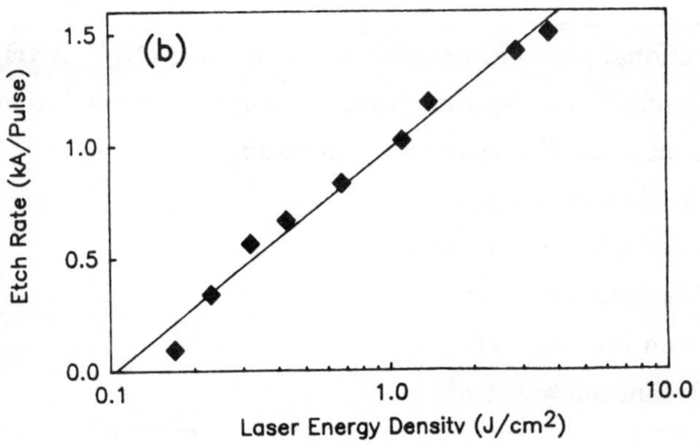

Figure 19. Pulsed excimer laser (248 nm, 30 ns) etching of high T_c thin films: (a) etch rate versus laser energy density on a linear graph; (b) same data on a semi log graph and the slope of the curve is the inverse absorption length of the light in the film (ref. 36).

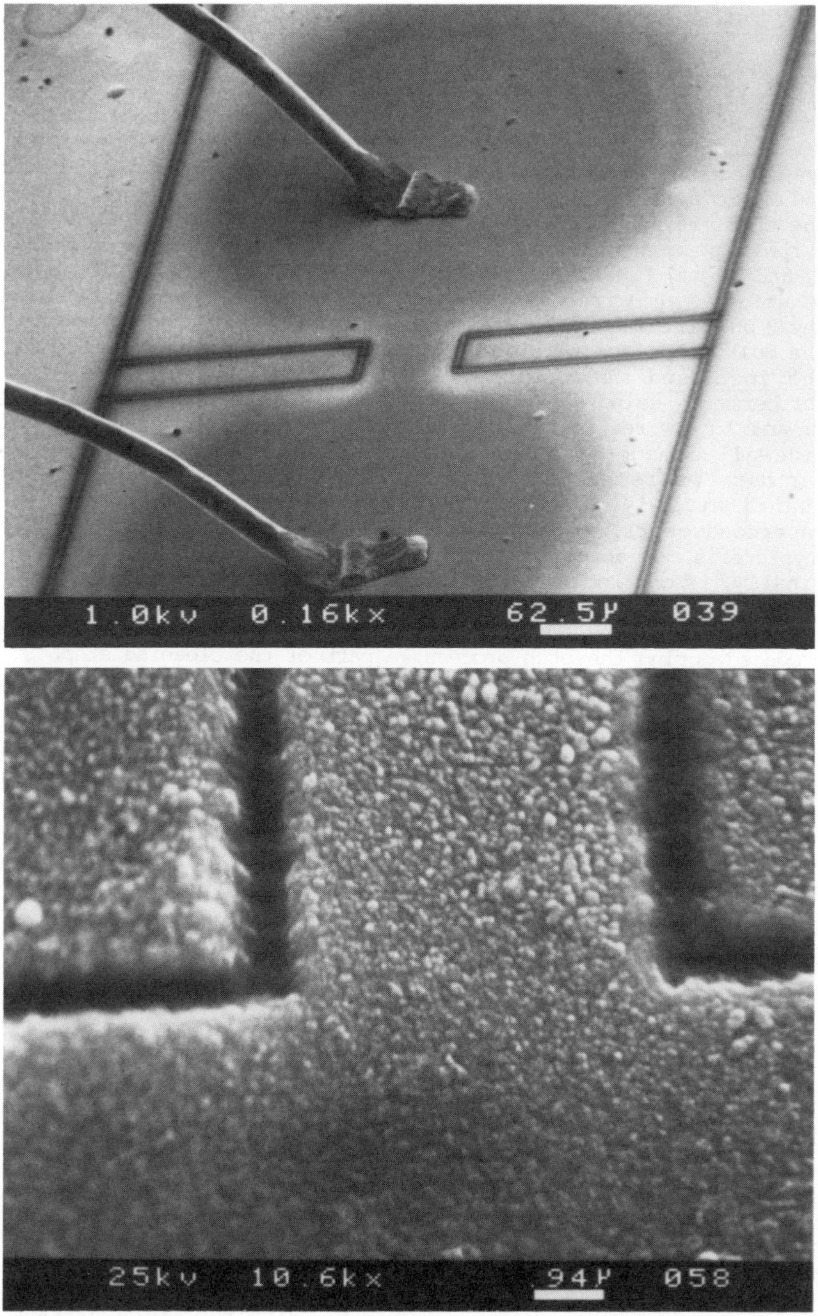

a

b

Figure 20. Argon laser etched patterns on the order of 1 um width by direct scanning. A low magnification image is shown in (a) and a high magnification image in (b). The surface roughness is produced by the debris deposited on the surface during the etching process.

MAGNETOTRANSPORT MEASUREMENT ON MICRON-SIZE WIRES

Four probe measurements on micron-size wires, formed by coventional optical lithography and argon-ion milling from the (in-situ deposited) films made at low temperature, were performed. Contacts were made using InSn solder to large area pads. The contact resistance limited the maximum critical current we could record to ~30 mA to avoid sample heating. To ensure that processing damage of the films is not significant, we have studied a sample of similar morphology processed into a set of wires of different widths (2, 4, 8, and 10 μm) (inset to fig. 21). The resistive transistion of the 10 μm and 2 μm wires are shown in fig. 21. The zero resistance in the 2 μm wire is depressed by less than 2 K from the 10 um one or the bulk film, and this difference may be partly accounted for by the increased measured current density in the smaller wire. Furthermore, using a 1 μV criterion, the critical current of the 10 μm and 2 μm wires is in the ratio 6.3:1. These findings suggest that sidewall damage of the wires during processing is not severe. Further, we remark that all the wires studied to date behave quantitatively and qualitatively like scaled down bulk superconductors, and do not show explicit weak link behaviour (hysteresis, or abrupt jumps in the current voltage charateristics). Finally, if one looks at the homogeneity of the film surface with respect to the smallest feature etched in the film, the results look indeed encouraging. The pit marks seen are voids on the substrate produced during the etch process, a part of the cleaning step.

Figure 22 shows the critical current as a function of temperature for a sequence of magnetic field applied perpendicular to the film. Three regions in the $I_c(T)$ charateristic separated by changes in slope can be identified; these are labelled I-III on B=0 T curve, but are also evident at higher fields. We will discuss these regions separately. In the inset are shown representative current-voltage characteristics for these three regions. At high temperature (region I), I_c does not reflect a true critical current, but is merely a consequence of our definition of the critical current, i.e., in this region I_c=10 μV/R_N, where the R_N is the normal state resistance of the film. In region II, the sample conductivity is rising by three orders of magnitude whilst remainig ohmic (inset to figure 22). This ohmic, and not fully superconducting region in the I-V characteristic has been observed in both thin film high T_c superconductors [19] and bulk polycrystalline material [38]. However, in a magnetic field the temperature range over which it occurs is increased. The orgin of this phase is still uncertain, but two interpretations of this behaviour are either that islolated regions in the film are acquiring bulk superconductivity, but do not yet form a sufficiently strongly coupled region or percolating path to support a bulk supercurrent. In this case, the transition at T=80 K (in zero magnetic field) represents a coupling transition of isolated superconducting regions which allows a bulk supercurrent to flow [19]. Alternatively, a homogeneous state with strong correlations of the order parameter exists. However, the measurements to be described, support the view that weakly coupled regions exist within the sample.

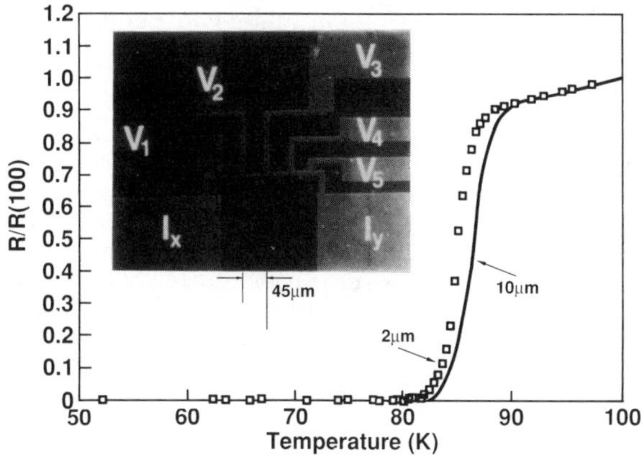

Figure 21. Resistive transition of a 10-μm-wide wire (solid line) and a 2-μm-wide wire (symbols). Inset is the wire geometry used for these comparative measurements. The wire widths in this case are 2, 4, 8, and 10 μm.

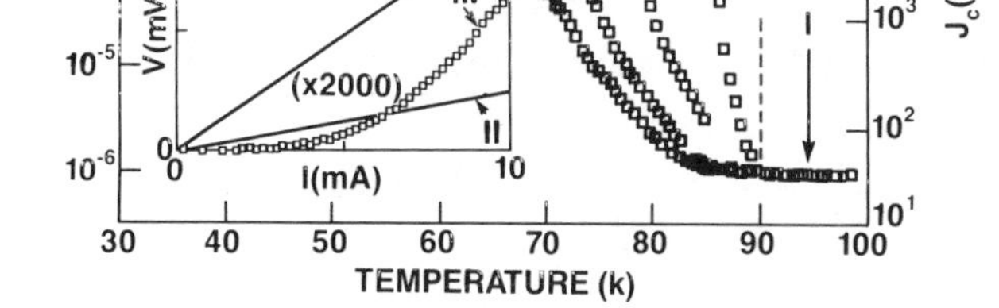

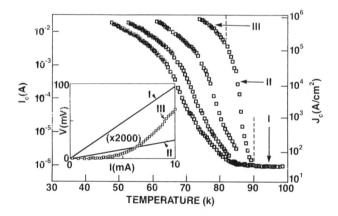

Figure 22. Critical current as a function of temperature in magnetic field (from the right) 0, 5, 10, and 14 T. Also indicated are the normal state (I), low-resistance ohmic state (II), and fully superconducting state (III). Inset are the current voltage characteristics representative of these regions. (Curve III has been expanded vertically 2000 times.)

In region III a true bulk supercurrent with associated non-zero critical current density exists. The critical current density in zero magnetic field is 0.7×10^6 A/cm^2 at 77 K, and rising according to the power law $I_c \sim (1-T/T_c)^{3/2}$ (not shown in the figure). The 3/2 power law in zero field has been observed repeatedly in granular material [18] and it is perhaps surprising that it should occur in this very homogenous material.

As the magnetic field is increased, the various transitions are displaced downwards in temperature, but remain well defined. We may linearly extraplated the critical filed at low temperature from the high field behaviour close to T_c. For a critical current density of 10^6 A/cm^2, we estimate a critical field H_c larger than 16 T at 4.2 K. Note that this is a usable critical field with a finite critical current, as opposed to the upper critical field which signifies the complete suppression of superconductivity.

The magnetotransport measurements have been made on films with dimensions comparable to those formed on state-of-the-art superconducting devices. The number for the critical current density as well as the critical field are quite impressive for films processed at these low temperatures.

CONCLUSION

In conclusion, how does the laser processing of high T_c thin films meet with the needs for device fabrication? The results are shown in table 5, where the laser deposition process seems to be

Table 5. Performance of laser deposition in meeting the needs of device fabrication

1. High T_c(R=0)	89 K
2. Small ΔT (transition width)	≤ 1 K
3. High J_c (critical current density)	$\sim 10^6$ A/cm^2 @ 77 K
4. Surface smoothness	< 10 nm
5. Stable film on substrates such as Si	YES!
6. Sharp interfaces	< 15 nm

close to producing viable films. The reproducibility of the process needs to be improved further, by the identification of critical parameters. A number of further in-situ processing schemes could be incorporated to enhance the crystallization of the films at lower temperature, the clear direction in which the film processing must head to improve reproducibility. The laser deposition process is demonstrated to have a number of inherent advantages over other techniques. It is fast, efficient, relatively inexpensive, versatile, allows for other processing tools to be incorporated independently and is capable of scale up. The technique is a serious contender for the production of high quality superconducting films as well as insulators and multiple interfaces. Eventually the technique may be extended to layer by layer synthesis of high T_c superconducting structures.

ACKNOWLEDGMENTS

The authors would like to thank J. M. Tarascon, P. Barboux, L. Nazar, L. H. Greene, B. G. Bagley, P. F. Miceli and J. M. Rowell of Bellcore, and M. Croft of Rutgers University for their help and for helpful discussions.

REFERENCES

1. Hand Book of Materials Science, Vol III, Non-matallic Materials and Applications, ed. C. T. Lynch (CRC Press, OH)
2. C. W. White, L. A. Boatner, P. S. Sklad, C. J. McHargue, J. Rankin, G. C. Farlow, and M. J. Aziz, Proceeding of the Radiation Effects in Insulators, (Lyon, France, July 6-10,1987).
3. P. Chaudhari, R. H. Roch, R. B, Laibowitz, T. R. McGuire, and R. J. Gambino, Phys. Rev. Lett, 58, 2684(1987).
4. M. Naito, R. H. Hammond, B. Oh, M. R. Hahn, J. W. P. Hsu, P. Rosenthal, A. F. Mashall, M. R. Beasley, T. H. Geballe, and A. Kapitulnik, J. Mater. Res. 2, 713(1987).
5. P. M. Mankiewich, J. H. Scofiels, W. J. Skocpol, R. E. Howard, A. H. Dayem, and E. Good, Appl. Phys. Lett. 51, 1753(1987).
6. Y. Enomoto, T. Murakami, M. Suzuki, and K. Moriwaki, Jpn. J. Appl. Phys.26, L1248(1987).
7. K. Char, A.D. Kent, A. Kapitulnik, M.R. Beasley, and T. Geballe, Appl. Phys. Lett. 51, 1370(1987).
8. M. Hong, S. H. liou, J. Kwo, and B. A. Davidson, Appl. Phys. Lett. 51, 694(1987).
9. C. Webb, S.-L. Weng, J. N. Eckstein, N. Missert, K. Char, D. S. Schlom, E. S. Hellman, M. R. Beasley, A. Kapitulnik, and J. S. Harris, Jr., Appl. Phys. Lett. 51, 1191(1987).
10. J. Kwo, T. C. Hsieh, R. H. Fleming, M. Hong, S. H. Liou, B. A. Davidson, and L. C. Feldman, Phys. Rev. B36, 4039(1987).
11. P. Barboux, J. M. Tarascon, B. G. Bagley, L. H. Greene, G. W. Hull, B. W. Meafger and C. B. Eom, MRS Proc. Vol. 99, eds. M. B. Brodsky, R. C. Dynes, K. Kitazawa, and H. L. Tuller, (MRS, Pittsburgh, 1988), p49.
12. D. Dijkkamp, T. Venkatesan, X. D. Wu, S. A. Shaheen, N. Jisrawi, Y. H. Min-Lee, W. L. McLean, and M. Croft, Appl. Phys. Lett. 51, 619(1987).
13. X. D. Wu, A. Inam, T. Venkatesan, C. C. Chang, E. W. Chase, P. Barboux, J. M. Tarascon, and B. Wilkens, Appl. Phys. Lett. 52, 754(1988).
14. C. C. Chang, X. D. Wu, A. Inam, D. M. Hwang, T. Venkatesan, P. Barboux, and J. M. Tarascon, Appl. Phys. Lett. (in press).
15. D. M. Hwang, L. Nazar, T. Venkatesan, and X. D. Wu, Appl. Phys. Lett. 52, 1834(1988).
16. X. D. Wu, T. Venkatesan, A. Inam, E. W. Chase, C. C. Chang, Y. Jeon, M. Croft, C. Magee, R. W. Odom, and F. Radicati,

High T_c Superconductivity: Thin Films and Devices, eds. C. C. Chi and R. B. Van Dover, (SPIE, Bellingham, Washington, 1988).
17. T. Venkatesan, E. W. Chase, X. D. Wu, A. Inam, C. C. Chang, and F. K. Shokoohi, Appl. Phys. Lett. (July, 1988)
18. S. B. Ogale, D. Dijkkamp, T. Venkatesan, X. D. Wu, and A. Inam, Phys. Rev. B36, 7210(1987);
19. P. England, T. Venkatesan, X. D. Wu, and A. Inam, Phys. Rev.
20. H. Adachi, K. Hirochi, K. Setsune, M. Kitabake, and K. Wasa, Appl. Phys. Lett. 51, 2263(1987).
21. D. K. Lathrop, S. E. Russek, and R. A. Buhrman, Appl. Phys. Lett. 51, 1554(1987).
22. T. Terashima, K. Iijima, K. Yamamoto, Y. Bando, and H. Mazaki, Jpn. J. Appl. Phys.27, L91(1988).
23. R. M. Silver, A. B. Berezin, M. Wendman, and A. L. de Lozanne, Appl. Phys. Lett. 52, 2174(19880.
24. S. Witanachchi, H.S. Kwok, X.W. Wang, and D.T. Shaw (preprint).
25. L. A. Farrow, A. Inam, X. D. Wu, M. S. Hegde, and T. Venkatesan, (unpublished)
26. W. J. Gallagher, T. K. Worthington, T. R. Dinger, F. Holtberg, D. L. Kaiser and R. L. Sandstrom, Superconductivity in Highly Correlated Fermion System, eds. M. Tachiki, Y. Muto, and S. Maekawa (North-Holland, Amsterdam, 1987), pp. 228-232.
27. L. C. Feldman, J. W. Mayer, and S. T. Picraux, Materials Analysis by Ion Channeling, (Academic Press, New York, 1982).
28. N. G. Stoffel, P. A. Morris, W. A. Bonner, and B. J. Wilkens, Phys. Rev. **B37**, 2297(1988).
29. J. Halbritter, P. Walk, H.-J. Mathes, B. Haeuser, and H.Rogalla, High Temperature Superconductors: Materials and Mechanishs of Superconductivity, (Interlaken, Switzerland, Feb. 29 - Mar. 4, 1988).
30. L. R. Doolittle, Nucl. Instrum. Methods. **B9**, 344(1985).
31. X. D. Wu, A. Inam, M. S. Hegde, T. Venkatesan, E. W. Chase, C. C. Chang, B. Wilkens, and J. M. Tarascon, (unpublished).
32. W. K. Ford, C. T. Chen, J. Anderson, J. Kwo, S. H. Liou, M. Hong, G. V. Rubenacker, and J. E. Drumheller, Phys. Rev **B37**, 7924(1988).
33. M. Ayyoob and M. S. Hegde, Surface Science, **147**, 361(1984).
34. M. S. Hedge, (unpublished).
35. S. Tanuma, C. J. Powell and D. R. Penn, Surface Sci. **192**, L849(1987).
36. A. Inam, X. D. Wu, T. Venkatesan, S. B. Ogale, C. C. Chang, and D. Dijkkamp, Appl. Phys. Lett. 51, 1112(1987).
37. A. Gupta and G. Koren, Appl. Phys. Lett. 52, 655(1988).
38. M. A. Dubson, S. T. Herbert, J. J. Calbrese, D. C. Harris, B. R. Patton, and J. C. Garland, Phys. Rev. Lett. 60, 1061(1988).

RECEIVED July 13, 1988

Chapter 20

Preparation of Superconducting Oxide Films from Metal Trifluoroacetate Solution Precursors

A. Gupta, E. I. Cooper, R. Jagannathan, and E. A. Giess

IBM Research Division, T. J. Watson Research Center, Yorktown Heights, NY 10598

Superconducting thin films of $Y_1Ba_2Cu_3O_{7-\delta}$ have been prepared on single-crystal yttria-stabilized zirconia (YSZ) and $SrTiO_3$ substrates using metal trifluoroacetate spin-on precursors. The films exhibit resistive transition widths of 1.5-1.7 K with zero resistance at temperatures as high as 91-94 K. The superconducting phase is formed by first decomposing the trifluroacetate film to the fluorides and then converting the fluorides by reaction with water vapor. The properties of the films prepared on YSZ substrates depend on the amount of conversion of the fluorides by reaction with water. Films which show the the presence of some unreacted barium fluoride have strong c-axis normal preferred orientation, with a sharp resistive transition. When all the barium fluoride is converted, the films are more randomly oriented, and have a broader transition to zero resistance. Films prepared on $SrTiO_3$ show different preferential growth directions depending on substrate orientation. Unlike films grown on YSZ substrates, these films exhibit the best superconducting properties when all the fluoride phases in the films are converted.

Various vacuum and non-vacuum techniques have recently been reported for depositing thin films of copper oxide perovskite superconductors (1). Non-vacuum techniques for producing films from solution precursors, by spin, spray or dip-coating, have the obvious advantage of low cost and simplicity. In addition, solution precursors allow accurate control of composition and intimate mixing of the constituents, which is important for multicomponent systems like the superconducting oxides. Solution precursors of Y, Ba and Cu like nitrates (2,3), alkoxides (4), and different carboxylates (5-8), have been used for preparation of superconducting films of $Y_1Ba_2Cu_3O_{7-\delta}$ (hereafter referred to as 1-2-3). Typically, a mixture of the precursors with the right cation stoichiometry is deposited on the substrate and is converted to the oxides by thermal decomposition at temperatures

0097–6156/88/0377–0265$06.00/0

of 300-500°C. The superconducting 1-2-3 phase is formed by subsequent annealing in oxygen at a higher temperature.

In spite of good control of stoichiometry , the superconducting properties of films prepared using these solution precursors have not compared favorably with films prepared using conventional vacuum techniques like sputtering or coevaporation. While good quality films of 1-2-3, which exhibit zero resistance in the 80-90 K range, can now be routinely prepared using vacuum techniques, it has been much more difficult to prepare films with such high transition temperatures from solutions. The critical current density of films prepared from solution precursors has also been much lower. The primary reason for the inferior properties is believed to be due to carbon contamination during decomposition . The intermediate barium oxide reacts readily with carbon dioxide evolved during decomposition (of metallo-organics) and also from the atmosphere. The resulting barium carbonate is difficult to decompose at low temperatures, and causes formation of secondary phases when decomposed at high temperatures. The evolution of large amounts of gases during decomposition can also lead to porosity in the films, which can also degrade the superconducting properties of the films.

We have used a spin-on solution containing a mixture of Y, Ba and Cu trifluoroacetates (TFA) to prepare 1-2-3 films which exhibit much improved properties compared to other reported solution precursors. Unlike other precursors, the trifluoroacetates decompose primarily to the fluorides which are much less sensitive to carbon contamination. The advantage of using BaF_2 as a Ba source for preparing superconducting films with reproducible properties by coevaporation was first reported by Mankiewich et al. (9). It has also been recently used to prepare good quality films by rf magnetron sputtering from composite targets (10). By using precursors of Y, Ba and Cu, which provide fluoride intermediates, we have been able to prepare 1-2-3 films on yttria-stabilized zirconia (YSZ) and $SrTiO_3$ substrates which exhibit an extremely sharp resistive-to-superconducting transition (1.5-1.7 K transition width), and have zero resistance at temperatures near 90 K. Preliminary results on films prepared on YSZ substrates was reported by us in a recent communication (11).

Experimental

Most metal trifluoroacetates can be readily synthesized by reacting corresponding metal oxides, carbonates, hydroxides, etc. with trifluoroacetic acid (12). We prepared yttrium and barium trifluoroacetates ($Y(TFA)_3$ and $Ba(TFA)_2$) by reacting Y_2O_3 and $BaCO_3$, respectively, with aqueous trifluoroacetic acid (CF_3COOH), and evaporating to dryness. Copper trifluoroacetate ($Cu(TFA)_2$) was prepared by reacting copper metal with trifluoroacetic acid in the presence of hydrogen peroxide. The anhydrous trifluoroacetates absorb between 2-3 moles of water per mole when left exposed to air. The fluorine substitution in the carboxylate makes all the three trifluoroacetates readily soluble in a variety of organic solvents, including alcohols. The precursor solution was prepared by dissolving a stoichiometric mixture of the trifluoroacetates in methanol. The concentration of metal ions in solution was confirmed by inductively coupled plasma emission spectroscopy (ICP), which showed the relative concentration of Y:Ba:Cu to be within 5% of 1:2:3. Films were spin-coated on (100) oriented and randomly oriented single crystals of yttria-stabilized zirconia (YSZ). Single-crystal $SrTiO_3$ substrates ((100) and (110) orientations) were also used for the study. The concentration of the 1-2-3 TFA mixture was adjusted so as to produce ~ 3 micron thick films at a spin speed of 2000 rpm. The spun-on films were baked at 200°C to remove the water of crystallization and any residual solvent. The dried films were amorphous, as determined by x-ray diffraction, and were featureless.

The superconducting phase was formed by first decomposing the films in air at 400°C. The film thickness after decomposition was 1.0-1.1 μm. Thicker films were prepared by spinning a second layer and decomposing it to build up the thickness to ~2 μm. In the next step, the films were heated in an oven at 850°C for 30-120 min under flowing helium saturated with water vapor. Helium was bubbled through water at room temperature to introduce the water vapor into the oven. For some films, this step was followed by a short anneal (5-10 min) in dry helium at temperatures of 900-920°C. The helium flow was then stopped and replaced by dry oxygen flow and the oven was shut off. The samples were then allowed to cool in the presence of flowing oxygen over a period of 3 hours to below 200°C before being removed from the oven. The thickness of the films decreased from 2 μm to 0.8-1.0 μm after the reaction and annealing steps.

For measuring the resistivity of the films as a function of temperature, a continuous flow cryostat (Oxford Inst. Model CF-1204) with a temperature controller was used. The details of the sample preparation and measurement have been described previously (3). A temperature controlled silicon diode (±0.1 K) was mounted next to the sample for accurate temperature measurement. X-ray powder diffraction patterns were obtained using a computerized Phillips powder diffractometer (Type APD 3520) with Ni-filtered CuKα radiation.

Results and Discussion

Thermogravimetric and differential calorimetric analysis (TGA & DSC) of the individual trifluoroacetate precursors, and their 1-2-3 mixture, was carried out in different atmospheres to study the thermal decomposition under slow heating conditions (4°C/min). Thermogravimetric analysis of the individual TFA's and 1-2-3 mixture, carried out in argon atmosphere, are shown in Fig. 1. The TFA's slowly lose water of crystallization between 50 and 200°C, and then all of them decompose readily between 250 and 350°C. The gaseous products which are evolved have not been identified. Decomposition of $Cu(TFA)_2$ also results in the loss of a small amount of copper through formation of a volatile solid which is probably Cu(I)(TFA). The decomposition temperatures are lowered by as much as 30-70°C when the decomposition is carried out in an oxygen atmosphere. Complete decomposition of the 1-2-3 mixture occurs around 310°C. There is also less loss of copper through volatilization. Differential calorimetric analysis of decompositions, done in air, are shown in Fig. 2. Endothermic peaks due to loss of water are seen at temperatures below 200°C. Multiple exothermic peaks are observed during decomposition of the the individual TFA's, indicating the complex nature of the decomposition process. Almost all the peaks observed during decomposition of the 1-2-3 mixture correspond closely to the peaks of the individual components. An

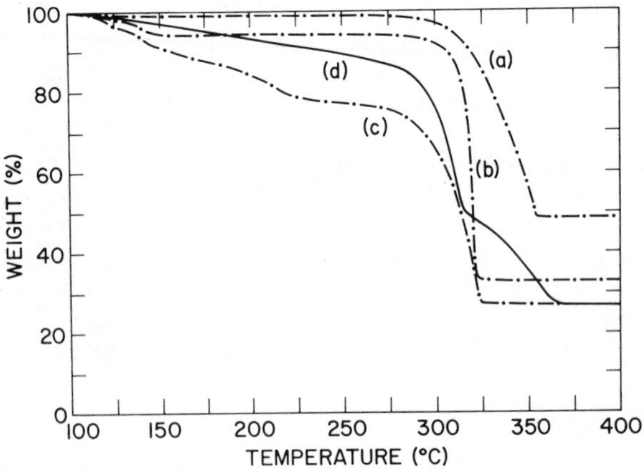

Figure 1. Thermogravimetric analysis of trifluoroacetates in argon at 4 °C/min. (a) Ba, (b) Y, (c) Cu, and (d) 1-2-3 mixture of trifluoroacetates.

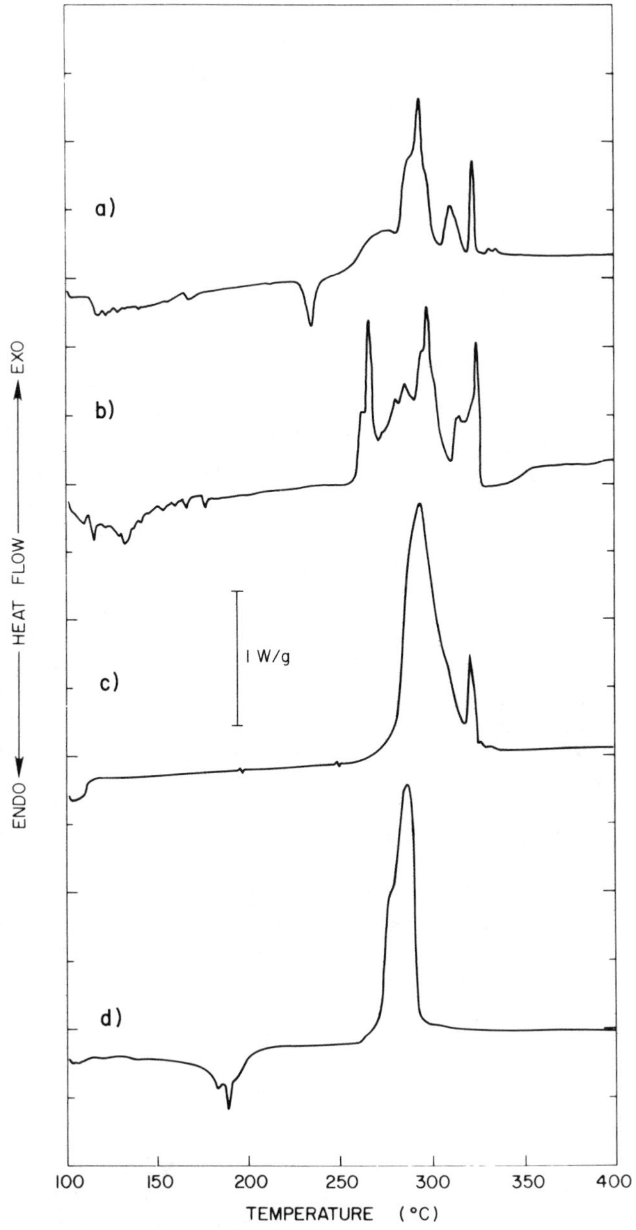

Figure 2. Differential scanning calorimetric analysis of trifluoroacetates in air at 4 °C/min. (a) 1-2-3 mixture, (b) Cu, (c) Ba, and (d) Y trifluoroacetate.

additional endothermic peak is observed at 235°C, just prior to decomposition, which is not observed in any of the individual components. The reason for the occurrence of this additional peak is not understood at present.

Weight loss during thermal analysis and x-ray patterns of the residues (after heating to 550°C) have been used to identify the products formed after decomposition. $Y(TFA)_3$ and $Ba(TFA)_2$ are observed to decompose to the respective fluorides, YF_3 and BaF_2, when the decomposition is done in argon. $Cu(TFA)_2$ forms a mixture of CuF_2, Cu_2O and CuO. Fluoride decomposition products are also observed for $Y(TFA)_3$ and $Ba(TFA)_2$ when the decomposition is carried out in an oxygen atmosphere, whereas $Cu(TFA)_2$ decomposes to form CuF_2 and CuO. Formation of fluorides by decomposition of rare-earth trifluoroacetates has been reported in an earlier study (13). The x-ray pattern of spun-on 1-2-3 films show only small diffraction peaks of BaF_2 after decomposition, suggesting that the products are mostly amorphous. Peaks corresponding to mixed fluoride phase are observed to appear when the films are heated to higher temperatures.

The oxides are formed by reacting the intermediate fluorides with water vapor. The relevant reactions and their standard free energies per mole of reactant fluoride at 298 K are given below.

$$BaF_2 + H_2O \rightarrow BaO + 2HF \qquad \Delta G^0_{298} = 73.4 \text{ Kcal/mol},$$
$$YF_3 + 3/2H_2O \rightarrow 1/2Y_2O_3 + 3HF \qquad \Delta G^0_{298} = 61.6 \text{ Kcal/mol},$$
$$CuF_2 + H_2O \rightarrow CuO + 2HF \qquad \Delta G^0_{298} = 21.5 \text{ Kcal/mol}.$$

The oxides formed in the above reactions combine to form the superconducting 1-2-3 oxide phase. The superconducting phase can also be formed through other intermediate or metastable oxide phases produced during conversion of the fluorides. The higher thermodynamic stability of the fluorides relative to the oxides, particularly for Y and Ba, makes it necessary to run the reaction at high temperatures under a continuous flow of water vapor. The free energies for the above reactions as a function of temperature are shown in Fig. 3 for two different ratios of partial pressures, P_{HF}/P_{H_2O}. These were calculated using the standard free energy values of the reactants and products at different temperatures given in Ref. 14. It is clear from Fig. 3 that the reactions become more favorable at lower temperatures as the vapor pressure of water is increased and/or the pressure of HF is lowered. Operating under a continuous flow of water vapor allows the product gas to be flushed away from the reaction zone. The HF gas is also effectively gettered by the quartz reaction tube. Under our operating conditions at 850°C, YF_3 and CuF_2 in the films are easily converted to oxides, even for reaction periods of a few minutes. Complete conversion of BaF_2 is observed after 45-60 min of reaction time.

X-ray diffraction patterns of films prepared on YSZ substrates, which have been reacted at 850°C for 30 and 60 min, respectively, are shown in Fig. 4(a) and 4(b). The films are annealed in helium at 920°C for 5 min after the reaction step before slow cooling in oxygen. The major peaks of the pattern correspond to the reflections of the superconducting oxide phase. Unreacted barium fluoride is observed in Fig. 4(a), whereas it is completely absent in Fig. 4(b). An unidentified phase with Bragg

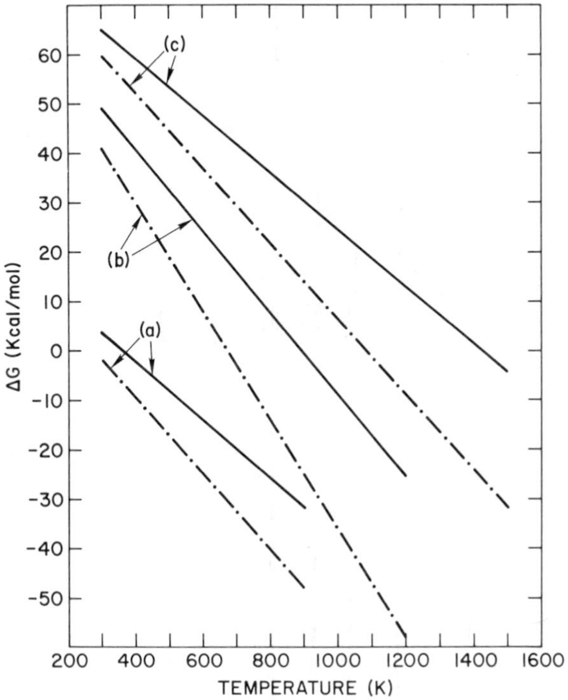

Figure 3. Gibbs free energies for reactions involving conversion of fluorides to oxides as a function of temperature for two different ratios of P_{HF}/P_{H_2O}. Solid line: $P_{HF}/P_{H_2O} = 10^{-3}$, $P_{H_2O} = 1$; dash-dot line: $P_{HF}/P_{H_2O} = 10^{-5}$, $P_{H_2O} = 1$. (a) $CuF_2 \rightarrow CuO$; (b) $YF_3 \rightarrow Y_2O_3$; (c) $BaF_2 \rightarrow BaO$.

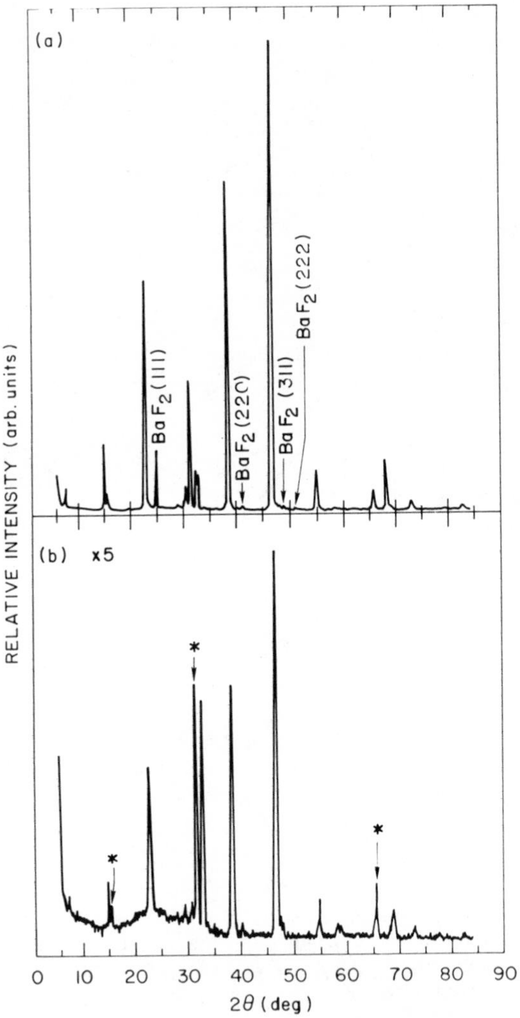

Figure 4. X-ray diffraction patterns of films on YSZ substrates. Process conditions: (a) water reaction at 85O °C for 30 min, annealed in helium at 920 °C for 5 min; (b) water reaction at 85O °C for 60 min, annealed in helium at 920 °C for 5 min. The unlabeled peaks belong to the 1-2-3 phase, BaF_2 peaks are labeled in (a), unidentified peaks marked by "*" in (b). Note that the vertical scale has been expanded x5 in (b) relative to (a).

peaks occurring at 2Θ ~ 15.66, 31.53 and 65.77, is present in both the patterns. It should be noted that the peaks corresponding to the superconducting oxide phase in Fig. 4(a) are primarily the (00l) reflections, indicating a strong c-axis normal preferred orientation. The barium fluoride is also textured with enhanced (111) reflection. The [111] long diagonal in the BaF_2 unit cell is only 1.5% shorter than twice the [110] diagonal in the 1-2-3 unit cell, which may lead to mutual preferred orientation of the two phases. When the film is reacted with water vapor for longer periods so as to convert all the barium fluoride, the preferential orientation of the 1-2-3 phase is significantly reduced, as seen in Fig. 4(b). The results obtained using (100) oriented and randomly cut YSZ substrates are very similar, suggesting that texturing of the film is not related to the substrate alone.

Scanning electron micrographs of superconducting films formed after annealing at two different temperatures (890 and 920°C) are shown in Fig. 5. Both the films were reacted at 850°C for 30 min in presence of water vapor prior to annealing, which leaves some unreacted BaF_2. The superconducting oxide grains are seen to grow mostly as large interconnected platelets which lie parallel to the surface. The film annealed at lower temperature (Fig. 5(a)) also has a significant fraction of needle-like grains growing perpendicular to the surface, suggesting that this film is less textured than the film annealed at higher temperature (Fig. 5(b)). Smaller grains with spheroidal morphology decorate the surface of both the films. Microprobe analysis show that these are barium fluoride grains. These grains are absent in films that are reacted for longer periods to convert all the BaF_2. The superconducting grains in these films are also much more randomly oriented, which is consistent with the x-ray results.

The preferentially oriented films on YSZ substrates exhibit extremely sharp resistive transition with zero resistance obtained at temperatures of 90 K and above. Typical resistivity vs temperature results are shown in Fig. 6 for a 1 μm thick film. The 10-90% transition width is 1.5 K and zero resistance is obtained at 91.4 K. On some films we have observed zero resistance at temperature as high as 94 K, which is one of the highest values reported for films of 1-2-3. When the films are reacted for a longer period to convert all the BaF_2, the transition gets broader, with zero resistance obtained at 80-85 K. The critical current density (J_c) of films with $T_c(R=0)$ higher than 90 K is about 10^3 A/cm^2 at 77 K and 10^4 A/cm^2 at 4 K. Films with broader transitions, which are also less textured, have critical current densities an order of magnitude lower.

We have also prepared films using TFA precursors on $SrTiO_3$ substrates. The x-ray diffraction pattern of a film prepared on (110) oriented $SrTiO_3$ substrate is shown in Fig. 7. Preferred orientation of the film with respect to the substrate, with Bragg peaks near the (110) and (220) reflections is observed. Films grown on (100) oriented $SrTiO_3$ show both a-axis and c-axis preferred orientation (not shown). Unlike the behavior observed on YSZ substrates, the orientation of the film is not dependent on the presence of unreacted barium fluoride. In fact, films on $SrTiO_3$ substrates display better superconducting properties when all the fluorides are reacted. Figure 8 shows the resistance vs temperature of a film grown on (100) $SrTiO_3$. The resistive transition is very sharp, similar to what is observed for films on YSZ substrates. The onset and completion temperatures are, however, a few

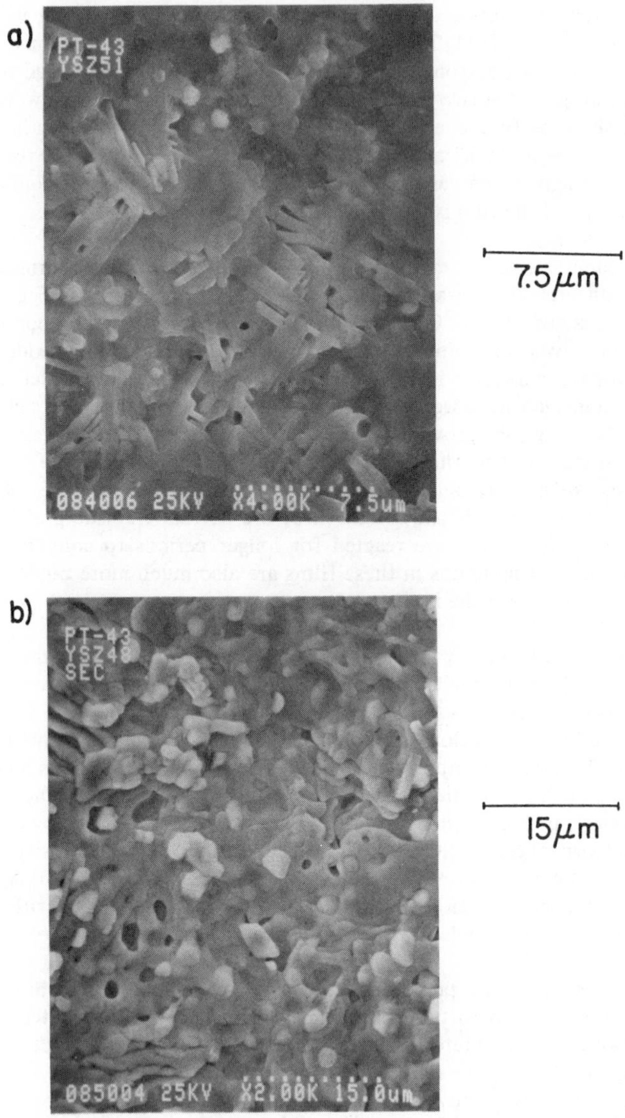

Figure 5. Scanning electron micrographs of films on YSZ substrates. Process conditions: (a) water reaction at 85O °C for 30 min, annealed in helium at 890 °C for 10 min; (b) water reaction at 85O °C for 30 min, annealed in helium at 920 °C for 5 min.

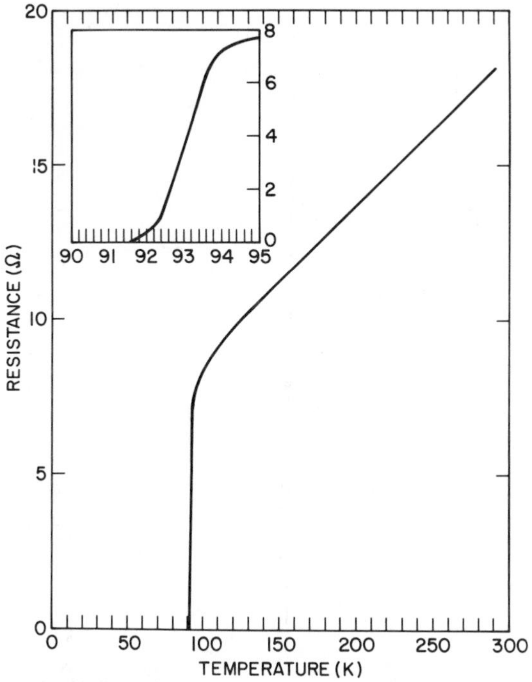

Figure 6. Resistance vs temperature for the same film as in Figure 5(b).

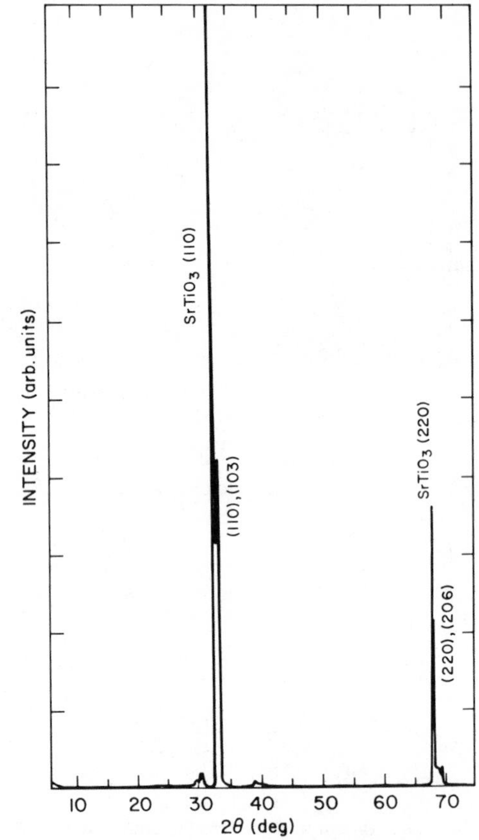

Figure 7. X-ray diffraction pattern of superconducting film on (100) $SrTiO_3$. The superconducting phase was formed by reacting in presence of water vapor at 850 °C for 120 min.

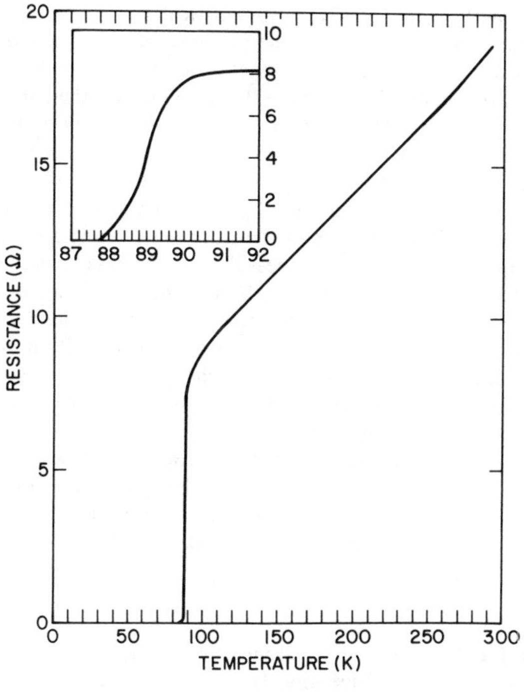

Figure 8. Resistance vs temperature of superconducting film on (100) $SrTiO_3$. Process condition was the same as given for Figure 7.

degrees lower than films on YSZ substates. On the other hand, the critical current density of films on $SrTiO_3$ are significantly better than oriented films on YSZ substrates. Values of J_c around $5x10^5$ A/cm^2 at 4 K (calculated using Bean's formula from magnetization measurements) have been obtained.

The superior properties obtained using fluoride precursors are consistent with results obtained by Mankiewich et al. for films deposited by coevaporation using BaF_2 (9). Our results show that films prepared on YSZ substrates show preferential c-axis normal orientation due to the presence of some unreacted BaF_2. These films also exhibit extremely sharp transitions with zero resistance temperatures higher than 90 K. It is interesting to note that sharp transitions are observed in these films in spite of the presence of secondary phases, unlike films prepared from metal or oxide sources. However, the relatively low critical current densities observed for these oriented films is probably due to the presence of unreacted secondary phases. Unlike YSZ substrates, the preferential orientation obtained on $SrTiO_3$ substrates is due to the close lattice match between the 1-2-3 oxide and the substrate. For films prepared on both substrates, the use of fluoride precursors avoids formation of intermediate carbonates which can degrade the superconducting properties of the films. Another possible reason for the improved properties could be due to fluorine substitution in the superconducting phase, particularly for films on YSZ substrates which show enhanced T_c's when some of the fluorides are left unreacted. An earlier claim of superconductivity in fluorine-substituted bulk samples of 1-2-3 at temperatures as high as 155 K (15) have been largely discounted by a number of recent publications (16-18). There is, however, some evidence that low level fluorine substitution in bulk samples does sharpen the transition and increases the zero resistance temperature to 93 K (18), which is very similar to our results. We have looked for fluorine in the 1-2-3 phase of the films using x-ray microprobe analysis and have not been able to detect it. Any fluorine in the superconducting phase, if present, is below the detection limit of 1 at.%. We are presently studying the properties of the films by systematically changing the amount of unreacted fluoride phase present in the films. This will help to better understand the role of fluorine in improving the superconducting properties of the films.

Acknowledgments

We thank J. I. Landman for electrical measurements, P. J. Bailey for the scanning electron micrographs, T. R. McGuire for magnetic measurements, and K. H. Kellener for microprobe analysis.

Literature Cited

1. Thin Film Processing and Characterization of High-Temperature Superconductors; Harper, J. M. E.; Colton, R. J.; Feldman, L. C., Eds.; American Vacuum Society Series 3, American Physical Society Proceedings No. 165, American Physical Society: New York, NY, 1988; pp 2-211.

2. Kawai, M.; Kawai, T.; Masuhira, H.; Takahasi, M. Jpn. J. Appl. Phys. 1987, 26, L1740.

3. Gupta, A.; Koren, G.; Giess, E. A.; Moore, N. R.; O'Sullivan, E. J. M.; Cooper, E. I. Appl. Phys. Lett. 1988, 52, 163.

4. Kramer, S.; Wu, K.; and Kordas, G. In High-Temperature Superconductors; Brodsky, M. B.; Dynes, R. C.; Kitazawa, K.; Tuller, H. L., Eds.; Materials Research Society Symposium Proceedings Vol. 99; Materials Research Society: Pittsburgh, PA, 1988; p 323.

5. Kumagai, T.; Yokota, H.; Kawaguchi, K.; Kondo, W.; Mizuta, S. Chem. Lett. 1987, 1987, 1645.

6. Rice, C. E.; van Dover, R. B.; Fisanick, G. J. Appl. Phys. Lett. 1987, 51, 1842.

7. Hamdi, A. H.; Mantese, J. V.; Micheli, A. L.; Laugal, R. C. O.; Dungan, D. F.; Zhang, Z. H.; Padmanabhan, K. R. Appl. Phys. Lett. 1987, 51, 2152.

8. Gross, M. E.; Hong, M.; Liou, S. H.; Gallagher, P. K.; Kwo, J. Appl. Phys. Lett. 1988, 52, 160.

9. Mankiewich, P. M.; Scofield, J. H.; Skocpol, W. J.; Howard, R. E.; Dayem, A. H.; Good, E. Appl. Phys. Lett. 1987, 51, 1753.

10. Liou, S. H.; Hong, M.; Kwo, J.; Davidson, B. A.; Chen, H. S.; Nakahara, S.; Boone, T.; Felder, R. J. Appl. Phys. Lett. 1988, 52, 1735.

11. Gupta, A.; Jagannathan, R.; Cooper, E. I.; Giess, E. A.; Landman, J. I.; Hussey, B. W. Appl. Phys. Lett. 1988, 52, 2077.

12. Garner, C. D.; Hughes, B. In Advances in Inorganic Chemistry and Radiochemistry; Emeléus, H. J.; Sharpe, A. G., Eds.; Academic: New York, 1975; Vol. 17, p 5.

13. Rillings, K. W.; Roberts, J. F. Thermochimia Acta 1974, 10, 285.

14. Barin, I.; Knacke, O.; Kubaschewski, O. Thermochemical Properties of Inorganic Substances; Springer-Verlag: New York, 1977.

15. Ovshinsky, S. R.; Young, R. T.; Allred, D. D.; DeMaggio, G.; Van der Leeden, G. A. Phys. Rev. Lett. 1987, 58, 2579.

16. Davies, P. K.; Stuart, J. A.; White, D.; Lee, C.; Chaikin, P. M.; Naughton, M. J.; Yu, R. C.; Ehrenkaufer, R. L. Solid State Commun. 1987, 64, 1441.

17. Wang, H. H.; Kini, A. M.; Kao, H. I.; Appelman, E. H.; Thompson, A. R.; Botto, R. E.; Carlson, K. D.; Williams, J. M.; Chen, M. Y.; Schlueter, J. A.; Gates, B. D.; Hallenbeck, S. L.; Despotes, A. M. Inorg. Chem. 1988, 27, 5.

18. Bansal, N. P.; Sandhuki, A. L.; Farrell, D. E. Appl. Phys. Lett. 1988, 52, 838.

RECEIVED July 13, 1988

Chapter 21

Surface and Interface Properties of High-Temperature Superconductors

H. M. Meyer, III[1], D. M. Hill[1], J. H. Weaver[1], and David L. Nelson[2]

[1]Department of Chemical Engineering and Materials Science, University of Minnesota, Minneapolis, MN 55455

[2]Chemistry Division, Office of Naval Research, Arlington, VA 22217

Electron spectroscopy has been used to identify occupied and unoccupied electronic states of the superconductors $La_{1.85}Sr_{0.15}CuO_4$, $YBa_2Cu_3O_{6.9}$, $Bi_2Ca_{1+x}Sr_{2-x}Cu_2O_{8+y}$, and related cuprate compounds. Results from polycrystalline and single crystalline materials reveal that the valence band emission for the 2-1-4, 1-2-3, and 2-1-2-2 superconductors are remarkably similar, with a low density of states near E_F, a central feature at ~3.4 eV, and shoulders at 2.1 and 5.4 eV that are derived primarily from Cu-O hybrid states. Comparison with calculated densities of states shows the importance of correlation effects. Structure in the empty states can be related to La 5d and 4f, Ba 5d and 4f, and Y 4d empty states. Core level results show the Cu $2p_{3/2}$ main line and satellite structure associated with formal Cu^{2+} configuration. The O 1s emission reveals inequivalent chemical environments. Interfaces studies show substrate disruption, contact formation, and passivation, depending on the overlayer. Ag and Au overlayers have a minimal effect (inert contacts) while the deposition of oxygen-scavenging atoms (Ti, Fe, Cu, Pd, La, Al, In, Bi, and Ge) results in oxygen removal and surface disruption. Deposition of the reactive metal Bi in an activated oxygen atmosphere leads to a passivating overlayer. CaF_2 also serves as a passivating layer since it has no effect on the superconductor surface.

In the eighteen months since the announcement of record breaking superconducting transition temperatures, the scientific community has witnessed unprecedented activity directed at a single set of fundamental and materials challenges. While the focus has shifted rapidly from $La_{2-x}Sr_xCuO_4$ (2-1-4, T_c ~35 K) (Ref. 1) to $YBa_2Cu_3O_{7-x}$ (1-2-3, ~90 K) (Ref. 2) to $Bi_2Ca_{1+x}Sr_{2-x}Cu_2O_{8+y}$ and to $Tl_2Ca_{1+x}Sr_{2-x}Cu_2O_{8+y}$ (2-1-2-2, 85-125 K) (Refs. 3 and 4), common challenges include enumeration of physical and chemical properties, identification of the underlying mechanism of high temperature superconductivity, and integration of these materials with new and existing technologies. The synthesis of high quality samples has allowed the identification of many of the intrinsic characteristics of the superconductor, superceding much of the early work done with less than ideal samples.

0097–6156/88/0377–0280$06.00/0

Photoemission and inverse photoemission spectroscopies have been used to probe the valence band spectra for comparison to theoretical models and have been used extensively to examine chemical changes during interface formation, surface degradation, and passivation (5). In this paper, we first present photoemission and inverse photoemission results for a number of copper-oxide-based polycrystalline and single crystalline 2-1-4, 1-2-3, and 2-1-2-2 compounds. These results have allowed us to identify those parts of the valence band and core level emission features which are intrinsic to the superconductor and those which are due to grain boundary phases. We then examine a number of overlayer/superconductor interface problems, reviewing the effects of deposition of adatoms of reactive metals and semiconductors (Ti, Fe, Cu, Pd, La, Al, In, Bi, and Ge). These reactive systems form surface oxides by withdrawing oxygen and inducing substantial atomic reorganization. Other overlayers do not disrupt the surface, and studies of the formation of Ag or Au overlayers have demonstrated their potential as ohmic contacts. We then demonstrate that Bi_2O_3 and CaF_2 can be used to form layers for passivation and encapsulation.

Experimental

Polycrystalline $La_{1.85}Sr_{0.15}CuO_4$ samples were synthesized by co-precipitation of the oxalate salts of La, Sr, and Cu with the proper stoichiometry, followed by sintering in air at 1100 °C for ~12 hours (6). These samples were 95% dense, were shown by neutron diffraction to be single phase, and had superconducting transition temperatures of ~35 K. Polycrystalline $YBa_2Cu_3O_{6.9}$ was formed from a fine stoichiometric powder obtained from Rhone-Poulenc (France) and subjected to a step-wise heating schedule in O_2, achieving a maximum temperature of 950 °C after 12 hours and returning to room temperature over an additional 19 hours. The measured density was 93%. Both $La_{1.85}Sr_{0.15}CuO_4$ and $YBa_2Cu_3O_{6.9}$ were obtained from D.W. Capone II of Argonne National Laboratory.

Single crystals of Pr_2CuO_4, Eu_2CuO_4, Gd_2CuO_4, and $NdBa_2Cu_3O_x$ were grown in air from stoichiometric melts using PbO-based fluxes. $GdBa_2Cu_3O_x$ single crystals were grown by slow cooling a flux containing 10 at.% Gd_2O_3, 30 at.% BaO_2, and 60 at.% CuO in flowing oxygen. The Pr_2CuO_4, Eu_2CuO_4, and Gd_2CuO_4 crystals were undoped and crystallized in a tetragonal structure related to the K_2NiF_4 structure but with oxygen atoms not in the Cu-O planes rotated 45° about the z-axis. The $NdBa_2Cu_3O_x$ and $GdBa_2Cu_3O_x$ single crystals were tetragonal with x ~ 6 and were semiconducting. The 2-1-4 samples were undoped and were also semiconducting. These samples were grown by D. Peterson and Z. Fisk of Los Alamos National Laboratory.

Samples of polycrystalline and single crystalline $Bi_2Ca_{1+x}Sr_{2-x}Cu_2O_{8+y}$ were examined. The sintered material (received from Argonne National Laboratory) was synthesized from a melt of Bi_2O_3, $SrCO_3$, CaO, and CuO powders, was splat quenched on a cooled copper plate, and was annealed at 780 °C in O_2/air. Resistivity measurements showed this to produce material with a transition temperature of 85 K. The single crystals were grown by C.F. Gallo of 3M from a eutectic melt of the parent powders, followed by annealing in O_2/air. Magnetic susceptibility measurements indicated that the single crystals were superconducting at 85 K.

X-ray photoemission (XPS) studies were performed in an ultrahigh vacuum system described in detail elsewhere (7). A monochromatized photon beam (Al K_α, $h\nu = 1486.6$ eV) was focussed onto the sample surface and the emitted electrons were energy dispersed with a Surface Science Instruments hemispherical analyzer onto a position-sensitive detector. The size of the x-ray beam was 300 μm and the pass energy of the analyzer was 50 eV, giving an instrumental resolution of 0.6 eV. The small spot size was particularly important for these studies because the single crystal samples were often little more than ~1 mm in diameter. Samples were prepared for study by either fracturing or cleaving *in situ* at operating pressures of $5\text{-}10 \times 10^{-11}$

Torr. For the inverse photoemission (IPES) measurements, a monochromatic low energy electron beam was directed onto a sample at normal incidence and the emitted photons were monochromatized and then energy analyzed with a position sensitive detector. The size of the electron beam was ~1x5 mm and the overall resolution was 0.3-0.6 eV for photon energies 12-44 eV (8). For interface studies, overlayers were deposited from well outgassed thermal evaporation sources at pressures below 5 x 10^{-10} Torr with evaporation rates of ~1 Å/min.

Results and Discussion

Clean Surfaces. In Fig. 1 we summarize our valence band and shallow core level results for seven single crystalline and polycrystalline rare earth copper oxides. The four curves at the bottom are for sintered, Sr-doped La_2CuO_4 and single crystals of undoped, semiconducting Pr_2CuO_4, Eu_2CuO_4, and Gd_2CuO_4. The top three are for sintered $YBa_2Cu_3O_{6.9}$ and single crystals of $NdBa_2Cu_3O_x$ and $GdBa_2Cu_3O_x$ (x ~ 6 for these semiconducting samples). To simplify visual comparison, the valence band spectrum of $La_{1.85}Sr_{0.15}CuO_4$ is superimposed on the other spectra, shifted to align their leading edges.

The valence band results can be represented by a relatively invariant Cu-O manifold derived from Cu 3d and O 2p levels with small variations in energy due to the movement of E_F. The process of photoemission alters the distribution of states from the ground state because of the creation of a photohole and the response of the correlated electrons to this hole. This complicates comparisons with ground state band calculations, to be discussed shortly. In Fig. 1, it is also possible to detect the relatively complex emission from the rare-earth-derived 4f electrons. The approximate energy of the O 2s emission is identified by the vertical line at 20 eV. Other features include the Ba $5p_{1/2,3/2}$ doublet (at 12.5 and 14.2 eV for $YBa_2Cu_3O_{6.9}$) and the Ba 5s emission centered at 28 eV for the 1-2-3 samples. The rare earth 5p levels exhibit complex lineshapes because of 5p-4f multiplet coupling and shift to greater binding energy with increasing atomic number (9).

In each of the XPS valence band spectra of Fig. 2, we have identified features at ~12 eV (labeled A) and at ~9 eV (labeled B). The 12 eV feature is derived from the excitation of a d electron from the Cu $3d^9$ ground state configuration, leaving two correlated 3d holes on a Cu site, denoted d^8. This feature has been discussed extensively in the literature (10). The feature at ~9 eV has been the source of more controversy. Two extrinsic origins for this feature have been proposed: contamination-related C 2s states and surface hydroxyl species. They have been eliminated by studies of carbon-free samples (1) and by failure to enhance this feature via room temperature water adsorption (9). The appearance of the 9 eV feature in all of the rare-earth-substituted single crystals and in the carbon free polycrystalline samples suggests that it is intrinsic and associated with either copper or oxygen. The absence of Cu 3p-3d resonance enhancement eliminates copper and leaves only oxygen (11). Indeed, Wendin (12) has suggested that the 9 eV feature might reflect two holes localized on an oxygen site in a photoemission final state akin to the Cu d^8 satellite at ~12 eV.

In Fig. 2 we emphasize the similarities in the valence bands for the 2-1-4, 1-2-3, and 2-1-2-2 superconductors by showing the occupied and unoccupied electronic states extending ~14 eV on either side of E_F for $La_{1.85}Sr_{0.15}CuO_4$ and $YBa_2Cu_3O_{6.9}$ and the occupied states for $Bi_2Ca_{1+x}Sr_{2-x}Cu_2O_{8+y}$. These spectra were obtained from nearly-carbon-free samples that were fractured or cleaved *in situ* in ultrahigh vacuum just prior to analysis (no C 1s emission for polycrystalline $La_{1.85}Sr_{0.15}CuO_4$ and less than 2 at.% for polycrystalline $YBa_2Cu_3O_{6.9}$ and single crystal $Bi_2Ca_{1+x}Sr_{2-x}Cu_2O_{8+y}$). The valence band similarities are striking with low state density near E_F, central features at ~3.4 eV, and shoulders at 2.1 and 5.4 eV. The d^8 satellite at 12 eV and the oxygen related feature at 9 eV are clearly seen for the

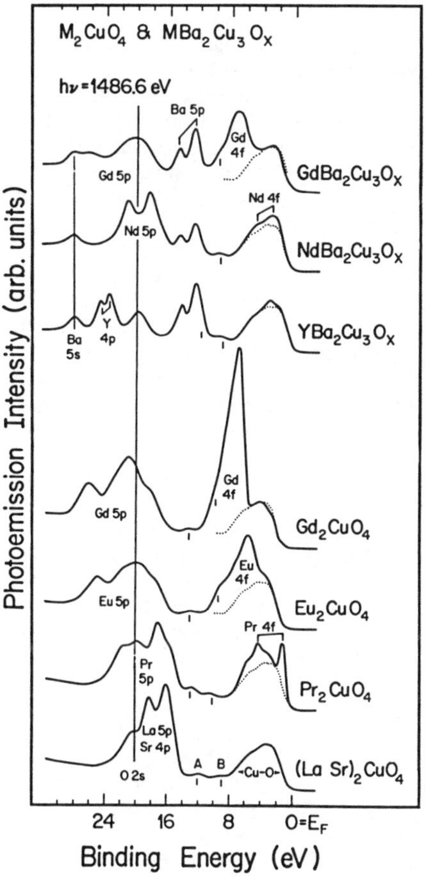

Figure 1. X-ray photoemission spectra for 1–2–3 and 2–1–4 copper oxides. The valence bands near the Fermi level are dominated by Cu–O hybrid states. These are not affected by the substitution of rare earth ions, but new 4f related features can be identified by subtraction of the $La_{1.85}Sr_{0.15}CuO_4$ baseline curve (dotted line).

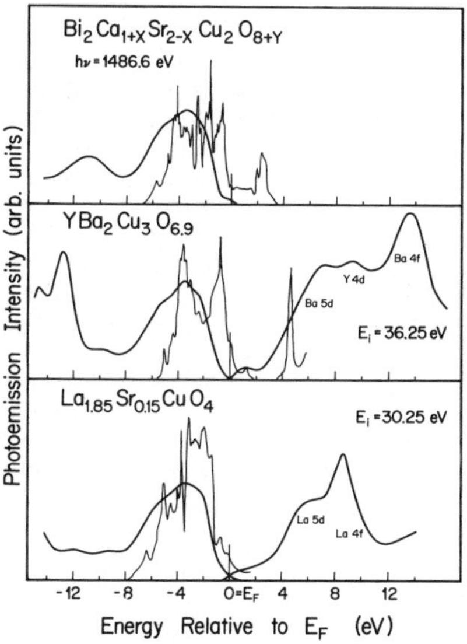

Figure 2. Comparison of calculated densities of states (from refs. 14, 15, and 15, top to bottom) and X-ray photoemission results for the 1–2–3, 2–1–4, and 2–1–2–2 superconductors. The offset of 1–1.5 eV in the centroids of the experimental and theoretical results is taken to reflect the photoemission process of removing an electron from a correlated system. The empty state features have been obtained with inverse photoemission, and their origin is identified.

$La_{1.85}Sr_{0.15}CuO_4$, but the d^8 satellite is obscured by Ba 5p emission for $YBa_2Cu_3O_{6.9}$ (it can be distinguished easily in synchrotron radiation photoemission studies at the Cu 3p-3d resonance[11]). Both the 12 and the 9 eV features are masked by Bi 6s emission for $Bi_2Ca_{1+x}Sr_{2-x}Cu_2O_{8+y}$. The unoccupied states for $La_{1.85}Sr_{0.15}CuO_4$ and $YBa_2Cu_3O_{6.9}$ have been discussed elsewhere (13). For comparison with theory, we have superimposed the calculated density of states (DOS) from Refs. 14 and 15. Inspection shows that the calculated state densities near E_F are consistently higher than those observed experimentally. Interestingly, the calculations for $Bi_2Ca_{1+x}Sr_{2-x}Cu_2O_{8+y}$ show the fewest states at E_F, while the 2-1-2-2 exhibits the largest emission and the best defined Fermi cutoff of the three sample types. Moreover, there is an overall shift in the DOS centroid by 1-1.5 eV relative to experiment. This shift can be understood in terms of the correlation between the photohole and the remaining electrons, as noted above. Fujimori *et al.* (16) and Shen *et al.* (17) have estimated the Coulomb (Hubbard) energy to be ~6 eV.

The O 1s and Cu $2p_{3/2}$ spectra for these same 2-1-4, 1-2-3, and 2-1-2-2 samples are shown in Fig. 3. The Cu $2p_{3/2}$ lineshape indicates a nominal 2+ valence state, analogous to that discussed by Sawatzky and coworkers for a number of divalent Cu halides (18). The broad doublet at higher binding energy is derived from 8 multiplets which reflect the interaction of the $\underline{2p}_{3/2}$ core hole (denoted by underlining) with nine screening 3d electrons. This satellite is centered at 942 eV for each of the samples, consistent with the argument that the $\underline{2p}_{3/2}3d^9$ final state energies should be independent of ligand influences (18). The main line, centered near 933 eV, is derived from ligand screening of the 2p core hole, namely $\underline{2p}_{3/2}d^{10}\underline{L}$ where L denotes a suitably symmetrized ligand p orbital. The width of these Cu core level features is believed to be due to mixing of the d^9 and $d^{10}\underline{L}$ ground state configurations since they are close in energy (18). Experimentally, we find the full widths at half maximum of the main lines for $La_{1.85}Sr_{0.15}CuO_4$ and $YBa_2Cu_3O_{6.9}$ to be nearly equal. In contrast, the main line for $Bi_2Ca_{1+x}Sr_{2-x}Cu_2O_{8+y}$ is ~0.5 eV wider and there is a distinct asymmetry at both lower and higher energy. We have no explanation for this as yet, but note that it has been observed in ten single crystal and polycrystalline $Bi_2Ca_{1+x}Sr_{2-x}Cu_2O_{8+y}$ samples. From Fig. 3, it can also be seen that the $YBa_2Cu_3O_{6.9}$ main line is shifted 0.2 eV to lower binding energy relative to that in $La_{1.85}Sr_{0.15}CuO_4$ while that in $Bi_2Ca_{1+x}Sr_{2-x}Cu_2O_{8+y}$ is shifted 0.3 eV to higher binding energy. This can be accounted for by slight differences in ligand screening, possibly due different bond lengths or wave function overlap.

At the right of Fig. 3, we show the O 1s spectra for these 2-1-4, 1-2-3, and 2-1-2-2 samples. In all cases, the spectra exhibit a wide, asymmetric main peak, indicating the presence of more than one chemical environment. These can be identified more readily after lineshape analysis which produces the decomposition shown by the dashed lines. These fits were guided by examination of single crystals of CuO, which provided the full width at half maximum fitting parameter. Moreover, they were done self-consistently for the O 1s core levels for all of the samples summarized by Fig. 1.

For the 2-1-4 samples, the O 1s emission is made up of two approximately equal main peaks split by ~0.7 eV. The shallower peak is derived from oxygen atoms in the Cu-O planes and can be written (O $\underline{1s}$ $2p^6$) $3d^9$ where the 1s hole is screened by $2p^6$ orbitals and $3d^9$ denotes one of the two possible ground state configurations. The small peak at ~1.5 eV higher binding energy than the Cu-O oxygen feature is related to the second configuration, (O $\underline{1s}$ $2p^6$) $3d^{10}\underline{L}$. The deeper of the two large peaks arises from oxygen atoms bound to La (and Sr) in the more ionic arrangement of the off-planes of the unit cell. For more details, see Ref. 9

For the 1-2-3 materials, the O 1s spectrum is narrower than for the 2-1-4's. It is made up of a dominant component at ~529 eV derived from Cu-O planes (accounting for 4 of 7 oxygen atoms in a formula unit) and a shallower component that we attribute to Cu-O chains (accounting for the remainder oxygen atoms). Again, we identify the

(O 1s $2p^6$) $3d^{10}$L feature at ~530 eV, although it overlaps with what is probably contamination-phase oxygen emission at ~531 eV.

For the 2-1-2-2 structure, the assignment of the peak components is less straightforward because we currently have less information and fewer systematics. Based on previous assignments and using equivalent linewidths for fitting, we can make tentative assignments as follows. The shallowest component is probably associated with oxygen atoms of the Cu-O planes since these are the least ionic. The next higher binding energy component belongs to oxygen atoms bound to alkali earth atoms and oxygen atoms bound to Bi atoms in off-plane sites. Oxygen atoms in the Bi-O planes are responsible for the third component. We speculate that the small, highest binding energy component is due to the (O 1s $2p^6$) $3d^{10}$L satellite, although this feature is not as intense as in the spectra for the other crystal types.

Interface Studies. The integration of high temperature superconductors with existing or new technologies will require the solution of a large number of materials-related problems. Major efforts must consider the formation of stable contacts and passivating layers. Indeed, early work by ourselves and others has shown that the surfaces of these high T_c ceramics are highly reactive. In particular, adatoms which are reactive with respect to oxide formation cause disruption of the 2-1-4, 1-2-3, and 2-1-2-2 surfaces by withdrawing oxygen to form oxide overlayers with thicknesses that are probably kinetically limited (5,19,20). The loss of oxygen (and superconductivity) from the surface region can be readily observed through changes in the Cu $2p_{3/2}$ satellite feature and sharpening of the main line (conversion from nominal 2+ to 1+ valence state). We have previously shown that Ag and Au are chemically inert, producing a metal layer that leaves the intrinsic Cu $2p_{3/2}$ and O 1s features unchanged (21). In the following, we use Cu reduction to characterize overlayer reactivity and focus on reactive overlayers (Cu and Bi) and passivating layers (CaF_2 and Bi_2O_3).

Examination of the Cu $2p_{3/2}$ spectra following Cu deposition on $La_{1.85}Sr_{0.15}CuO_4$ and $YBa_2Cu_3O_{6.9}$ (Refs. 19 and 5) shows that the extent of reactivity is approximately the same, as judged from the fact that the satellite intensity disappears after the deposition of the same amount of Cu in both cases. In particular, the deposition of the equivalent to 4 Å is sufficient to reduce all Cu^{2+} to Cu^{1+} within the probe depth of our XPS measurements (50-60 Å), implying that the crystal structures of the 2-1-4 and 1-2-3 materials are prone to extensive disruption/rearrangement as oxygen is withdrawn and reacts with the adatoms. In contrast, the analogous results for Cu overlayer growth on polycrystalline $Bi_2Ca_{1+x}Sr_{2-x}Cu_2O_{8+y}$ indicate the persistence of significant $2p_{3/2}$ satellite intensity after the deposition of 16 Å, although there is evidence for limited surface reaction at lower coverage. Since the attenuation of the $Bi_2Ca_{1+x}Sr_{2-x}Cu_2O_{8+y}$ substrate emission by Cu adatoms is consistent with approximately layer-by-layer growth of the Cu, we conclude that it is more stable than the 2-1-4's or 1-2-3's (22).

In Fig. 4 we present Cu $2p_{3/2}$ EDC's which show the effects of the deposition of Bi adatoms on polycrystalline $YBa_2Cu_3O_{6.9}$ and $Bi_2Ca_{1+x}Sr_{2-x}Cu_2O_{8+y}$. Bismuth was chosen because of its relatively low heat of oxide formation (~145 kcal/mole) and its promise as a passivation or metallization layer. Moreover, it is a constitutent of the 2-1-2-2 compound, and studies have shown the tendency of the 2-1-2-2 material to be more stable than the 1-2-3's or 2-1-4's. This stability may be a consequence of the intrinsic Bi-O double layer structure. The spectra shown of Fig. 4 have been normalized to constant height to emphasize lineshape changes. The results for Bi/$YBa_2Cu_3O_{6.9}$ (lower half) indicate a reduction in the satellite intensity and a sharpening of the main line with increasing Bi coverage. However, there is still evidence for Cu^{2+} after the deposition of 50 Å of Bi . This is consistent with limited conversion of Cu^{2+} to Cu^{1+}, and we conclude that oxygen withdrawal is much less than is observed when equivalent amounts of Cu or Ti are deposited (5). Although the

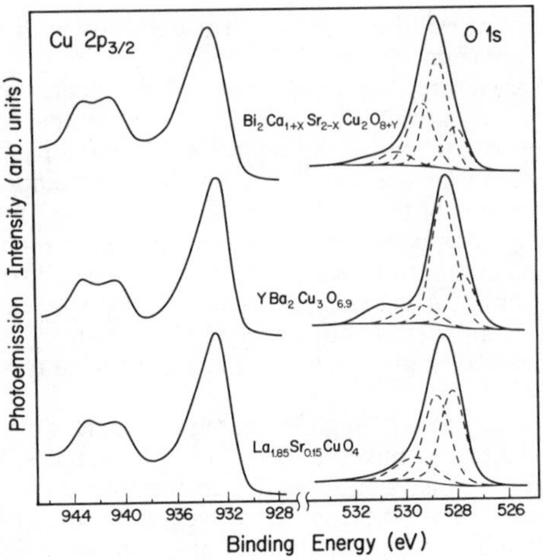

Figure 3. Cu $2p_{3/2}$ and O 1s emission for the 1-2-3, 2-1-4, and 2-1-2-2 superconductors. For Cu, the satellite feature indicates nominal 2+ bonding configurations. The background is different for $Bi_2Ca_{1+x}Sr_{2-x}Cu_2O_{8+y}$ because of superposition with Bi 4s emission at ~938 eV. Lineshape analysis for the O 1s emission reveals inequivalent chemical configurations corresponding to oxygen in Cu-O planes, off-planes, chains, and Bi-O planes for the various materials, as discussed in the text.

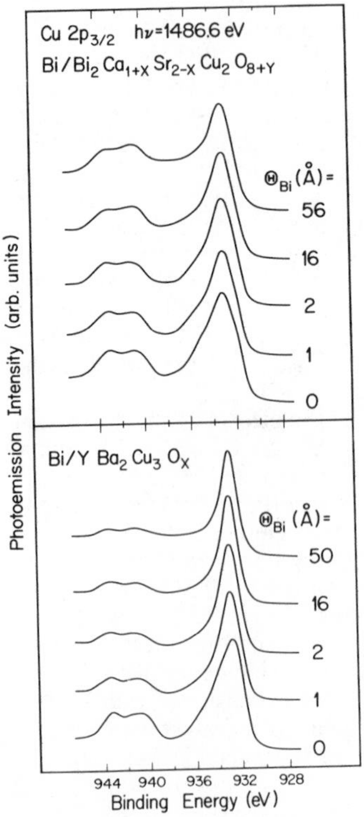

Figure 4. Changes in Cu $2p_{3/2}$ core level emission for Bi deposition onto $YBa_2Cu_3O_{6.9}$ and $Bi_2Ca_{1+x}Sr_{2-x}$-Cu_2O_{8+y}. The loss of the satellite doublet-structure reflects chemical conversion of Cu atoms from 2+ to 1+ configurations. In both cases, the reaction is much less than for adatoms of Ti, Fe, Cu, Pd, La, Al, In, and Ge. The spectra have been normalized to emphasize lineshape changes; the total emission decreases during interface formation. Its persistence to ~50 Å reveals incomplete covering because of the complex morphology of fractured polycrystalline surfaces.

persistence of Cu emission at this coverage suggests that there has been shadowing of the substrate because of its irregular, fractured-surface morphology, it is possible to estimate the extent of oxygen withdrawal from changes in the Cu $2p_{3/2}$ satellite-to-main-line ratio. (Note that the background changes with Bi deposition due to the growth of the Bi 4s emission at ~ 938 eV.) This ratio decreases from 0.35 for the clean surface to 0.18 for 8 Å. Further deposition of Bi does not change this ratio, indicating that the reaction is complete.

In the right panel of Fig. 5 we show the corresponding growth of the Bi 4f emission for Bi/$YBa_2Cu_3O_{6.9}$. These spectra are again normalized to constant height to emphasize chemical changes. The deposition of Bi leads to the growth of two separate Bi $4f_{5/2,7/2}$ doublets although the $4f_{7/2}$ peaks overlap with the Y 3d emission from the substrate (compare to bottom EDC). After the deposition of 1 Å of Bi, there is a prominent feature at ~165 eV and a smaller feature at 3 eV lower binding energy. These two components correspond to Bi-O and Bi metal, respectively. Lineshape decomposition and normalized area analysis show that the oxide peak grows in absolute magnitude until a total Bi deposition of 8 Å, corresponding to the point where the Cu $2p_{3/2}$ satellite/main line intensity ratio stabilizes. Concurrently, the metallic component grows more rapidly, as can be seen by visual inspection of Fig. 5. After 8 Å, the metallic peak continues to grow and attenuates the substrate and BiO_x interlayer. We conclude that the reaction of Bi with the substrate is probably initiated at nucleation sites and that these sites, followed by lateral growth of BiO_x across the surface until a continuous layer is formed. The fact that we observe the simultaneous formation of Bi metal indicates that the thickness of this oxide interlayer is very small and that it impedes the movement of oxygen into the region where the unreacted Bi atoms are deposited (kinetically limited reaction since the Bi-O reaction would be thermodynamically favored). Such nonuniform reaction and interlayer evolution has been observed during the equivalent growth of metal overlayers on semiconductor surfaces (23).

In the upper panel of Fig. 4 and the left panel of Fig. 5, we show the Cu $2p_{3/2}$ and Bi 4f results for the Bi/$Bi_2Ca_{1+x}Sr_{2-x}Cu_2O_{8+y}$ interface. Inspection reveals that the Cu satellite/main intensity ratio changes very little for Bi depositions up to 56 Å, although it is difficult to ascertain this ratio precisely because of the changing background (underlying Bi 4s from Bi of the superconductor plus growth due to deposition of Bi metal at ~938 eV). From the sharpening of the main line for depositions to ~2 Å, however, it can be inferred that some reduction of Cu^{2+} has occurred. The Bi 4f emission shows an interesting shift to 0.37 eV greater binding energy for the substrate Bi emission, suggesting that there has been a redistribution of oxygen within the near-surface Bi-O planes. This is consistent with the limited disruption of Cu-O planes, as inferred from the Cu emission. This shift is complete by 2 Å deposition and, in fact, the Bi 4f emission shows the presence of non-oxidized Bi (component 2 in Fig. 5). Analysis indicates a nearly constant intensity for the Bi oxide emission up to 2 Å Bi deposition, but a continuous reduction thereafter. We conclude that new Bi-O bonding configurations are being formed during the first few Ångstroms of Bi deposition, but that Bi metal attenuates this oxide emission. Comparison with the results for Bi deposited onto $YBa_2Cu_3O_{6.9}$ indicates that the 2-1-2-2 compound is substantially less reactive, and we speculate that this may be due to the relative stability of the bilayers of Bi-O.

To address the need for passivating layers that will preserve the integrity of the superconductor, we undertook another series of interface studies, again using the Cu $2p_{3/2}$ lineshape as an indicator of reaction and evaluating the emission ratio of the satellite to the main line. In Fig. 6, we show the results for a range of overlayers on $YBa_2Cu_3O_{6.9}$. The spectrum at the top demonstrates that 2.5 Å of Cu leads to almost-complete loss of the satellite and, therefore, the conversion of Cu^{2+} within the probe depth of 50-60 Å of the surface. In contrast, the deposition of 8 Å of Bi leads to much less disruption, and 10 Å of Au does not alter the Cu $2p_{3/2}$ lineshape. The

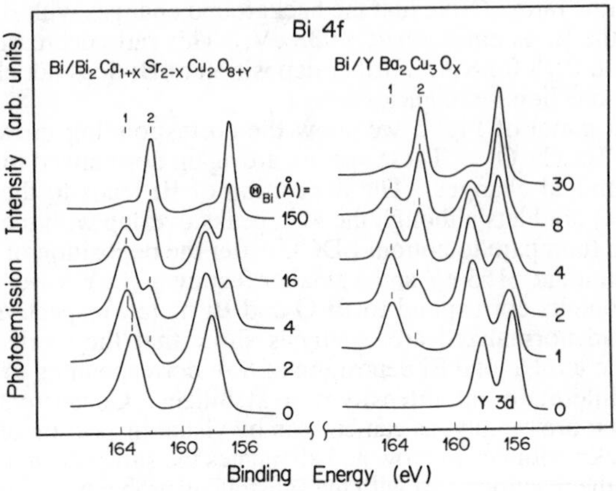

Figure 5. Comparison of the evolution of Bi overlayers on $Bi_2Ca_{1+x}Sr_{2-x}Cu_2O_{8+y}$ (left) and $YBa_2Cu_3O_{6.9}$ (right). The Bi $4f_{5/2}$ component #1 reflects Bi-O bonds of either the substrate or the overlayer. Component #2 is attributed to Bi metal which is seen to be present after the deposition of very small amounts of material. As discussed in the text, we conclude that Bi is a relatively unreactive metal on these superconductors.

deposition of 10 Å of CaF_2 also leaves the substrate completely intact with no evidence of disruptive interactions. Moreover, CaF_2 appears to cover the surface uniformly, as judged by the rapid attenuation of the substrate emission. In contrast to the other overlayers considered above, CaF_2 is a large bandgap insulator with a high static dielectric constant and it may be useful as an insulating layer in device fabrication (it is used as a dielectric grown epitaxially on GaAs or Si).

As part of these passivation studies, we examined the deposition of metals in an activated-oxygen environment at 1 x 10^{-6} Torr partial pressures of oxygen. In this way, it was possible to provide activated oxygen from the gas phase to form metal oxide precursors that would not react with oxygen from the superconductor. In Fig. 6 we compare the results for Bi deposited onto $YBa_2Cu_3O_{6.9}$ both with and without activated oxygen. As shown, the Cu 2p lineshape did not change following the deposition of 6 Å of Bi in activated oxygen. We conclude that it is possible to form metal oxide passivating overlayers on the superconductor without affecting the substrate, but that the metal adatoms must be able to combine with oxygen from a source other than the substrate.

These results have implications regarding contact formation with copper-oxide-based superconductors. Of the reactive metals studied to date, Bi appears to be the least disruptive. Ag and Au are completely nonreactive. Comparisons of the 1-2-3 and 2-1-4 ceramics indicate comparable reactivity for a given metal overlayer. The 2-1-2-2 Bi superconductor is more stable with respect to the loss of oxygen, but it also exhibits reaction, as was demonstrated by studies of Cu overlayers (22). This reactivity can be altered by providing oxygen from an external source during deposition of the metal atoms, as our results for Bi deposition demonstrate. Finally, CaF_2 forms a passivating layer without inducing substrate modification.

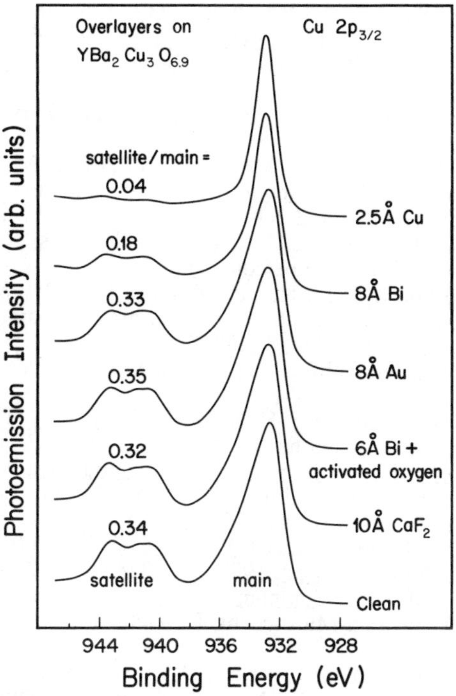

Figure 6. Comparison of changes in Cu $2p_{3/2}$ lineshape for highly reactive Cu adatoms, less reactive Bi adatoms, nonreactive CaF_2, and Bi that had been evaporated in an activated oxygen environment. As shown, this deposition process for Bi produces an overlayer which has not altered the bonding configurations of the underlying superconductor. The ratio of satellite to main line emission is a direct measure of the reduction of Cu from its 2+ configuration.

Acknowledgments

This work was supported by the Office of Naval Research under ONR N00014-87-K-0029 and DARPA. Stimulating discussions with T.J. Wagener are gratefully acknowledged.

Literature Cited

1. Bednorz, J.G.; Müller, K.A. Z. Phys. B 1986, 64, 189.
2. Chu, C.W.; Hor, P.H.; Meng, R.L.; Gao, L.; Huang, Z.J.; Wang, Y.Q. Phys. Rev. Lett. 1987, 58, 405.
3. Maeda, H.; Tanaka, Y.; Fukutomi, M.; Asano, T. Jpn. J. Appl. Phys. 1988, 27, L209.
4. Sheng, Z.Z.; Hermann, A.M. Nature 1988, 332, 55.
5. An exhaustive list of experimental and theoretical results for 2-1-4 and 1-2-3 type superconductors can be found in Meyer III, H.M.; Hill, D.M.; Wagener, T.J.; Gao, Y.; Weaver, J.H.; Capone II, D.W.; Goretta, K.C. Phys. Rev. B (in press). The number of such papers is now very large.

6. Jorgensen, J.D.; Schuttler, H.B.; Hinks, D.G.; Capone II, D.W.; Zhang, K.; Brodsky, M.B.; Scalapino, D.J. Phys. Rev. Lett. 1987, 58, 1024.
7. Chambers, S.A.; Hill, D.M.; Xu, F.; Weaver, J.H. Phys. Rev. B 1987, 34, 634.
8. Gao, Y.; Grioni, M.; Smandek, B.; Weaver, J.H.; Tyrie, T. J. Phys. E (in press).
9. Weaver, J.H.; Meyer III, H.M.; Wagener, T.J.; Hill, D.M.; Gao, Y.; Peterson, D.; Fisk, Z.; Arko, A.J. Phys. Rev. B (submitted - photoemission investigations of single crystals of 1-2-3 and 2-1-4 superconductors). Meyer III, H.M.; Hill, D.M.; Weaver, J.H.; Nelson, D.L.; Gallo, C.F. Phys. Rev. B (submitted - single crystals of Bi 2-1-2-2 superconductors).
10. Thuler, M.R.; Benbow, R.L.; Hurych, Z. Phys. Rev. B 1982, 26, 669.
11. Kurtz, R.L.; Stockbauer, R.L.; Mueller, D.; Shih, A.; Toth, L.E.; Osofsky, M.; Wolf, S.A. Phys. Rev. B 1987, 36, 8818.
12. Wendin, G.; 14th Inter. Conf. X-Ray and Inner-Shell Processes, J. Physique (in press).
13. Meyer III, H.M.; Gao, Y.; Hill, D.M.; Wagener, T.J.; Weaver, J.H.; Flandermeyer, B.K.; Capone II, D.W. in Thin Film Processing and Characterization of High Temperature Superconductors, American Institute of Physics Conference Proceedings No. 165, Harper, J.M.E; Colton, R.J. Feldman, L.C., Eds.: New York, NY, 1988, p. 254.
14. Massidda, S.; Yu, J.; Freeman, A.J. Physica C (in press) for Bi 2-1-2-2.
15. Mattheiss, L.F. Phys. Rev. Lett. 1987, 58, 1028 for La 2-1-4. Mattheiss, L.F.; Hamann, D.R. Solid State Commun. 1987, 63, 395 for Y 1-2-3.
16. Fujimori, A.; Takayama-Muromachi, E.; Uchida, Y.; Okai, B. Phys. Rev. B 1987, 35, 8814.
17. Shen, Z.; Allen, J.W.; Yeh, J.J.; Kang, J.S.; Ellis, W.; Spicer, W.; Lindau, I.; Maple, M. B.; Dalichaouch, Y. D.; Torikachvili, M.S.; Sun, J.Z.; Geballe, T.H. Phys. Rev. B 1987, 36, 8414.
18. van der Laan, G.; Westra, C.; Haas, C.; Sawatzky, G.A. Phys. Rev. B 1981, 23, 4369.
19. Gao, Y.; Meyer III, H.M.; Wagener, T.J.; Hill, D.M.; Anderson, S.G.; Weaver, J.H.; Flandermeyer, B. K.; Capone II, D.W. in Thin Film Processing and Characterization of High Temperature Superconductors, American Institute of Physics Conference Proceedings No. 165, Harper, J.M.E; Colton, R.J. Feldman, L.C., Eds.: New York, NY, 1988, p. 358.
20. Hill, D.M.; Meyer III, H.M.; Weaver, J.H.; Flandermeyer, B.; Capone II, D.W. Phys. Rev. B 1987, 36, 3979; Hill, D.M.; Gao, Y.; Meyer III, H.M.; Wagener, T.J.; Weaver, J.H.; Capone II, D.W. Phys. Rev. B 1988, 37, 511; Meyer III, H.M.; Hill, D.M.; Anderson, S.G.; Weaver, J.H.; Capone II, D.W. Appl. Phys. Lett. 1987, 51, 1750.
21. Meyer III, H.M.; Hill, D.M.; Anderson, S.G.; Weaver, J.H.; Capone II, D.W. Appl. Phys. Lett. 1987, 51, 1118; Wagener, T.J.; Gao, Y.; Vitomirov, I.M.; Joyce, J.J.; Capasso, C.; Weaver, J.H.; Capone II, D.W. Phys. Rev. B (in press -- noble metal and Pd overlayers on $YBa_2Cu_3O_{6.9}$).
22. Meyer III, H.M.; Hill, D.M.; Weaver, J.H.; Gallo, C.F. Phys. Rev. B (Cu overlayer growth on Bi-based 2-1-2-2 superconductors, in preparation).
23. del Giudice, M.; Joyce, J.J.; Ruckman, M.W.; Weaver, J.H. Phys. Rev. B 1987, 35, 6213.

RECEIVED July 7, 1988

Chapter 22

Chemical Compatibility of High-Temperature Superconductors with Other Materials

R. Stanley Williams[1,2] and Shiladitya Chaudhury[2,3]

[1]Department of Chemistry and Biochemistry, University of California, Los Angeles, CA 90024–1569
[2]Solid State Science Center, University of California, Los Angeles, CA 90024–1569
[3] Department of Physics, University of California, Los Angeles, CA 90024–1569

In any application, the new oxide superconductors will have to coexist in intimate contact with other materials. However, the copper-oxygen bonds common to all presently known oxide superconductors with T_c above 40K are relatively weak. This means that elements that form much stronger bonds with oxygen, such as Si, will be chemically unstable when in contact with the copper-oxide superconductors. In this paper, we will establish procedures and guidelines that will enable workers to choose materials that will not react chemically with the oxide super-conductors. These should enable researchers working in all applications areas to avoid unnecessary empirical searches for suitable substrate or host materials by eliminating the most thermodynamically unfavorable choices. In particular, we have determined which of the elemental metals should be most stable in contact with the copper-oxide semi-conductors, and examined schemes for integrating these superconductors into Si devices.

There has been a great deal of research over the past year on the new oxide superconductors. Justifiably, most of that work has concentrated on the preparation of materials with optimized properties; i.e., maximized T_c and critical currents. This is the proper time to think in broad terms about how to successfully encorporate the new superconductors into structures that will serve useful purposes. Particular attention must be paid to the processing of these structures, especially if any of the steps involve sintering or annealing steps at reasonably high temperatures (i.e., 700K or above).

0097–6156/88/0377–0291$06.00/0

The superconducting 1-2-3 phases are known to be chemically sensitive, and in fact are thought to be metastable compounds under all conditions of temperature and oxygen partial pressure (1). Thus, it is imperative that any material in intimate contact with 1-2-3 phases does not react with the component oxides to form more stable compounds, since such a reaction will destroy the superconducting material. This is especially important for thin film applications, since the amount of superconductor is small compared to that of the substrate, and the diffusion path for potential solid state reactions, i.e., the film thickness, is very short. In this paper we will concentrate on searching for materials that should be most stable in contact with the 1-2-3 superconductors, since they appear to be more sensitive to chemical disruption than the more recently discovered Bi- and Tl-based phases. The principles and much of the data presented here can be applied to any oxide superconductor, whether it is presently known or yet to be discovered.

The reason for the chemical vulnerability of phases based on copper oxides is shown in Fig. 1, which is a display of the heat of formation per gram-atom of the most stable solid oxides formed by the metallic and semiconducting elements in the periodic table. To generate this table, compilations of thermochemical data were consulted (2,3), and these sources provided the enthalpies of formation used for the calculations presented later. The reason for expressing the data in the manner shown in Fig.1 is to get an idea of the relative strengths of the M-O bonds that form the solid oxides. One can see immediately that very few elements form weaker bonds with O than does Cu. Thus, most elements will reduce CuO to elemental Cu, and are very likely to react chemically with any of the copper-oxide superconductors as well. Although a positive heat of reaction, determined for interactions of the 1-2-3 superconductors with another material, is not a sufficient condition for chemical stability, it is almost certainly necessary. Consequently, researchers interested in the processing of copper-oxide superconductors into structures should consider thermochemical properties carefully.

In this investigation, we have collected the relevant heats of formation and phase diagrams to analyze the stability of 1-2-3 compounds with respect to the solid elements and several oxides. To check that the thermochemical predictions are indeed correct, we have made use of reports of chemical reactivity in the literature, and have also performed simple experiments to determine whether or not 1-2-3 will react with another material and identify the reaction products when it does. The next section describes the experimental procedure followed in this study; the results and conclusions to be drawn from this paper are presented in the succeeding sections.

Experimental Procedure

The reactivities of various materials with respect to $YBa_2Cu_3O_{7-x}$ were determined by heating mixtures of carefully weighed powders sealed in

ΔH_f (heat of formation) in kcal/gm atom for stable solid oxides

IA	IIA	IIIA	IVA	VA	VIA	VIIA	VIIIA	VIIIA	VIIIA	IB	IIB	IIIB	IVB	VB	VIB	VIIB	VIIIB
H —																	He —
Li -47.5	Be -72.7											B -60.8	C —	N —	O —	F —	Ne —
Na -33.1	Mg -72.8											Al -80.2	Si -72.5	P -50.9	S -27.1	Cl —	Ar —
K -28.9	Ca -75.8	Sc -91.1	Ti -72.7	V -58.2	Cr -54.0	Mn -47.3	Fe -39.2	Co -30.9	Ni -28.8	Cu **-18.6**	Zn -41.9	Ga -51.7	Ge -46.2	As -31.2	Se -18.0	Br —	Kr —
Rb -26.3	Sr -70.7	Y **-91.1**	Zr -87.7	Nb -64.9	Mo -46.8	Tc -34.5	Ru -24.3	Rh -18.3	Pd -13.4	Ag -2.4	Cd -31.0	In -44.2	Sn -46.2	Sb -34.4	Te -25.8	I —	Xe —
Cs -27.0	Ba **-61.1**	La **-85.7**	Hf -88.7	Ta -69.8	W -50.3	Re -33.3	Os -23.5	Ir -19.3	Pt -5.6	Au > 0	Hg -10.9	Tl -18.6	Pb -26.2	Bi -27.3	Po ?	At —	Rn —
Fr ?	Ra -62.5	Ac ?															

Ce -87.1	Pr -87.2	Nd **-86.4**	Pm ?	Sm -87.6	Eu -82.5	Gd **-87.1**	Tb -87.3	Dy -89.1	Ho **-89.9**	Er -90.9	Tm -89.9	Yb **-86.7**	Lu -89.9
Th -97.7	Pa -97.5	U ?	Np ?	Pu ?	Am ?	Cm ?	Bk ?	Cf ?	Es ?	Fm ?	Md ?	No ?	Lr ?

— No relevant oxide
? No reliable data

Figure 1. Periodic table of the elements showing the heat of formation (in kcal/gram-atom) of the most stable solid oxide of each element. The more negative the value of ΔH_f, the more strongly the element bonds to oxygen. (Data were taken from Refs. 2 and 3.)

quartz ampoules. The 1-2-3 material, prepared by the citrate precipitation process (4), was obtained from the research group of Prof. Richard Kaner at UCLA. The other materials were reagent grade chemicals obtained from our stockroom. Small amounts of the mixed powders were reserved for x-ray powder diffraction analysis so that the phases present before and after heating could be compared. The ampoules were placed in a furnace at 1073K for four hours to initiate any reactions. The intent here was not to determine the equilibrium products of the multi-element systems, but just to see if indeed a reaction did occur. Both heated and unheated mixtures, as well as control samples that were single compounds sealed in ampoules and heated, were examined by x-ray powder diffractometry. The diffractometer was computer controlled, and data collection times as long as 12 hours were used to obtain a reasonable signal-to-noise ratio. These studies showed that the 1-2-3 material by itself did not react with the quartz tubes during the heating cycle, but mixtures of 1-2-3 with other materials often did attack the quartz.

Chemical Stability of 1-2-3 with Respect to Elements

There are only nine elemental metals that will not reduce CuO: Ru, Rh, Pd, Ag, Os, Ir, Pt, Au, and Hg. These are grouped with a bold border in the periodic table of Fig. 1. No solid oxide of Au is stable, and the oxides of Pd, Ag, Pt, and Hg are just barely stable. The chemical formula of the oxides of Ru, Os, and Ir is MO_2, so two equivalents of CuO would be required to oxidize one equivalent of these metals by the reaction $M + 2CuO \rightarrow MO_2 + 2Cu$. For Rh, the oxide is Rh_2O_3, so three equivalents of CuO would be required to produce one equivalent of Rh_2O_3. Thus, CuO is also stable with respect to these last four metals because the heat of formation of two (or three) equivalents of CuO is larger in magnitude than a single equivalent of the oxides of Ru, Os, and Ir (or Rh). Therefore, these nine metals are the only elements that may not react chemically with 1-2-3 superconductors, although each individual system should be checked to ensure that they are indeed stable.

We can actually use the reactivity or lack thereof of the 1-2-3 compounds with different elements to determine roughly the heat of formation of the superconductor. For the purposes of this paper, we will assume that 1-2-3 is a stable compound and that it has the ideal formula $YBa_2Cu_3O_{6.5}$. Thus, the reaction

$$1/2Y_2O_3 + 2BaO + 3CuO \rightarrow YBa_2Cu_3O_{6.5} \quad (1)$$

has an unknown but negative heat of reaction, ΔH_f, or the heat of formation with respect to the component oxides. We can place a lower limit on this value by considering a reaction between the 1-2-3 material and an element that is just barely above the threshold for reaction with CuO. The most suitable element for this purpose is Tl, since

$$2Tl + CuO \rightarrow Tl_2O + Cu \ ; \qquad \Delta H_1 = -2.9(\pm 2.6) \text{ kcal} \quad (2)$$

However, because of our lack of facilities to handle toxic materials, we chose instead to try the next best choice, which is Pb:

$$Pb + CuO \rightarrow PbO + Cu \quad ; \quad \Delta H_2 = -15.3\ (\pm 1.0)\ \text{kcal} \tag{3}$$

If we consider the idealized reaction between Pb and 1-2-3,

$$3Pb + YBa_2Cu_3O_{6.5} \rightarrow 1/2Y_2O_3 + 2BaO + 3PbO + 3Cu \quad ; \quad \Delta H_3 = ? \tag{4}$$

and this reaction is exothermic, then we can obtain limits for ΔH_f as follows:

$$0 > \Delta H_f = 3\Delta H_2 - \Delta H_3 > 3\Delta H_2 = -46\ \text{kcal} \tag{5}$$

where the second inequality holds because minus ΔH_3 is a positive number. We have carried out the reaction indicated by Eq.(4), and we do observe the appearance of elemental Cu in the reaction products. Although the uncertainty in the estimate that we obtain appears to be large, the limits that we place on ΔH_f do allow us to make some quantitative predictions with respect to the reactions of 1-2-3 with other elements and compounds. If we predict that 1-2-3 will react based on the above lower limit for ΔH_f, we can be very confident in our prediction since ΔH_f may even be slightly positive (1).

Chemical Stability of 1-2-3 with Respect to Si and Insulators

For many applications, a noble or near-noble metal cladding for copper-oxide superconductors may be perfectly acceptable. However, one of the primary uses envisioned for high-T_c superconductors is as a conductor on semiconductor devices. It is obvious that 1-2-3 will not be stable as a thin film on a bare Si substrate, since

$$Si + 2CuO \rightarrow SiO_2 + 2Cu \quad ; \quad \Delta H_4 = -143\ \text{kcal} \tag{6}$$

Thus, one may predict

$$3/2Si + YBa_2Cu_3O_{6.5} \rightarrow 1/2Y_2O_3 + 2BaO + 3/2SiO_2 + 3Cu \tag{7}$$

where

$$-97\ \text{kcal} > \Delta H_5 > -143\ \text{kcal} \tag{8}$$

In fact, mixing Si and 1-2-3 in the proportions shown in Eq.(7) and heating immediately yields a brown powder, which contains elemental Cu and various silicates.

An obvious idea is to use a layer of inert insulating material between the Si and the superconductor, such as SiO_2. However, here we run into the problem of silicate formation via the following reaction:

$$SiO_2 + 2BaO \rightarrow Ba_2SiO_4 \quad ; \quad \Delta H_6 = -64.5 \text{ kcal} \qquad (9)$$

which is more exothermic than the estimate of Eq.(5) for the heat of formation of 1-2-3 from its component oxides. A stabilizing layer of aluminum oxide does not appear to be much more useful, since

$$2Al_2O_3 + 2BaO \rightarrow 2BaAl_2O_4 \quad ; \quad \Delta H_7 = -48.0 \text{ kcal} \qquad (10)$$

which means that 1-2-3 probably reacts with Al_2O_3 as well. Thus, any processing procedure that will require a high temperature annealing of the superconductor after it has been deposited onto the Si wafer will require careful thought to design a suitable buffer layer.

Another possibility that comes to mind is to use Y_2O_3. However, this is also not a reasonable choice. Examination of the quasi-ternary phase diagram in Fig.2 that has been generated for the CuO-BaO-$YO_{1.5}$ system (5) reveals that there is no tie-line connecting 1-2-3 with Y_2O_3. This means that these two compounds will react with each other to produce the 2-1-1 compound and either CuO or $Y_2Cu_2O_5$, depending on the relative amounts of the starting materials. The three compounds that do form tie-lines to 1-2-3 are 2-1-1, CuO, and $BaCuO_2$. Neither CuO nor $BaCuO_2$ could be effective barriers, since both contain fragile Cu-O bonds that will be the focus of any chemical attack by other materials.

One might consider growing a multilayer insulating film with the structure SiO_2/Y_2O_3/2-1-1/1-2-3. This may also lead to problems, since the SiO_2 and the Y_2O_3 can react to form two different silicates, Y_2SiO_5 and $Y_2Si_2O_7$. Thus, a barrier layer involving Y_2O_3 would involve as many as five different films to separate 1-2-3 from Si and ensure that the materials in contact at each interface are chemically stable with respect to each other. Such complex structures should be avoided, since they will be difficult and expensive to produce.

Probably the best strategy is to search for a compound that is stable with respect to both 1-2-3 and to SiO_2. In principle, this requires that the phase diagram in Fig. 2 be generalized to a pseudo-quaternary diagram by including all the interactions of SiO_2 with the other phases. At present, only the compounds on the edges of the resulting tetrahedron that share the SiO_2 apex are known, and they are shown on the three pseudo-ternary diagrams in Figs. 3-5.

It is possible to make a good guess at the topology of both the pseudo-ternary phase diagrams in the SiO_2-Y_2O_3-CuO (Fig.3) and the SiO_2-BaO-CuO (Fig.4) systems. Since there are no copper silicates, it is unlikely that there are any pseudo-ternary compounds in either phase diagram. Moreover, the yttrium and barium silicates are very stable, so SiO_2 is likely to replace CuO in compounds with BaO or Y_2O_3. With these assumptions, the tie-lines in Figs.3 and 4 are the most logical ways of

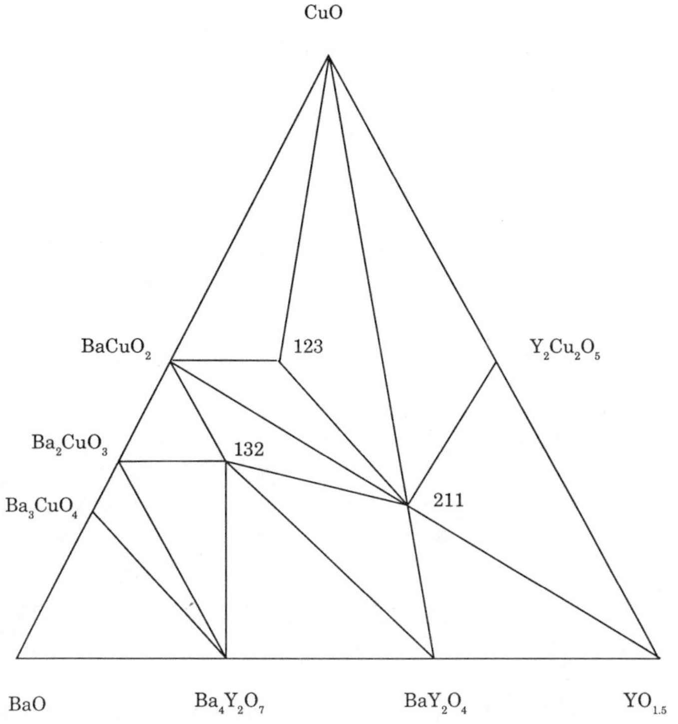

Figure 2. Idealized pseudo-ternary phase diagram for the Y_2O_3-BaO-CuO system, showing only the solid phases. (Adapted from Ref. 5.)

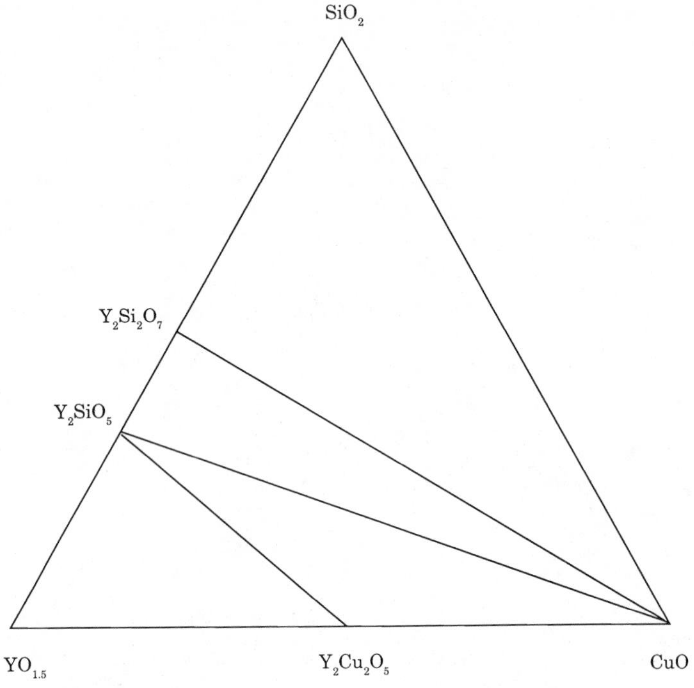

Figure 3. Idealized pseudo-ternary phase diagram for the SiO_2-Y_2O_3-CuO system, showing only the solid phases. The tie-lines were drawn based on the assumption that no pseudo-ternary compounds form.

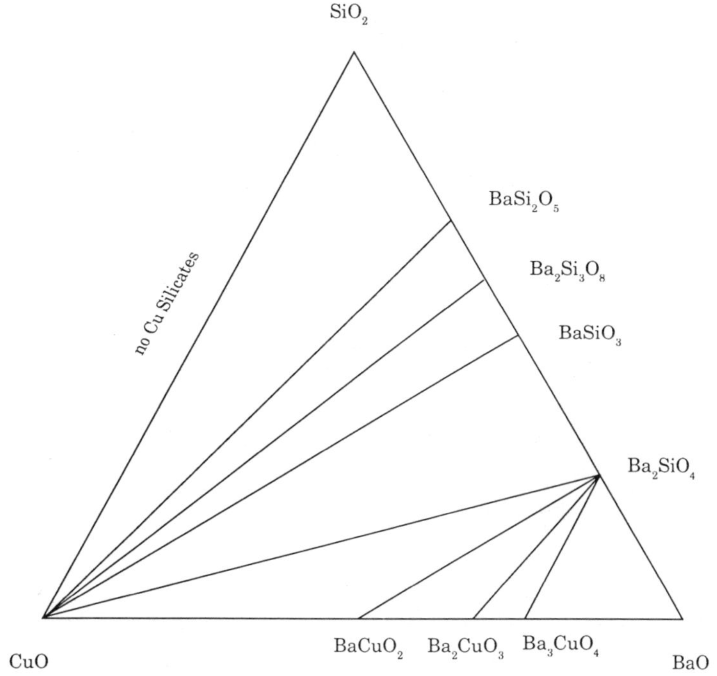

Figure 4. Idealized pseudo-ternary phase diagram for the SiO_2-CuO-BaO system, showing only the solid phases. The tie-lines were drawn based on the assumption that no pseudo-ternary compounds form.

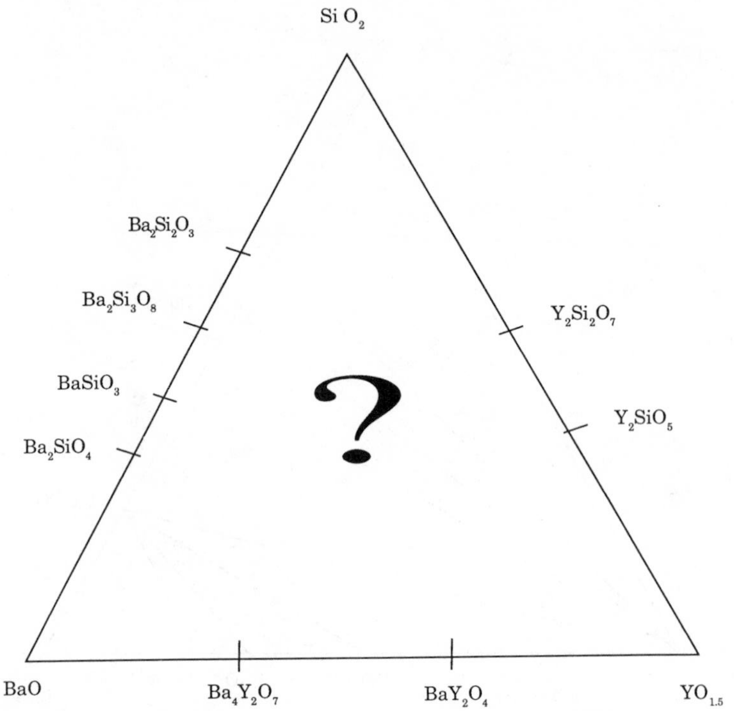

Figure 5. The boundary of the pseudo-ternary phase diagram for the SiO_2-BaO-Y_2O_3 system. Since it is almost certain that pseudo-ternary compounds exist in this system, no attempt has been made to estimate the positions of the tie-lines.

connecting the pseudo-binary compounds across the phase diagrams. The SiO_2-Y_2O_3-BaO system outlined in Fig.5, however, is much too complex to allow a phase diagram to be determined by guesswork. There are almost certainly pseudo-ternary compounds in this system, and there are no thermo-chemical data for the pseudo-binary compounds on the boundaries of the phase diagram. Thus, this system will have to be investigated in detail experimentally.

It would be fortunate if there were a tie-line cutting through the pseudo-quaternary system to connect 1-2-3 with either $Y_2Si_2O_7$ or $BaSi_2O_5$. The former is not very likely, since the CuO-211-BaY_2O_4 tie-lines shown in Fig.2 are probably part of a plane that effectively isolates 1-2-3 from Y_2O_3-rich species in the pseudo-quaternary system. The latter case was a possibility that we investigated. No reaction was observed between powders of 1-2-3 and $BaSi_2O_5$ heated to 973 K for 24 hours, which means that these materials do terminate a tie line. The *caveat* is that the silicate is apparently suitable as a buffer layer in terms of thermodynamic properties, but it may have other undesirable qualities, such as poor mechanical properties, rough film morphology, or severe lattice mismatch. These properties are the subject of further research.

Chemical Stability of 1-2-3 in Air

The 1-2-3 compounds are notoriously unstable with respect to CO_2 and H_2O in the air. This is the result of the large enthalpies of formation of $BaCO_3$ and $Ba(OH)_2$ from the following reactions:

$$BaO + H_2O \rightarrow Ba(OH)_2 \;;\; \Delta H_8 = -35.4 \text{ kcal} \qquad (11)$$

and

$$BaO + CO_2 \rightarrow BaCO_3 \;;\; \Delta H_9 = -64.4 \text{ kcal} \qquad (12)$$

Thus, we see that the reactivity of BaO with other oxides is a major factor in the chemical sensitivity of 1-2-3. The high-T_c superconductors that do not contain BaO should be significantly more stable than 1-2-3, and most likely much easier to process in the presence of other materials.

Acknowledgments

We thank B. Dunn and R.B. Kaner for their advice and help. This work was supported in part by the Office of Naval Research.

Literature Cited

1. Sleight, A.W. In Chemistry of High-Temperature Superconductors; Nelson, D.L.; Whittingham, M.S.; George, T.F., Eds.; ACS Symposium Series No. 351; American Chemical Society: Washington, DC, 1987; pp 2-12.

2. Kubaschewski, O; Alcock, C.G. Metallurgical Thermochemistry; Pergamon: New York, 1979; pp 267-323.
3. Wagman, D.D.; Evans, W.H.; Parker, V.B.; Schumm, R.H.; Halow, I.; Bailey, S.M.; Churney, K.L.; Nuttall, R.L. J. Phys. and Chem. Data 1982, 11, Supp. 2.
4. Chu, C.; Dunn, B. J. Am. Ceram. Soc. 1987, 70, C-375.
5. Frase, K.G.; Clarke, D.R. Advan. Cer. Mat. 1987, 2, No.3B, 295.

RECEIVED July 8, 1988

PROCESSING

Chapter 23

A New Layered Copper Oxide: $LaSrCuAlO_5$

J. B. Wiley[1], L. M. Markham[1], J. T. Vaughey[1], T. J. McCarthy[1], M. Sabat[1], S.-J. Hwu[1], S. N. Song[2], J. B. Ketterson[2], and K. R. Poeppelmeier[1]

[1]Department of Chemistry, Northwestern University, Evanston, IL 60208
[2]Department of Physics and Astronomy and Materials Research Center, Northwestern University, Evanston, IL 60208

Square-pyramidal coordination of copper and Cu-O planes are common structural features in many of the high T_c oxide superconductors that support this state. These poorly screened (highly ionic) compounds have occupied Cu-O dpπ bonding orbitals and empty Cu-O dpσ anti-bonding orbital partners. The combination of four and six coordination leads to similar layered structures and Cu-O σ and π bonding. Our recent efforts preparing quaternary oxides with strongly basic counter cations, small tetrahedrally coordinated cations and copper have led to a new series of layered copper oxides.

Following the report of Bednorz and Müller (1) on possible high T_c (>30K) superconductivity in the Ba-La-Cu-O mixed phase system, the compound $La_{2-x}M_xCuO_{4-x/2+\delta}$ (M = Ba^{2+}, Sr^{2+}, or Ca^{2+}) with the tetragonal K_2NiF_4 structure was identified (2,3) to be the superconducting phase. Soon thereafter Chu et al. (4,5) reported superconductivity above 90K in the Y-Ba-Cu-O system. The compound has been identified by numerous groups (6–9) and has the composition $YBa_2Cu_3O_{7-y}$ and a structure (10) similar to the tetragonal structure reported for $La_3Ba_3Cu_6O_{13.5+y}$ (11).

Since the high T_c oxide superconductors were discovered in 1986, their completely unexpected properties have challenged our collective scientific understanding. Even after two years this challenge largely remains. But despite the empirical nature associated with the synthesis of $YBa_2Cu_3O_{7-y}$ (4), and later with the bismuth (12,13) and thallium (14) compounds, there is an emerging understanding (15) that is derived from a fundamental theoretical and experimental basis of how metallic oxides transport charge in the normal and superconducting states.

The concept that a significant fraction of oxygen atoms may be removed by reduction from metallic oxide lattices with little or no structural change was first studied (16,17) and debated (18) over twenty-five years ago. In particular the perovskite lattice has

0097–6156/88/0377–0304$06.00/0

been shown to persist over the limits ABO_{3-x}; $0 \le x \le 0.5$ for many transition and some main group metal cations. The first compounds with the composition $ABO_{2.5}$ and ordered oxide ion vacancies that were recognized (19,20) to be related to perovskite were $Ca_2Fe_2O_5$ (21) and the mineral brownmillerite, Ca_2AlFeO_5 (22). In general the superstructure formed depends upon the electronic configuration (d^n) and preferred coordination of the smaller B-site cation and the ionic radius of the larger, electropositive A-cation (23,24). Many oxygen-deficient structures based on perovskite are known for manganese (25), iron (26), cobalt (27), nickel (28) and copper (29). Oxygen deficient compounds based on the related K_2NiF_4 structure, e.g. $Ca_2MnO_{3.5}$, are also known (30) although they have not been as thoroughly investigated.

In this paper we describe the synthesis, crystal structure, and conductivity of $LaSrCuAlO_5$. The crystal structure of this compound is similar to $Ca_2Fe_2O_5$ and Ca_2AlFeO_5, but there are also some interesting differences.

Experimental

Polycrystalline samples of $LaSrAlCuO_5$ were prepared by solid state reaction of Aldrich cupric oxide (99.999%), strontium carbonate (99.999%), lanthanum oxide (99.999%) and aluminum nitrate (99.999%). Powders were ground with a mortar and pestle and calcined in air at 950°C for 5 days with daily grindings. The product was dark black. Thermogravimetric studies with a Du Pont Thermal Analysis System by reduction in hydrogen were used to determine the oxygen composition.

For resistivity measurements disc-shaped specimens 1.25 cm in diameter and 1.5 mm thick were isostatically pressed at 7.2 kbar at room temperature. An oxygen treatment from 900°C to room temperature (100°C/hr) was used to anneal the pellet. Discs were cut into rectangular specimens with cross sections of 1.5 x 3 mm^2 and four leads were attached with silver paint for 4-point resistivity measurements.

X-ray diffraction (XRD) measurements were carried out on polycrystalline materials. XRD powder patterns were recorded on a Rigaku diffractometer with Cu $K\alpha$ radiation and a Ni filter. For reference all patterns were recorded with an internal NBS Si standard. The correct d-spacings and intensities were calculated from crystal data by the use of the program, LAZY PULVERIX (31). After correcting the peak positions based on the silicon line positions, the observed powder pattern matched the calculated one.

Single crystals of $LaSrCuAlO_5$ were grown from a CuO flux. A small plate-like crystal (0.2 x 0.08 x 0.08 mm) was used for the structure determination. The structure was solved by the use of a full-matrix least-squares program in the TEXSAN program package (32). A summary of the crystal data and refinement results is given in Table I. Further details of the structure determination will appear elsewhere (Wiley, J. B.; Poeppelmeier, K. R. J. Solid State Chem., submitted)

Table I. Crystallographic Data for $LaSrCuAlO_5$

Z	4
Space group	Pbcm
Lattice constants (Å)	
a	7.9219(6)
b	11.0200(10)
c	5.4235(4)
V (Å^3)	473.47(11)
Refinement	
R	0.056
R_w	0.084

Positional Parameters

Atom	Site	Mult.	x	y	z
La1	4d	0.345	0.1950(2)	0.1358(1)	0.2500
La2	4d	0.155	-0.2527(3)	0.0978(2)	0.2500
Sr2	4d	0.345	-0.2527(3)	0.0978(2)	0.2500
Sr1	4d	0.155	0.1950(2)	0.1358(1)	0.2500
Cu	4d	0.500	-0.0234(2)	0.1261(1)	-0.2500
Al	4d	0.500	0.5013(6)	0.1574(4)	-0.2500
01	4a	0.500	0.0000	0.0000	0.0000
02	4c	0.500	-0.0436(14)	0.2500	0.5000
03	4d	0.500	0.2840(12)	0.1386(9)	-0.2500
04	4c	0.500	-0.4460(15)	0.2500	0.5000
05	4d	0.500	0.3709(21)	-0.0331(13)	0.2500

Results and Discussion

The following generalizations are commonly believed to be important constraints on the chemical and physical structures that can support the superconducting state. First, mixed valency will be required, associated with a unusually high oxidation state for metals (e.g. Cu) to the right in the transition series (oxidation) or a low oxidation state for metals (e.g. Ti) to the left in the transition series (reduction), to favorably place the Fermi energy so that oxygen $p\pi$ and transition metal d-bands overlap. Second, electropositive cations such as Ba^{2+}, Sr^{2+} or La^{3+} will be required to stabilize the oxide. Another important feature appears to be that the narrow d-band can potentially exhibit polaronic (33) versus itinerant electron behavior at elevated temperature. And finally with copper the possibility for a $p\pi$-bonding hole associated with $p\pi$ orbitals perpendicular to strongly bonded CuO_2 planes should exist.

Our work has focused on the synthesis of new materials possessing metallic properties and those features, especially electron transfer between copper sites, that potentially affect high-temperature superconductivity. Metallic behavior in copper oxides results from a competition between covalent and short Cu-O bonds (~1.9 Å) and ionic and longer (~2.0 Å) bonds. The importance of electropositive cations such as Ba^{2+}, Sr^{2+} and Ca^{2+} in promoting covalent Cu-O bonds is now a well-documented feature (12,13,14) of the high-temperature oxide superconductors including the more recent $A_2Ba_2CuO_6$ (A = Bi^{3+}, Tl^{3+}), $A_2Ba_2CaCu_2O_{8+y}$ (A = Bi^{3+}, Tl^{3+}) and $A_2Ba_2Ca_2Cu_3O_{10+y}$ (A = Tl^{3+}) series of compounds some of which have $T_c > 100$ K.

In addition to large electropositive cations such as Ba^{2+}, other cations that are compatible with covalent Cu-O bonds are known. This stabilization can occur with small cations in their highest oxidation state (no outer s, p or d electrons) from displacement of oxygen towards the cations stabilizing the long Cu-O bond(s) of the Cu^{2+} ion. For example Cu_6PbO_8 readily forms in air (34) and is stable at temperatures where CuO (tenorite) would transform to Cu_2O (cuprite). In the 90 K superconductor, $YBa_2Cu_3O_7$, the Cu^{3+} (chain site) d^8-ion also forms such a bond.

We have begun a study of the stabilization of semi-conducting and metallic oxides with other metal cations that will form covalent metal-O-Cu bonds and a two level electronic band structure. These materials will be essentially semiconductors where the conductivity arises from doping to produce mixed-valence compounds. We chose to begin our study with cations that adopt tetrahedral coordination and focus on how to create structures that incorporate distorted octahedral, square pyramidal and square planar coordination of copper compatible with still other electropositive (ionic) cations. The mixed valency introduced by doping can then be accommodated on the copper metal and adjacent oxygen atom sites by an accompanying bond polarization around the cation with tetrahedral coordination.

The prototype compound we have studied is $LaSrCuAlO_5$. It has not been synthesized before and should be viewed as an oxygen deficient perovskite. The copper ions in the structure have very distorted $CuO_{6/2}$ octahedra, so much so that four short (≈1.9 Å) CuO bonds condense to $CuO_{4/2}$ planes. These planes are slightly modulated along the b-axis (see Figure 1). The aluminum cations adopt tetrahedral coordination and these "double" tetrahedral strings run throughout the structure parallel to the c-axis. The connectivity in the $AlO_{2/2}$ layer is broken by the missing oxygen atoms that would, if present, give the normal perovskite structure and composition ABO_3. The La and Sr atoms (not shown) are distributed unequally (69:31) over two nonequivalent A-sites. The sites are eight and nine coordinate, respectively, with the larger Sr ion preferring the more highly coordinated site.

Simple oxygen deficient perovskites, $ABO_{2.5}$, accommodate oxygen vacancies by different coordination schemes. In $CaMnO_{2.5}$, for example, square pyramidal coordination is observed for manganese (Fig. 2). In $CaFeO_{2.5}$ and brownmillerite (Ca_2FeAlO_5), the B-cations prefer a combination of octahedral and tetrahedral coordinations (Fig.3). $LaSrCuAlO_5$ is most closely related to brownmillerite in that it can also be viewed as a combination of these two coordinations. Figure 4 shows an idealized representation of the real structure with Cu and Al in octahedral and tetrahedral sites, respectively.

An important difference between $LaSrCuAlO_5$ and $Ca_2Fe_2O_5$ is seen in the ordering of oxygen vacancies. Both compounds have vacancies that order in alternate (001) $BO_{4/2}$ layers, with alternate [110] rows of anions removed. As shown in Figures 3 and 4 it is the relative arrangement of these [110] rows that distinguishes the two compounds. In brownmillerite-like $Ca_2Fe_2O_5$ the rows of vacancies are staggered in each tetrahedral $FeO_{4/2}$ layer, while in $LaSrCuAlO_5$ the vacancies occur in the same row in each $AlO_{4/2}$ layer. As would be expected, the difference in the ordering of the oxygen vacancies in the iron and copper compounds is reflected in the relationship between lattice parameters. (see Table II). Compared with the brownmillerite orientation, the

Table II. Comparison of Lattice Parameters

$Ca_2Fe_2O_5$	$LaSrCuAlO_5$
$a_{Fe} \simeq \sqrt{2}a_c$	$a_{Cu} \simeq b_{Fe}/2$
$b_{Fe} \simeq 4a_c$	$b_{Cu} \simeq 2a_{Fe}$
$c_{Fe} \simeq \sqrt{2}a_c$	$c_{Cu} \simeq c_{Fe}$

difference in ordering observed in $LaSrCuAlO_5$ causes a halving of the b_{Fe} lattice parameter while the a_{Fe} parameter is doubled.

$LaSrCuAlO_5$ is semiconducting as prepared with a resistivity at room temperature (ρ_{300K}) of 0.065Ω cm. Doping studies to introduce mixed valency (Cu^{3+}/Cu^{2+}) are underway. Many compounds with a variety of large and small trivalent and bivalent cations have been prepared that adopt this and closely related structures.

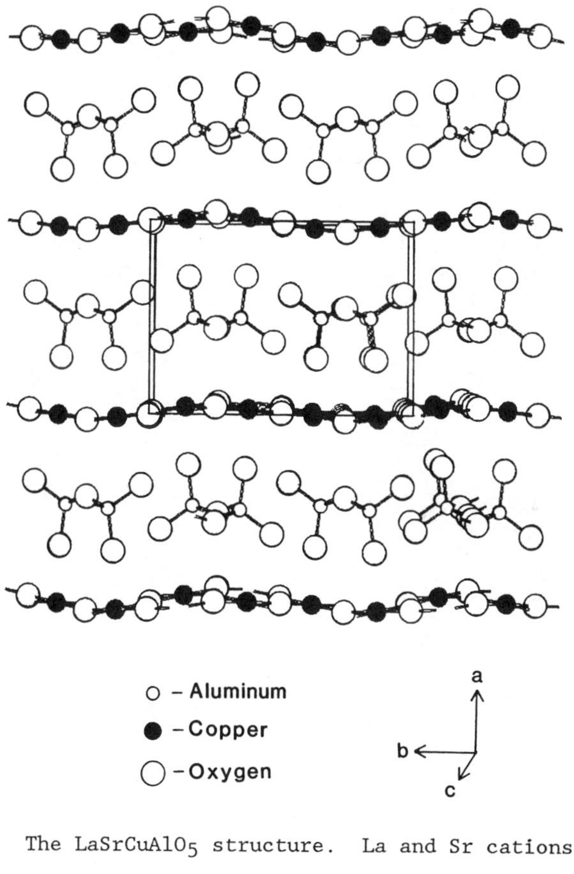

Figure 1. The $LaSrCuAlO_5$ structure. La and Sr cations are not shown.

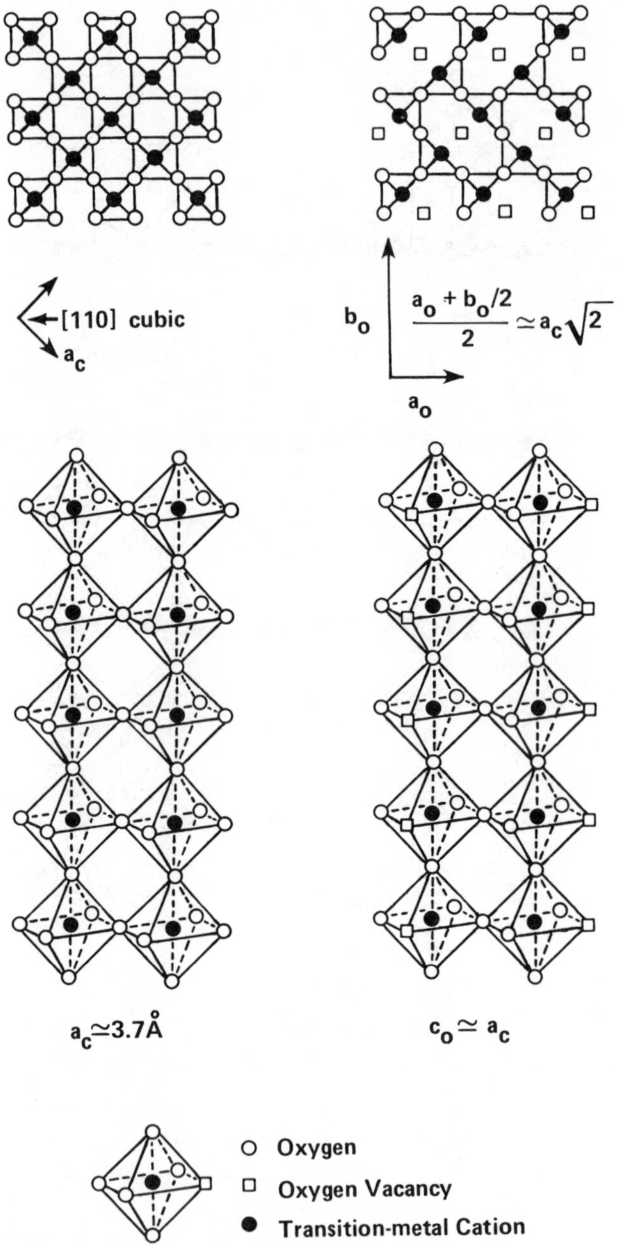

Figure 2. Structure of stoichiometric ABO_3 perovskite (left) and oxygen deficient $CaMnO_{2.5}$ (right). The A-cations are not shown.

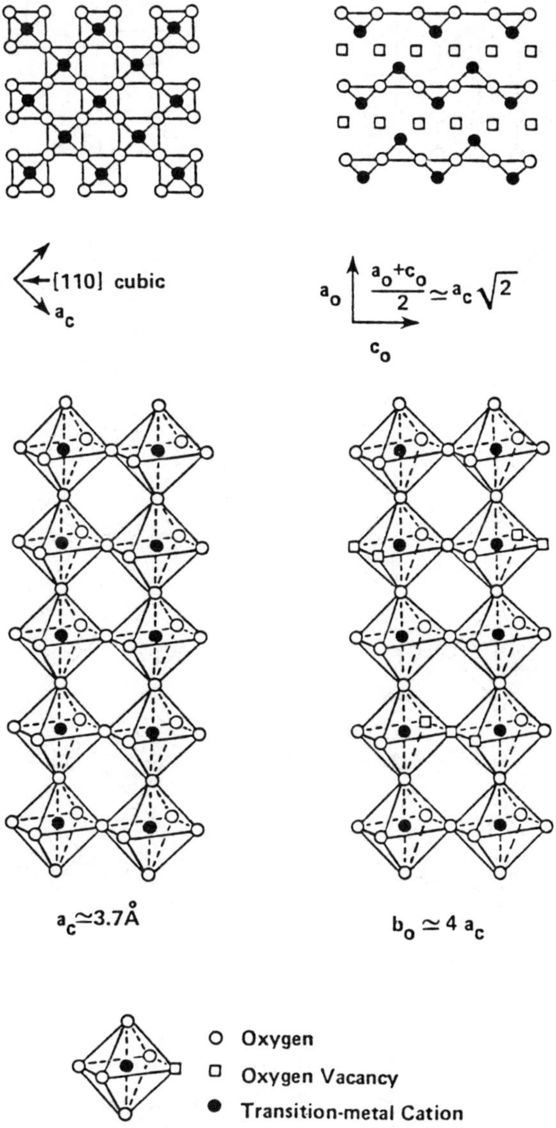

Figure 3. Structure of ABO_3 perovskite (left) and brownmillerite-like $Ca_2Fe_2O_5$ (right).

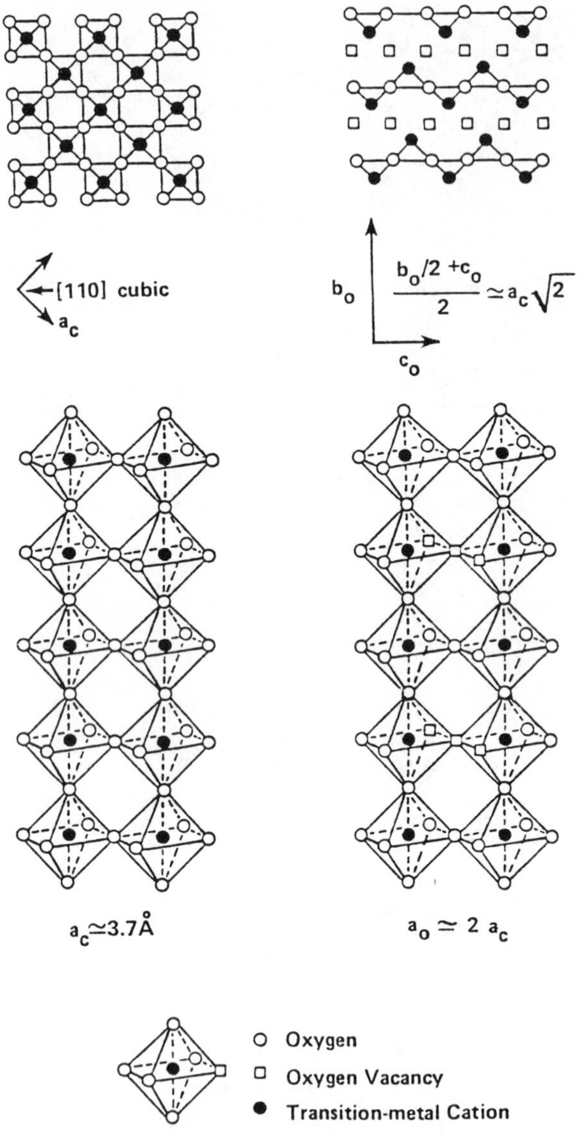

Figure 4. ABO_3 perovskite structure (left) and an idealized representation of $LaSrCuAlO_5$ (right).

Acknowledgments

We acknowledge support for this research from NSF (DMR-8610659) and the Materials Research Center of Northwestern University, NSF (MRL 8520280). We are grateful to S. Massidda, J. J. Yu and A. J. Freeman for helpful discussions.

Literature Cited

1. Bednorz, J. G.; Müller, K. A. Z. Phys. 1986, B64, 189.
2. Takagi, H.; Uchida, S.; Kitazawa, K.; Tanaka, S. Jpn. J. Appl. Phys. Lett. 1987, 26, L123.
3. Cava, R. J.; van Dover, R. B.; Batlogg, B.; Rietman, E. A. Phys. Rev. Lett. 1987, 58, 408.
4. Wu, M. K.; Ashburn, J. R.; Torng, C. J.; Hur, P. H.; Meng, R. L. Gao, L.; Huang, Z. J.; Wang, Y. Q.; Chu, C. W. Phys. Rev. Lett. 1987, 58, 908.
5. Hur, P. H.: Gao, L.; Meng, R. L.; Huang, Z. J.; Forster, K.; Vassilious, J.; Chu, C. W. Phys. Rev. Lett. 1987, 58, 911.
6. Hwu, S.-J.; Song, S. N.; Thiel, J. P.; Poeppelmeier, K. R.; Ketterson, J. B.; Freeman, A. J. Phys. Rev. B. 1987, 35, 7119.
7. Cava, R. J.; Batlogg, B.; van Dover, R. B.; Murphy, D. W.; Sunshine, S.; Siegrist, T.; Remcika, J. P.; Reitman, E. A.; Zahurak, S.; Espinosa, G. P. Phys. Rev. Lett. 1987, 58, 1676.
8. Stacy, A. M.; Badding, J. V.; Geselbracht, M. J.; Ham, W. K.; Holland, G. F.; Hoskins, R. L.; Keller, S. W.; Millikan, C. F.; zur Loye, H.-C. J. Am. Chem. Soc. 1987, 109, 2528.
9. Steinfink, H.; Swinnea, J. S.; Sui, Z. T.; Hsu, H. M.; Goodenough, J. B. J. Am. Chem. Soc. 1987, 109, 3348.
10. Beno, M. A.; Soderholm, L.; Capone, D. W.; Jorgensen, J. D.; Schuller, I. K.; Segre, C. U.; Zhang, K.; Grace, J. D. Appl. Phys. Lett. 1987, 51, 57.
11. Er-Rakho, L.; Michel, C.; Provost, J.; Raveau, B. J. Solid State Chem. 1981, 37, 151.
12. Michel, C.; Hervieu, M.; Borel, M. M.; Grandin, A.; Deslandes, F.; Provost, J.; Raveau, B. Z. Phys. 1987, B68, 421.
13. Maeda, H.; Tanaka, Y.; Fukutomi, M.; Asano, T. Jpn. Appl. Phys. 1988, 27, L209.
14. Sheng, Z. Z.; Hermann, A. M. Nature 332, 55, 1988; 1988, 332, 138.
15. Subramanian, M. A.; Torardi, C. C.; Calabrese, J. C.; Gopalakrishnan, J.; Morrissey, K. J.; Askew, T. R.; Flippen, R. B.; Chowdhry, U.; Sleight, A. W. Science, 1988, 239, 1015.
16. Jonker, G. H. Physica. 1954, 20, 1118.
17. Kestigian, M.; Dickenson, J. G.; Ward, R. J. Am. Chem. Soc. 1960, 79, 5598.
18. Anderson, S.; Wadsley, A. D. Nature 1960, 187, 499.
19. Watanabe, H.; Sugimoto, M.; Fukase, M.; Hirone, T. J. Appl. Phys. 1965, 36, 988.

20. Geller, S.; Grant, R. W.; Gonser, U.; Wiedersich, H.; Espinosa, G. P. Phys. Lett. 1966, 20, 115.
21. Bertaut, E. F.; Blum, P.; Sagnieres, A. Acta Crystallogr. 1959, 12, 149.
22. Hansen, W. C.; Brownmiller, L. T. Amer. J. Sci. 1928, 15, 224.
23. Grenier, J. C.; Parriet, J.; Pouchard, M. Mat. Res. Bull. 1976, 11, 1219.
24. Poeppelmeier, K. R.; Leonowicz, M. E.; Longo, J. M. J. Solid State Chem. 1982, 44, 89.
25. Poeppelmeier, K. R.; Leonowicz, M. E.; Scanlon, J. C.; Longo, J. M.; Yelon, W. B. J. Solid State Chem. 1982, 45, 71.
26. Grenier, J. C.; Pouchard, M.; Hagenmuller, P. Structure and Bonding 1981, 47, 1.
27. Vidyasagar, K.; Gopalakrishnan, J.; Rao, C. N. R. Inorg. Chem. 1984, 23, 1206.
28. Vidyasagar, K.; Reller, A.; Gopalaknishnan, J.; Rao, C. N. R. J. Chem. Soc., Chem. Commun. 1985, 7.
29. Michel, C.; Raveau, B. Revue de Chimie Minerale 1984, 21, 407.
30. Leonowicz, M. E.; Poeppelmeier, K. R.; Longo, J. M. J. Solid State Chem. 1985, 59, 71.
31. a) Yvon, K.; Jeitschko, W.; Parthe, E. LAZY PULVERIX Program Laboratoire de Cristallographie Aux Rayon-X, Univ. Geneve, Geneva Switzerland 1977.
    b) Keszler, D.; Ibers, J. A. Modified LAZY PULVERIX Program, Northwestern University, 1984.
32. "TEXSAN"; Molecular Structure Corp.: College Station, TX 77843.
33. De Jongh, L. J. Physica C, 152, 171-216, 1988.
34. a) Kasper, J. S.; Crist, C. L. J. Chem. Phys. 1953, 21, 1897.
    b) Kasper, J. S.; Prener, J. S. Acta Crystallogr., 1954, 7, 246.

RECEIVED July 15, 1988

Chapter 24

Structure–Property Correlations in Superconducting Copper Oxides

D. W. Murphy, L. F. Schneemeyer, and J. V. Waszczak

AT&T Bell Laboratories, 600 Mountain Avenue, Murray Hill, NJ 07974

We present initial results of a survey correlating structural features of the cuprate superconductors with the occurrence of superconductivity. The best correlation is with the Cu–Cu distance in the plane of the two-dimensional compounds. Superconductivity has been shown only over a narrow region of Cu–Cu separations for both the A_2CuO_4 and square pyramidal based compounds. On either side of the superconducting region semiconductivity is found for both systems. It is suggested that the key difference between the classes is that oxidation of the A_2CuO_4 compounds results in removal of anti-bonding electrons, whereas in the square pyramidal compounds these electrons are non-bonding, which allows for greater carrier densities before chemical instabilities occur. The copper oxide based systems are compared to the bismuth based superconductors. Unsuccessful attempts to synthesize superconductors based on lead are described.

The pace of discovery of new superconducting copper oxides has been remarkable going from no known superconductors to T_c's as high as 125K in less than two years. The structures of several of the new superconducting materials have been determined by x-ray and neutron diffraction. While important questions of a structural nature remain, such as the nature of defects, superlattices, and twinning, there is now a large enough body of structural data to begin make broad structure-property correlations that could be useful in understanding the known materials and help predict where else to look for new superconductors. This paper presents our initial attempts at such correlations.

The structural feature common to each of the known copper oxide superconductors is the presence of $[CuO_2]_\infty$ planes. These two-dimensional layers are schematically illustrated in Figure 1. The layers are formally

0097–6156/88/0377–0315$06.00/0

partially oxidized (beyond Cu^{2+}) with charge compensation between the layers. Three modes of charge compensation have been employeed; 1) substitution of lower valent cations for the A ion (e.g., Sr^{2+} for La^{3+}), 2) a variable amount of interlayer oxygen (e.g., $Ba_2YCu_3O_{7-\delta}$ ($0 \leq \delta \leq 1$)), and 3) a fixed interlayer composition incommensurate with an integral charge on the layers (e.g., $Ba_2YCu_4O_8$).

We have attempted to correlate the electronic properties (particularly the occurrence and T_c of superconductivity) with structural parameters from both superconducting and non-superconducting compounds containing $[CuO_2]_\infty$ layers. Specific correlations have been presented in the literature for individual superconducting systems, including an increase in T_c with a decrease in *a* in or increase in the *c* lattice parameter for $La_{2-x}(Ba,Sr,Ca)_x CuO_4$,(1) a decrease in T_c for $Ba_2RECu_3O_x$ with an increase in the *c* lattice parameter(2), and in the same materials an increase in T_c with the length of the bond between the chain Cu and the apical oxygen.(3) Here we have attempted broader correlations and, most importantly, have included non-superconducting compositions in hopes of establishing a predictive basis to distinguish between superconductors and non-superconductors. We have attempted correlations with a number of possible structural parameters such as Cu-O distances (in and out of plane), separation between layers, lattice parameters, and planarity of the $[CuO_2]_\infty$ sheets. We recognize the multidimensionality of the problem and have not included in this first attempt important parameters such as the formal charge. Rather, we have hoped that such properties will be reflected in the structures.

The most notable correlation we have found is with the distance between copper atoms in the $[CuO_2]_\infty$ plane. The use of the Cu-Cu distance is not meant to imply a bond. Any correlation could be thought of as O-O or Cu-O. However, the Cu-Cu distance is easily derived from unit cell parameters because Cu is virtually always on a special position, whereas O generally is not making those distances less reliable if full crystallographic characterization is lacking.

A_2CuO_4 Compounds

Two A_2CuO_4 structure types exist with the La based materials having the K_2NiF_4 structure and all others having the T′ Nd_2CuO_4 structure in which the axial oxygens are bonded only to the RE and not to Cu.(4) Superconducting T_c's are plotted against Cu-Cu distances for the A_2CuO_4 are summarized in Figure 2. The striking feature of this data is that superconductivity has been found only in a narrow region near 3.78Å for the A_2CuO_4 compounds. Values are plotted from several sources (1,5-7) which can be misleading, since generally only resistive transitions are reported which could be due to a small portion of the sample. It does appear, however, that T_c is maximized near a = 3.777Å (x=0.15 for M=Sr). Superconductivity disappears and semiconducting behavior appears between x=0.3 (a = 3.766) and 0.4 (a = 3.761) for M=Sr.(5)

Note that electronic effects such as oxidation are reflected in the structural data. Even though La is the largest of the RE, it has a small *a*

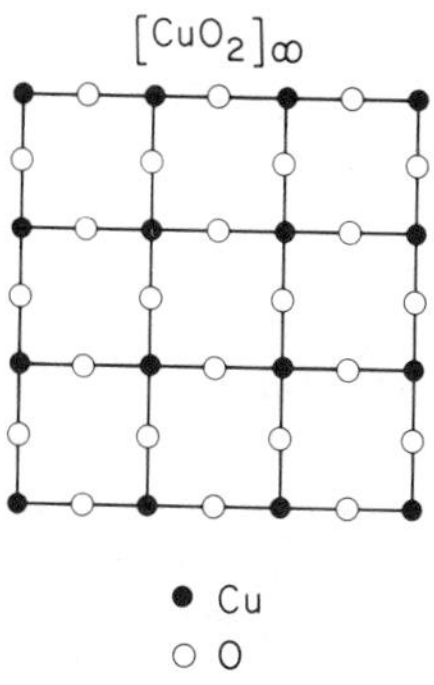

Figure 1: The square $[CuO_2]_\infty$ net common to all copper oxide superconductors.

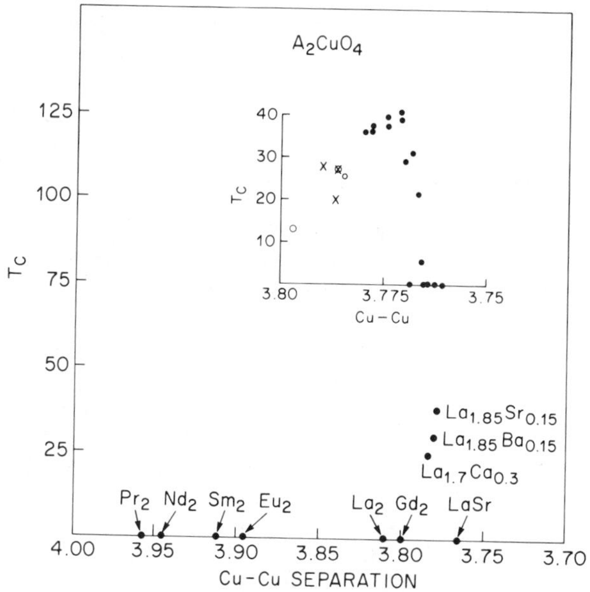

Figure 2: T_c vs. Cu - Cu separation for A_2CuO_4 compounds. The inset is an expanded region showing data for Ca (○), Ba (x), and Sr (•). Data taken from references 1,5-7.

lattice parameter. In addition, formal oxidation resulting from substitution of La with Ca, Sr or Ba leads to shorter Cu-Cu distances even in the case of Ba which is larger than La. The A_2CuO_4 phases with large Cu-Cu distances are insulating (green).

Square Pyramidal Layered Copper Oxides

The superconducting T_c is plotted against the intraplane Cu-Cu distance for square pyramidal layered compounds in Figure 3. A narrow window of superconductivity is found near 3.83-3.86Å. The plot is complicated by the fact that the orthorhombic splitting in $Ba_2YCu_3O_7$ is large (3.82 x 3.885Å). The average Cu-Cu distance is shown along with the magnitude of the orthorhombic splitting. Just as was the case for the A_2CuO_4 compounds, there are semiconducting square pyramidal Cu compounds with both larger (e.g., $La_3Ba_3Cu_6O_{14+x}$) and smaller ($La_2SrCu_2O_6$ and $Bi_2Sr_2CuO_6$) Cu-Cu separations.

The square pyramidal layer compounds do not show as clear a dependence of T_c on distance as the A_2CuO_4. For example, $Ba_2YCu_3O_x$ goes from a 93K superconductor to a semiconductor as x goes from 7.0 to 6.0, but the average Cu-Cu intraplane distance varies little with x. However, the orthorhombicity at x=7 spans a large range and decreases with x, making it unclear whether the average, shorter, or longer distances are most appropriate. Within this single structure type, additional parameters such as the previously noted correlations in *c* or in the apical Cu-O distance may be needed.

Comparisons Between Classes

The data presented above are generally supportive of the qualitative ideas of Goodenough,(8) which have been schematically addressed by Sleight (9) for copper oxides as shown in Figure 4. These authors have used the transfer energy and covalency of Cu-O bonds, respectively, as the variable. The directly measurable Cu-Cu separation may reflect these underlying properties, but we do not assume its origin. The A_2CuO_4 phases with large Cu-Cu distances are insulating, but little is known about the magnetic properties of these materials because of the difficulty of extracting the Cu contribution to the magnetic susceptibility in the presence of magnetic RE ions. Antiferromagnetism has, however, been confirmed in La_2CuO_4,(10) thus we believe that the region of large Cu-Cu distances is similar to that of low transfer energy in Goodenough's diagram. The origin of semiconductivity on the short Cu-Cu separation side needs further exploration, but appears to be of the Mott insulator type.

Formal oxidation of the layered cuprates beyond Cu^{2+} results in holes in the conduction band which is derived from Cu-3d and O-2p orbitals. Both theoretical and experimental evidence point to a large O contribution at the Fermi level. The contraction of La_2CuO_4 with oxidation indicates that the electrons removed are antibonding with respect to the $[CuO_2]_\infty$ planes. As the number of holes increases, the Cu-Cu separation decreases

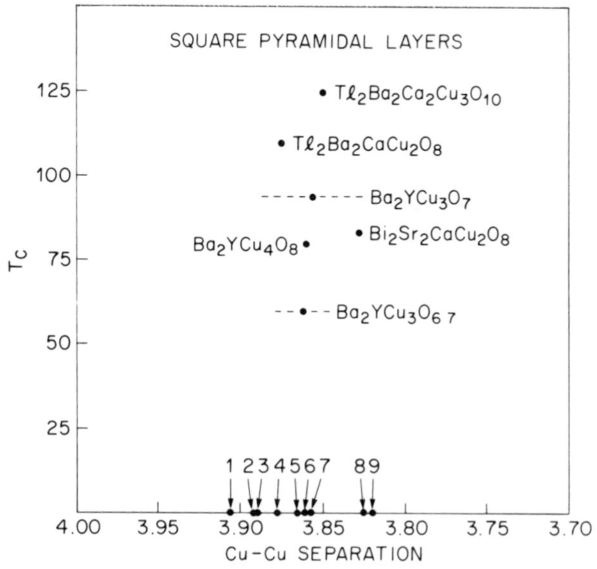

Figure 3: T_c vs. Cu - Cu separation for square pyramidal based copper oxides. The numbered points represent the oxides of $La_3Ba_3Cu_6$ (1), Ba_2YCu_2Co (2), $Pr_3Ba_3Cu_6$ (3), $Ba_2YCu_{2.5}Fe_{.5}$ (4), $Ba_2YCu_{2.8}Al_{.2}$ (5), La_2SrCu_2 (6), Ba_2YCu_3 (7), Bi_2Sr_2CaCu (8), La_2CaCu_2 (9), $Sr_2YCu_{3-x}Al_x$ (9).

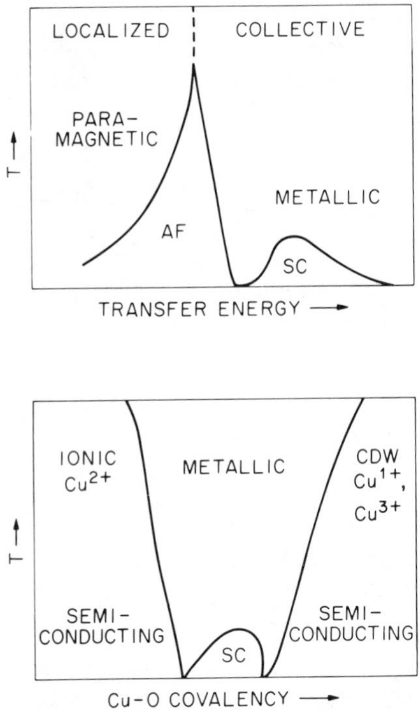

Figure 4: Simplified views of the evolution of electronic properties of oxides as the interaction increases (top - adapted from ref.8), and specifically for copper oxides (bottom - adapted from ref.9).

until a chemical instability occurs. That instability may be thought of as arising from a sufficiently high hole concentration on oxygen that oxygen atoms become favored resulting in oxygen loss and a decrease in the hole concentration. Thus, at $LaSrCuO_x$ under one atm. oxygen x=3.5 (11) and high oxygen pressure is needed to give x=4.0 (12), and at Sr_2CuO_x x=3.0 (13) with the oxygen vacancies ordered in the layers to form exclusively chains similar to those in $Ba_2YCu_3O_7$ rather than layers. It is important to realize that any oxygen vacancies occur in the layers rather than in the axial oxygens and thus are very effective at disrupting conductivity. In contrast to the *anti-bonding* nature of the conduction band in La_2CuO_4, the Cu-Cu separation in the square pyramidal compounds changes little with oxidation indicating that the band at the Fermi level is essentially *non-bonding* with respect to the layers. *An important consequence of the non-bonding character of the band is that larger hole concentrations are tolerated before a chemical instability occurs.* No examples are known of oxygen vacancies in the layers of the square pyramidal layer compounds. Toleration of a higher density of states in turn leads to higher T_c's.

The compounds $La_3Ba_3Cu_6O_{14}$, La_2CuO_4, and $Bi_2Sr_2CuO_6$ are on the borderline of the superconducting region and can be made superconducting with slight modification. In $La_{3-x}Ba_{3+x}Cu_6O_{14}$ this is accomplished by increasing x (to a maximum of x=1).(14) There is little change in lattice parameters with x and bulk superconductivity is difficult to obtain. Similarly La_2CuO_4, which tends to be oxygen deficient, can be made superconducting by annealing in oxygen with little resulting change in lattice parameters.(15) The single Cu layer compounds, $Bi_{2.3}(Sr,Ca)_{1.7}CuO_6$ (six coordinate Cu rather than five, but useful to plot here because of its relationship to the multi-Cu-layer structures), are semiconducting in our hands, but others have reported superconductivity near 10K. This case may be similar to $La_3Ba_3Cu_6O_{14}$ and La_2CuO_4 in requiring a subtle modification to produce superconductivity. Alternatively, superconductivity may simply be due to intergrowth regions.

The compounds $La_4BaCu_5O_{13}$ and $La_5SrCu_6O_{15}$ are metallic, but not superconducting. (16) They formally contain $Cu^{2.4+}$ and $Cu^{2.17+}$, respectively, but have incomplete two-dimensional O nets. Ternary copper oxides such as $CaCu_2O_3$ have condensed (edge shared) planes. The condensed layers contain oxygen bonded to three rather than two Cu atoms, making holes on oxygen unfavorable because of the need to maintain local charge neutrality. We have been unable to make substitutions to oxidize these condensed layer systems.

Potential New Non–Copper based Superconductors

It has been noted that the T_c's in copper oxides are anomalously high for the density of states when compared to other superconductors. (17) The T_c of 13K for $BaPb_{0.7}Bi_{0.3}O_3$ (18) is also unusually high, suggesting some common features with the copper oxides. This system is a mixed B ion perovskite. Recently, this sytem was modified by A ion substitution (19,20) to give T_c=29.8K at the composition $Ba_{0.6}K_{0.4}BiO_3$.(20) The formal oxidation state of Bi at this composition's $Bi^{4.4+}$. A common feature with the Cu systems is the possibility of oxidation on either Bi or O. The 13K superconductor, $BaPb_{0.7}Bi_{0.3}O_3$, contains more Pb than Bi. Therefore, we have explored the Pb only compounds and some of their derivatives because Pb shares several features displayed by Cu and Bi in the superconducting oxides including multiple oxidation states and the formation of perovskite-related phases. Band structure considerations have similarly suggested Pb based compounds. (21) Lead is typically found in the +2 or +4 state and mixed valent oxides such as Pb_2O_3 (equal amounts of Pb^{2+} and Pb^{4+}) are known. Among its ternary compounds, $BaPbO_3$ is a simple perovskite, Ba_2PbO_4 has the K_2NiF_4 structure and Sr_2PbO_4 has the the T′ version. We have carried out simple chemical modifications of these systems in an attempt to systematically alter their properties in hopes of producing new superconducting phases.

We consider first the perovskite, $BaPbO_3$. By substituting La^{+3} for Ba^{+2}, Pb is reduced. Up to 25% of the Ba can be substituted by La, with the unit cell volume increasing as the smaller La^{+3} ion replaces Ba^{+2} as shown in Figure 5, indicating that electrons are being removed from bonding levels. All members of the series, $Ba_{1-x}La_xPbO_3$, are metallic to 4K, but do not become superconducting. The static magnetic susceptibility of $Ba_{0.75}La_{0.25}PbO_3$ is temperature independent and paramagnetic ($53x10^6$ emu/mole) after correction for core diamagnetism. In terms of the Goodenough model, these phases may lie sufficiently far into the metallic regime that chemical modification does not approach the the region where superconductivity might occur.

In the K_2NiF_4 phase, lanthanum substitution again reduces Pb from Pb^{+4} in unsubstituted Ba_2PbO_4. A quarter of the Ba^{+2} can be replaced by La^{+3}, but with no effect on the lattice parameters. Ceramic samples show metallic resistivity with essentially linear ρ vs T to low temperature, but again do not superconduct. The T′ phase Sr_2PbO_4 phases are all yellow and insulating. Substitution of Y for Sr changes the color to brown, but does not make them metallic.

Summary

We have noted the existence of a narrow region of in-plane Cu-Cu separations in superconducting layered cuprates. For the A_2CuO_4 compounds, the distance is strongly reduced by oxidation whereas for the square pyramidal cuprates the distance is not strongly affected by oxidation. The fact that oxidation occurs in a more non-bonding band (with respect to the layers) in the latter system is likely responsible for greater chemical variations and higher hole concentrations that lead to higher T_c's. These ideas have been expressed using only traditional chemical viewpoints without the need to invoke any particular mechanistic model for superconductivity.

The occurrence of holes predominately on oxygen is unprecedented in other metal oxide systems, and thus it is tempting to assume they play a key role in the mechanism of superconductivity in the cuprates. Both magnetic and phonon based mechanisms have been proposed that could attribute a

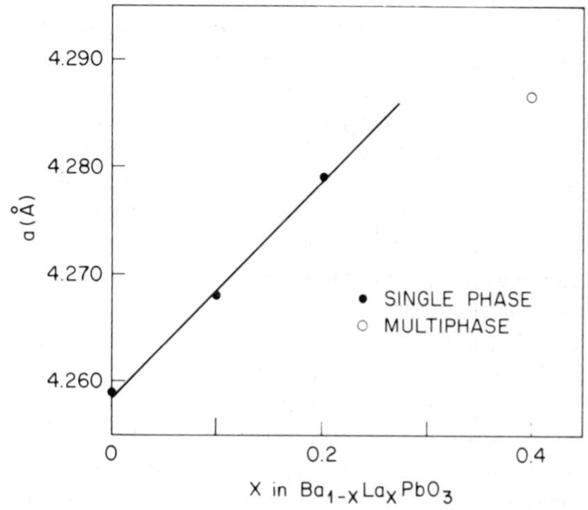

Figure 5: The cubic lattice parameter vs. composition for the series $Ba_{1-x}La_xPbO_3$.

key role to holes on oxygen. We have found no reason to consider either the magnetic state of Cu or the availability of three oxidation states of Cu in the above discussion. That does not mean that they are unimportant. Similarly the occurrence of antiferromagnetism in non-superconducting La_2CuO_4 and $Ba_2YCu_3O_6$ does not demonstrate the importance of such interactions in the superconducting state. It has long been known that localization and antiferromagnetism can occur close to superconductivity.

There appears to be a common basis for superconductivity in Cu and Bi based oxides. To date there is insufficient data to properly characterize many of the copper oxides, especially in those cases with shorter Cu-Cu separations than similar superconductors. More copper free superconductors would be of great value in testing and developing theoretical models.

Acknowledgments

We have benefited from helpful discussions with R. J. Cava, T. Siegrist, and M. L. Steigerwald.

Literature Cited

1. Kishio, K.; Kitizawa, K.; Sugii, N.; Kanbe, S.; Fueki, K.; Takagi, H.; Tanaka, S. *Chem. Lett.*, 1987, 635.
2. Cava, R. J.; et. al. Proceedings of the Intl. Conf. on Materials and Mechanisms of Superconductivity, Interlaaken, Switzerland, March 1988, to be published.
3. Miceli, P. F.; Tarascon, J. M.; Greene, L. H.; Barboux, P.; Rotella, F. J.; Jorgensen, J. D. *Phys. Rev. B.*, 1988, *37,* 5932.
4. For a review of these and other copper oxide structure types see Muller-Buschbaum, H. *Angew. Chem. Int. Ed.* 1977, *16,* 674.
5. Tarascon, J. M.; Greene, L. H.; McKinnon, W. R.; Hull, G. W.; Geballe, T. H. *Science,* 1987, *235,* 1373.
6. Jorgensen, J. D.; Schuttler, H. -B.; Hinks, D. G.; Capone, D. W.; Zhang, K.; Brodsky, M. B. *Phys. Rev. Lett.* 1987, *58,* 1024.
7. Cava, R. J.; Santoro, A.; Johnson, D. W.; Rhodes, W. W. *Phys. Rev. B.* 1987, *35* 6716.
8. Goodenough, J. B. *Progress in Solid State Chem.,* 1971, *5,* 145.
9. Sleight, A. W., In *Chemistry of High Temperature Superconductors,* ACS Symposium Series 351, American Chemical Society, Washington, DC, 1987, pp. 2-12.
10. Vaknin, D.; Sinha, S. K.; Moncton, D. E.; Johnston, D. C.; Newsam, J. M.; Sarfinya, C. R.; King, H. E. *Phys. Rev. Lett.* 1987, *58,* 2802.

11. Nguyen, N.; Studer,F.; Raveau, B. *J. Phys. Chem. Solids* 1983, *44,* 389.

12. Goodenough, J. B.; Demazeau, G.; Pouchard, M.; Hagenmuller, P. *J. Solid State Chem.* 1973, *3,* 325

13. Teske, C. L.; Muller-Buschbaum, H. *Z. Anorg. Allg. Chemie* 1969, *371,* 325.

14. Sunshine, S. A.; Schneemeyer, L. F.; Waszczak, J. V.; Murphy, D. W.; Miraglia, S.; Santoro, A.; Beech, F. *J. Crystal Growth* 1987, *85,* 632.

15. Beille, J.; Cabanel, R.; Chaillout, C.; Chevallier, B.; Demazeau, G.; Deslandes, F.; Etourneau, J.; Lejay, P.; Michel, C.; Provost, J.; Raveau, B.; Sulpice, A.; Tholence, J. L.; Tournier, R. *Compt. Rend. Acad. Sc.* 1987, *304,* 1097.

16. Torrance, J. B.; Tokura, Y.;Nazzal, A. *Chemtronics* 1987, *2,* 120.

17. Batlogg, B.; Ramirez, A. P.; Cava, R. J.; van Dover, R. B.; and Rietman, E. A. *Phys. Rev. B* 1987, *35,* 5340.

18. Sleight, A. W.; Gillson, J. L.; Bierstedt, P. E. *Solid State Commun.* 1975, *17,* 27.

19. Mattheiss, L. F.; Gyorgy, E. M.; Johnson, D. W. *Phys. Rev. B,* 1988, *37,* 3745.

20. Cava, R. J.; Batlogg, B.; Krajewski, J. J.; Farrow, R. C.; Rupp, L. W.; White, A. E.; Short, K. T.; Peck, W. F.; Kometani, T. Y. *Nature* 1988, *332,* 814.

21. Mattheiss, L. F.; Hamann, D. R. to be published.

RECEIVED July 15, 1988

INDEXES

Author Index

Affiliation Index

Subject Index

F

G

R

Production by Cheryl Shanks and Barbara J. Libengood
Indexing by Janet S. Dodd

Elements typeset by Hot Type Ltd., Washington, DC
Printed and bound by Maple Press, York, PA